THE CHAUVENET PAPERS

A Collection of Prize-Winning Expository Papers in Mathematics

VOLUME I

J. C. ABBOTT, editor

United States Naval Academy

Published and distributed by
THE MATHEMATICAL ASSOCIATION OF AMERICA

Library of Congress Catalog Card Number 78-53419.

Vol. I ISBN 0-88385-425-2.

Printed in the United States of America

Current printing (last digit)

10 9 8 7 6 5 4 3 2 1

PREFACE

The By-Laws of the Mathematical Association of America state that:

"Its object shall be to assist in promoting the interest of mathematics in America, *especially in the collegiate field*, by *holding meetings* in any part of the United States of America or Canada for the presentation and discussion of mathematical papers, by the publication of mathematical papers, journals, *books*, monographs, and reports, by conducting investigations for the purpose of *improving the teaching of mathematics*, by accumulating a mathematical library and by cooperating with other organizations whenever this may be desirable for attaining these or similar objects."[1] [Italics added.]

In accordance with these objectives, President Julian Lowell Coolidge proposed in March 1925 "that the Association establish a prize for *special merit in mathematical exposition*."[2] The prêsident spoke of the importance of emphasizing *excellence in exposition and presentation of papers*, mentioning the great improvement in papers presented at the Harvard Mathematics Club since small prizes had been offered.[3] The proposition was sanctioned by mail by a vote of the Board of Trustees, and a committee consisting of Professors A. J. Kempner, Chairman; Louise D. Cummings, and D. R. Curtis was appointed to formulate the details. A report of the Committee was made at the meeting of the Board of Trustees in September 1925 at Ithaca, N.Y. The report read:

I. We believe that the proposed prize *will exert a desirable influence on the production of high-grade expository articles*. The adoption of some scheme in this respect is recommended. Concerning the details, we take some pleasure in recommending, with only minor modifications, the unofficial suggestions of President Coolidge, as follows:

II. The name of the prize shall be the *Chauvenet Prize*.

This name, suggested by President Coolidge, seems in every respect appropriate. For a study of the influence and life of William Chauvenet, 1820–1870,... Professor of Mathematics in the U. S. Navy (1841–1859), and President of the Academic Board of the Naval Academy (1847–50), Professor of Mathematics and Natural Philosophy at Washington University, St. Louis, (1859–69), author of many works and treatises...see an article by W. H. Roever, Washington University Series, No. 2, 1925.

III. The amount of the prize shall be one hundred dollars. ...

V. The prize shall be conferred for the best article of an *expository character*, dealing with some mathematical topic, written by a member of the Association and published in English in a journal during the five calendar years preceding the year of the award. ...

---

[1]The MONTHLY, 32 (1925), 439–440. See also: Kenneth O. May, *The Mathematical Association of America: Its First Fifty Years*, The Mathematical Association of America, 1972.

[2]Minutes of Meetings of the Association, 1920–1930.

[3]*Ibid.*

VIII. The Association accepts the offer of Professor Coolidge to defray the expenses of the first award. Professor Coolidge asks that the gift be anonymous. ...

We feel that the scrutinizing committee should be restricted as little as possible. For this reason we have made no attempt to define in V the term "some mathematical topic" nor have we considered it desirable to suggest any definite rule against the division of the prize, or awarding the prize to the same author twice.[4]

After adopting the report unanimously, President Coolidge appointed the first committee, consisting of Professor E. B. Van Vleck, Wisconsin, Chairman; W. C. Graustein, Harvard; and Mrs. Anna Pell Wheeler, Bryn Mawr.

It is clear from these minutes that credit for the establishment of the Chauvenet Prize belongs to Julian Lowell Coolidge, the President of the Association and Professor at Harvard. Professor Coolidge was himself not only a well-recognized research mathematician, but also one of the great teachers of his time. His classes were a joy to attend and he inspired many students to take up careers in mathematics. He was a proper Bostonian who conducted his classes in the formal fashion current in those days. But at the same time he had a wry sense of humor behind the facade that made him a delight to know. His classes were large but every student felt personal contact. His course in freshman calculus covered many of the rudiments of real analysis long before it had crept into the undergraduate curriculum. He was able to bring the results of his own research in projective geometry and his love of the history of mathematics into the most elementary courses. His suggestion for the creation of the Chauvenet Prize was certainly made from personal concern and example.

As mentioned in the minutes, President Coolidge backed up his proposal by a personal contribution of $100. With characteristic modesty he requested that this contribution be kept anonymous and that the prize be dedicated to the honor of Professor Chauvenet. A detailed account of the career of Professor Chauvenet constitutes the first article of this volume.

The original committee selected an article by Professor G. A. Bliss for the first award. In a return letter Professor Bliss wrote: "I congratulate the Association on the inauguration of the Chauvenet Prize, and hope that it may give to many others in the future the same pleasant impetus which it gives to me."

During the following years the Association donated $20 annually toward the Prize, and subsequently gifts were received of $500 from W. B. Ford and $100 from Dunham Jackson. By 1940 the awards were reduced to $50 due to falling interest rates. The Association also transferred funds from the General Fund and eventually the Chauvenet Fund was merged with the Awards Fund. In 1965 the award was increased to $500.

In 1928 the period of the award was reduced to three years and the wording was changed to bring the prize more distinctly within the range of "younger American mathematicians" and "best expository paper" was changed to "noteworthy paper." Again in 1962 the Board voted to authorize the establishment of a standing

[4] *Ibid.*

committee on the Chauvenet Prize and the award was changed to an annual award "provided there is a suitable recipient."[5]

A list of the various committees on the Chauvenet Prize follows:

1929: A. J. Kempner, Chairman; D. R. Curtis, W. A. Hurwitz
1933: W. B. Ford, Chairman
1944: Philip Franklin, Chairman; Saunders Mac Lane, G. T. Whyburn
1948: R. P. Agnew, Chairman; R. W. Barbard, R. P. Boas, Jr.
1951: W. B. Carver, Chairman; R. E. Gillman, D. H. Holl
1954: C. B. Allendoerfer, Chairman; C. C. MacDuffee, Virgil Snyder
1956: P. A. White, Chairman; J. D. Mancill, J. G. Wendel
1959: H. G. Bohnenblust, Chairman; R. H. Bruck, Mark Kac
1959–61: Walter Rudin, Chairman; Ernst Snapper, R. L. Wilder
1962: R. L. Wilder (1962–63), Chairman; P. J. Davis (1962–65), Ernst Snapper (1962–64)
1964: P. J. Davis (1963–65), Chairman; E. F. Beckenbach (1964–66), Irving Kaplansky (1964)
1965: P. J. Davis, Chairman; E. F. Beckenbach, V. L. Klee
1966: E. F. Beckenbach (1964–66), Chairman; Samuel Eilenberg (1966–68), V. L. Klee (1965–67)
1967: Victor Klee (1965–67), Chairman; Samuel Eilenberg (1966–68), G. L. Weiss (1967–69)
1968: Samuel Eilenberg (1966–68), Chairman; Mark Kac (1968–70), G. L. Weiss (1967–69)
1969: G. L. Weiss (1967–69), Chairman; Mark Kac (1968–70), N. D. Kazarinoff (1968–70)
1970: Mark Kac (1968–71), Chairman; S. S. Chern (1970–72), N. D. Kazarinoff (1969–71)
1971: N. D. Kazarinoff (1969–71), Chairman; S. S. Chern (1970–72), Norman Levinson (1971–73)
1972: S. S. Chern (1970–72), Chairman; Norman Levinson (1971–73), François Trèves (1972–74)
1973: Norman Levinson (1971–73), Chairman; C. D. Olds (1973–75), François Trèves (1972–74)
1974: François Trèves (1972–74), Chairman; P. D. Lax (1974–76), C. D. Olds (1973–75)
1975: C. D. Olds (1973–75), Chairman; P. D. Lax (1974–76), M. D. Davis (1975–77)
1976: P. D. Lax (1974–76), Chairman; M. D. Davis (1975–77), L. A. Zalcman (1976–78)

[5]The MONTHLY, 33 (1926), 177–78.

Since its inception in 1925 there have been twenty-four awards for the Chauvenet Prize. The prize papers constitute the contents of these two volumes. They attest to the success of Coolidge's original concept of a prize for outstanding expository papers. A perusal of the list of winners together with their backgrounds as described in the accompanying biographies shows that all are not only expert expositors but top-flight research mathematicians. Furthermore, they have all shown deep concern for the welfare of their students and for the future of mathematics. They all recognize that without proper communication with the mathematical public as well as all users of mathematics in general, the survival of their work is not guaranteed. Their works cover the broad spectrum of all of modern mathematics from analysis to algebra, from geometry to foundations, from probability to applied mathematics. They may have conceived their original researches in the ivory towers of their individual minds, but they universally recognize that it is necessary to climb out of these towers to make known their results to their fellow mathematicians and students.

Their lists of achievements and honors is long and very respectable, and their positions held throughout the world show that their accomplishments are not simply one-time affairs, but their success is consistent. Their bographies show that they combine the best of the two worlds of pure research and sympathetic teaching. Their work serves as an inspiration to make the telling of mathematics on a par with its creation.

There is a myth that constantly creeps into the edges of the academic world that a good research scientist cannot also be a good teacher. He is too involved in the inner world of his personal research to be concerned that the outer world understands him. Good researchers do not have time for their students and they do not have patience. Certainly the winners of the Chauvenet Prize belay all such false myths. Many of the past winners have served as leaders of the mathematical world outside of pure research as high officers in both the Association as well as the American Mathematical Society. They have been the innovators in new programs such as CUPM designed to improve the teaching of mathematics, both on the undergraduate as well as high school levels. They have certainly lived up to the goals of the Association as stated in the By-Laws quoted above. The Association is proud to publish this collection of all past Chauvenet Prize papers as a symbol of its concern for the future of the teaching and exposition of American Mathematics.

The Subcommittee hopes that the publication of these papers will continue to inspire other mathematicians to strive toward goals of better exposition and continued concern for the future education of all undergraduates. The future of mathematics rests on the attainment of these goals.

Since Professor Chauvenet had played an outstanding role in the founding of the U.S. Naval Academy, the Naval Academy chose the name "Chauvenet Hall" when it dedicated a new mathematics building in 1969. In honor of this name and its connection with the Chauvenet Prize, a Chauvenet Symposium was held at the Academy at the time of the dedication ceremonies. Six of the past winners of the Chauvenet Prize were invited to give "repeat performances" at this symposium.

Since these papers are typical of the goals of the Chauvenet Prize we list them here as further readings:

P. R. Halmos, *Finite-Dimensional Hilbert Spaces*, THE MONTHLY, 77 (1970) 457-464.

Guido Weiss, *Complex Methods in Harmonic Analysis*, THE MONTHLY, 77 (1970) 465-474.

G. T. Whyburn, *Dynamic Topology*, THE MONTHLY, 77 (1970) 556-570.

Saunders Mac Lane, *Hamiltonian Mechanics and Geometry*, THE MONTHLY, 77 (1970) 570-586.

Mark Kac, *On Some Probabilistic Aspects of Classical Analysis*, THE MONTHLY, 77 (1970) 586-597.

Leon Henkin, *Mathematical Foundations for Mathematics*, THE MONTHLY, 78 (1971) 463-487.

In addition to this Symposium, the Association conducted a second Chauvenet Symposium on the occasion of the announcement of the publication of the present two volumes at the Annual Winter Meeting in Atlanta, Georgia, January 6, 1978. Two additional papers by past Chauvenet Prize winners were presented at this meeting. They were:

L. Zalcman, *Offbeat Integral Geometry*.

Martin Davis, *Boolean-Valued Models in Set Theory, Analysis, and Quantum Mechanics.*

These papers attest to the continued ability of the past winners to produce outstanding expository papers and to the success of the Chauvenet Prize program in promoting continued efforts to communicate the best of mathematics to the mathematical public.

The Subcommittee on the publications of the Chauvenet Papers wishes to thank the following publishers for permission to reproduce the following papers:

1. The Annals of Mathematics, Editor: Armand Borel, for the articles by G. A. Bliss: *Algebraic Functions and Their Divisors*, The Annals of Mathematics, 26 (1924) 95–124; Dunham Jackson: *Series of Orthogonal Polynomials* and *Orthogonal Trigonometric Sums*, The Annals of Mathematics, Series 2, 34 (1933) 527–545 and 799–814.
2. The Bulletin of the American Mathematical Society, for the articles by T. H. Hildebrandt: *The Borel Theorem and Its Generalizations*, Bull. AMS, 32 (1926) 423–474; G. H. Hardy: *An Introduction to the Theory of Numbers*, Bull. AMS, 35 (1929) 778–818; G. T. Whyburn: *On the Structure of Continua*, Bull. AMS, 42 (1936) 49–73; Jean François Trèves: *On Local Solvability of Linear Partial Differential Equations*, Bull. AMS, 76 (1970) 552–571.
3. The National Mathematics Magazine, for the article by R. H. Cameron: *Some Introductory Exercises in the Manipulation of Fourier Transforms*, Nat. Math. Mag. (1941) 331–356.

4. Science, for the article by Leon Henkin: *Are Logic and Mathematics Identical?*, Science, 138 (1962) 788–794. Copyright 1962 by the American Association for the Advancement of Science.

5. The Society for Industrial and Applied Mathematics, for the article by Jack K. Hale and Joseph P. LaSalle: *Differential Equations: Linearity vs. Nonlinearity*, SIAM Review, 5 (1963) 249–272. Copyright 1963 by the Society for Industrial and Applied Mathematics.

6. Scientific American, for the article by Martin D. Davis and Reuben Hersh: *Hilbert's Tenth Problem*, Scientific Am., 229, No. 5 (November 1973) 84–91.

The remaining articles appearing in these volumes, I and II, appeared under the auspices of the Association, either in the MONTHLY, as Slaught Papers, or in the MAA Studies in Mathematics.

The following biographies, with revisions, were taken from the MONTHLY: Mark Kac 75 (1968) p. 4; Philip Davis, 70 (1963) p. 2; Leon Henkin, 71 (1964) p. 3, and 78 (1971) p. 463; Jack Hale and Joseph P. LaSalle, 72 (1965) p. 3; Guido Weiss, 74 (1967) p. 3; Shiing-Shen Chern, 77 (1970) p. 117; Norman Levinson, 78 (1971) p. 112; Jean François Trèves, 79 (1972) p. 113; Carl Douglas Olds, 80 (1973) p. 120; Peter D. Lax, 81 (1974) p. 113; Martin Davis and Reuben Hersh, 82 (1975) p. 108; Lawrence Zalcman, 83 (1976) p. 84.

The Subcommittee also wishes to thank the following for their help in preparing appendices for some of the earlier papers: E. J. McShane, the Hildebrandt paper; Guido Weiss, the Jackson papers; F. B. Jones, the Whyburn paper; Colin Blyth, the Halmos paper; Mark Kac, the Kac paper; E. J. McShane, the McShane paper, and R. H. Bruck, the Bruck paper.

The Subcommittee of the Committee
on Publications for the Publication
of the Chauvenet Prize Papers

J. C. ABBOTT, Chairman
G. B. GALE
C. D. OLDS
J. A. TIERNEY
R. L. WILDER

WILLIAM CHAUVENET

Born in 1820 in Milford, Pennsylvania, William Chauvenet had the advantage of an excellent training in his youth. His parents gave him all their attention and saw to it that he attended the best schools and received the best instruction available. It soon became evident that he was unusually talented in mathematics and music, and that he possessed a high degree of mechanical aptitude. Though but sixteen years of age when he entered Yale College, William had mastered the entire four-year course in mathematics prior to his entrance to the college. He graduated with high honors in mathematics and the classics and in 1889 Florian Cajori, the eminent mathematical historian, referred to him as "the ablest mathematician and astronomer Yale has produced."

After participating in a series of magnetic experiments at Girard College in Philadelphia, he was appointed a professor of mathematics in the United States Navy in 1841. In this capacity he served briefly aboard the U.S. Steamer Mississippi, after which he became head of the shore Naval School in Philadelphia. His success at this school eventually led to the establishment of the present Naval Academy at Annapolis, Maryland. Often referred to as the "father of the Naval Academy," William Chauvenet was the first Head of the Department of Mathematics at Annapolis. During the next fourteen years he worked unceasingly to make the Academy the greatest institution of its kind in the world. Increasing awareness of his stature as a mathematician and a scientist notably enhanced the reputation of the Academy. Always the most prominent member of the faculty he was the guiding spirit of the institution.

In 1859 Professor Chauvenet was elected to the chair of mathematics at Washington University in St. Louis and in 1862 was chosen Chancellor of the University. His reputation, untiring devotion, and inspiration contributed in large measure to the growth and stature of that university. Illness forced him to reduce his activities in the last six years of his life and he died in St. Paul, Minnesota, in 1870.

William Chauvenet was one of the leaders in American science and one of the first of our mathematicians to gain recognition abroad. Cajori wrote in 1889 that "Professor William Chauvenet ranks among the coryphaei of science in America. He and Benjamin Peirce have done more for the advancement of mathematical and astronomical science, and for raising to a higher level the instruction in these subjects, than any other two Americans." He was a member of several scientific societies including the American Philosophical Society, the American Society of Arts and Sciences, and the National Academy of Sciences, of which he was a prominent charter member. At the time of his death he was president of the American Association for the Advancement of Science and vice president of the National Academy of Sciences. In 1860 St. John's College in Annapolis conferred upon him the degree LL.D.

Professor Chauvenet's international reputation was based upon the brilliant papers he presented to learned societies and upon his textbooks. His *Treatise on*

Plane and Spherical Trigonometry, written and published while he was at the Naval Academy, became a classic and for many years was used in the leading colleges of America. An 1850 review in the *Journal of the Franklin Institute* referred to this work as "the most complete treatise on trigonometry extant in the English language." Professor Chauvenet's greatest work, *A Manual of Spherical and Practical Astronomy*, evolved from his teaching of astronomy and navigation at Annapolis. It has been translated into many languages and its reputation was as great in Europe as in America. Professor Herman Struve, world-renowned director of the *Königliche Sternwarte* in Berlin, once remarked to an American astronomer that "you have in America the best work in existence on practical astronomy, that of Chauvenet, which is complete." In 1895 F. P. Matz, one of Professor Chauvenet's biographers, stated that "Few works of a scientific nature, by American authors, have been received with such universal favor."

Professor Chauvenet's works have exerted marked influence upon mathematical and scientific literature. The 1886 revision of Nathaniel Bowditch's *The New Practical Navigator* was due primarily to William Chauvenet's new methods. In 1904 Roberdeau Buchanan based his *Mathematical Theory of Eclipses* on the chapter on eclipses in Professor Chauvenet's treatise on astronomy, referring to this chapter as "the most thorough and exhaustive treatise on this subject that has yet been published."

Although Professor Chauvenet made many original contributions to knowledge, his fame is due as much to his writings as to his creations. Professor Peirce stood alone as a mathematician, but William Chauvenet had no equal as a writer. Thoroughly familiar with the works of the great French and German masters, he had the genius for exposition and lucid presentation which enabled him to present their methods and techniques in a form accessible to the scientists and teachers of America. The ease, grace of style, and lucidity of his texts set the standard in the colleges of the country. Professor William H. Roever, one of his biographers, stated in 1926 that he was gifted with "a power of expression and purity of language unexcelled in American scientific literature."

As a teacher he had an uncanny ability, so often lacking in men of genius, for imparting his knowledge to others. His biographer, J. H. Coffin, states: "His intellectual abilities, his thorough knowledge of the subjects of instruction, the wide range of his attainments, a just appreciation of merit, an unwavering integrity, a uniform disposition, never disturbed by passionate excitement, and a kindly interest in those with whom he was associated, gained the esteem and respect of all. In naval circles his memory is revered. Admiral S. R. Franklin, American naval hero and former midshipman, writes in his memoirs that "Professor Chauvenet, our instructor in Mathematics, had the faculty of imparting what he knew to others in a higher degree than any man I have ever known."

Few scientists of distinction have had more varied interests outside their specialties than Professor Chauvenet. He was an accomplished pianist and was highly regarded as a music critic. His knowledge of English literature, together with his vast acquaintance with the world of science, stood him in good stead as the

Naval Academy Librarian. Matz refers to him as a "man of wide and varied culture, and keen critical taste." His colleague, Professor C. W. Woodward, describes him as "a delightful gentleman in every respect; genial, polished, sympathetic, and always interesting," and states that he "moved about with dignity and grace, and a most distinguished air." He was deeply religious and extremely devoted to his parents, his wife and his five children.

In announcing the death of President Chauvenet to the members of the American Association for the Advancement of Science, acting President Thomas Sterry Hunt concluded his remarks with these words:

"In his assiduous devotion to scientific studies he did not neglect the more elegant arts, but was a skillful musician, and possessed of great general culture and refinement of taste. In his social and moral relations he was marked by rare elevation and purity of character, and has left to the world a standard of excellence in every relation of life which few can hope to attain."

J. A. Tierney

CONTENTS

VOLUME I

VOLUME II

1

GILBERT AMES BLISS

Gilbert Bliss was born in Chicago on May 9, 1876. He entered the University of Chicago in 1893 and obtained his B.S. in 1897, his M.S. in 1898, and his Ph.D. in 1900. He became an instructor in mathematics at the University of Minnesota, 1900-02, studied at Göttingen in 1902-03, and returned to the United States as assistant professor at the University of Missouri in 1904-05. He transferred to Princeton as assistant professor in 1905 and became associate professor at Chicago in 1908. He was promoted to professor in 1913 and in 1933 was named Distinguished Service Professor. He was department chairman from 1927-1941 and retired in 1941. He also gave courses at Wisconsin, Chicago, Princeton, and at Harvard during summer sessions in 1906-1911.

He was associate editor of the *Annals*, 1906-08 and of the *Transactions*, 1906-16 and Chairman of the Editorial Committee from 1929-41. He was on the Fellowship Board of the National Research Council and was a member of the National Academy of Sciences, a member of the American Philosophical Society, and a Fellow of the American Academy of Arts and Sciences as well as a member of the American Mathematical Society and the Mathematical Association of America. He was President of the Mathematical Society in 1921-22 and served as Vice-President of the American Association for the Advancement of Science. He was also a member of the Illinois Academy of Science, the London Mathematical Society, the Deutsche Mathematische Verein, and Circolo Mathematica di Palermo. He received an honorary Doctor of Science from Wisconsin in 1935.

His principal contributions to mathematics were in the field of the calculus of variations as leader of the Chicago School. He was under the influence of F. R. Moulton in astronomy during his early days, but switched to pure mathematics for his dissertation on geodesics on the anchor ring. Thereafter, influenced by Bolza's work on Weierstrass methods, he turned to the calculus of variations. During his tenure at Chicago he inspired many followers in his field and essentially dominated the area for some time. He made applications of boundary value problems to quantum mechanics and relativity as well as to the theory of ballistic trajectories during a stay at Aberdeen Proving Grounds in 1918.

Bliss took especial pride in his teaching, stimulating all those who worked under him. He always sought for simplicity, clarity, and comprehensiveness in his mathematical exposition. He was an appropriate choice as the first winner of the Chauvenet Prize in 1925. In his own words, "The real purpose of graduate work in mathematics, or in any other subject, is to train the student to recognize what men call the truth, and to give him what is usually his first experience in searching out the truth in some special field and recording his impressions. Such training is invaluable for teaching, or business, or whatever activity may claim the student's future interest." These words inspired over fifty doctoral candidates including such names as L. M. Graves, W. L. Duren, E. J. McShane, and M. R. Hestenes; i.e., most of the great leaders in the calculus of variations.

ALGEBRAIC FUNCTIONS AND THEIR DIVISORS

GILBERT AMES BLISS, University of Chicago

Introduction. The theory of algebraic functions and their integrals is one of the most impressive of the contributions of the mathematicians of the nineteenth century, and one of the most beautiful of all the chapters in the domain of modern mathematics. It had its origin in the effort to generalize the theory of elliptic integrals of the form

$$I = \int \eta(x,y)\,dx,$$

where $\eta(x,y)$ is a rational function of x and y, and y is defined as a function of x by an equation of the form

$$y^2 - A_0x^4 - A_1x^3 - A_2x^2 - A_3x - A_4 = 0.$$

Early in the nineteenth century Abel conceived the idea of studying integrals of this type for which the variable y is defined as a function of x by means of a more general equation $f(x,y)=0$ in which the first member may be any polynomial whatsoever in x and y instead of the relatively simple one in the equation just given for the elliptic case. His extension of the problem seemed at first almost too great a generalization, but his results, notably the famous theorem now called by his name, indicated clearly that the theory of such integrals had a richness of content which justified elaborate and searching study. A function $y(x)$ defined by an equation $f(x,y)=0$ such as has just been described is called an algebraic function and the integrals I associated with it are called Abelian integrals.

There have been three principal methods of attacking the theory of algebraic functions and their integrals*. The first of these is based largely upon the researches of Abel (1826–9) and Riemann (1857), and is called the transcendental theory because in it the Abelian integrals play the central role. A second is the geometrical method, closely interwoven with the theory of higher plane curves. It was inaugurated by Clebsch and Gordan (1863–6), continued by Brill and Noether (1871), and presented more recently in attractive form by Severi (1914). The third and last method has been called, not any too appropriately, the arithmetic method. In contrast with the transcendental method its emphasis is placed primarily upon the construction and theory of rational functions $\eta(x,y)$ such as those which occur in the integrands of the integrals I, and only secondarily upon the integrals themselves. One of the earliest suggestions of such a theory is found in a paper

* An excellent description and comparison of these method has been given in brief by Hensel and Landsberg, Theorie der algebraischen Funktionen, pp. 694–702. For more extensive summaries and references see Brill and Noether, Die Entwickelung der Theorie der algebraischen Funktionen, Deutsche Math. Ver., vol. 3, (1894), pp. 107–566; Wirtinger, Algebraische Funktionen und ihre Integrale, Encykl. d. math. Wiss., II B 2, pp. 115–175; Hensel, Arithmetische Theorie der algebraischen Funktionen, Encykl. d. math. Wiss., II C 5, pp. 533–674.

which Kronecker presented to the Berlin Academy in 1862 but published first in 1881. More elaborate theories, differing widely in detail, are those of Dedekind and Weber (1882), Weierstrass (1902)*, and Fields (1906). The methods of Dedekind and Weber have been elaborated and improved by Hensel and Landsberg (1902) in their book above referred to.

The transcendental and the various arithmetic theories are in reality identical in purpose, though differing markedly in the mechanisms by means of which this purpose is carried out. One of the undesirable features of both the transcendental and geometrical theories has been the necessity of applying preliminary birational transformations in order to simplify the singular points of the curve represented by the fundamental algebraic equation $f(x,y)=0$. In both cases the development of the theory is much easier if only multiple points with distinct tangents, or better only double points with distinct tangents, are present. The arithmeticists have shown that one may succeed in attaining the same, and in many cases even more general, results without the assistance of such transformations. This has been done by Hensel and Landsberg in most interesting fashion in their treatise, and the purpose of the present paper is to give a relatively elementary introduction to their theory. The goals which it is planned to attain here are two, the construction of the three types of elementary integrals upon which the superstructure of the theory of Abelian integrals is based, and the proof of the Riemann-Roch Theorem which is fundamental for the theory of the rational functions $\eta(x,y)$.

It has always seemed to the writer to be unfortunate to have results of the importance of those which have just been mentioned buried so deeply in the text of the book of Hensel and Landsberg.† Before reaching them one must read through some three hundred pages dealing with these and many other matters. In the following sections only those theorems have been collected which form links in the chain of reasoning leading up to the principal results described in the last two sections of this paper. The theorems themselves will be familiar to those who have read Hensel and Landsberg's book, but the proofs are frequently fundamentally different and it is hoped simpler. This is most noticeably so perhaps in the establishment of Theorem 3, page 10, concerning the existence of a basis for every divisor, where the properties of the determinant Δ of page 9 turn out to be of great assistance, and in the discussion of the properties of complementary bases.

It is of course not possible to discuss completely any large portion of the theory of algebraic functions in a short paper of this sort. For this reason the properties of the fundamental expansions for an algebraic function have been described without proofs in Section 1, but the remainder of the paper is intended to be logically complete within itself. The expansions are familiar ones to those who have studied the theory of algebraic functions. It is hoped that they are here described with

---

* The date is that of the publication of volume IV of Weierstrass' Werke. The theory there presented was developed in his lectures during the years 1875–6.

† This feeling was emphasized by the necessity of referring to the Riemann-Roch Theorem in the footnote on page 277 of the author's paper, Birational transformations simplifying singularities of algebraic curves, Trans. Amer. Math. Soc., vol. 24 (1922), p. 274.

sufficient clearness, so that others may, after accepting them, gain some idea of the content of this very beautiful theory.

1. Expansions for an algebraic function. Let $f(x,y)$ be an irreducible polynomial in x and y with coefficients which are either real or complex. By an irreducible polynomial is meant one which is not decomposable into a product of factors of the same sort. If the degree of the function $f(x,y)$ in y is n, then the n values of y which satisfy the equation $f(x,y)=0$ form what is called an n-valued algebraic function of x.

The theory of such algebraic functions is based upon the fact that near every value $x=x_0$ the roots of the equation $f=0$ are expressible by a number of series of the form*

$$y=b(x-x_0)^{\frac{\mu}{r}}+b'(x-x_0)^{\frac{\mu'}{r}}+\cdots \tag{1}$$

where r is a positive integer and the increasing integers $\mu<\mu'<\cdots$ are either positive or negative. The numbers $r,\mu,\mu',\ldots,$ are relatively prime so that the particular series (1) furnishes r distinct values of y when the r values of the root $(x-x_0)^{\frac{1}{r}}$ are substituted. The number of series corresponding to different values $x=x_0$ may be different, but the sum of the integers r for each $x=x_0$ is exactly n, corresponding to the fact that for each value of x near $x=x_0$ the equation $f=0$ has n roots.

Near $x=\infty$ the expansions giving the values of y are in powers of $1/x$ instead of $x-x_0$. We may include both cases in the formula

$$y=bt^{\mu}+b't^{\mu'}+\cdots \tag{2}$$

where t is now either $(x-x_0)^{\frac{1}{r}}$ or $(1/x)^{\frac{1}{r}}$. Such an expansion is called a cycle and its order is defined to be the number $r-1$. There are only a finite number of cycles for which $r>1$ and these are called branch cycles. By a transformation $x'=1/(x-x_1)$ we can always bring it about that the cycles for $x=\infty$ have all values $r=1$, and we shall in the following pages always suppose that this has been done unless expressly indicated otherwise.

The cycles (2) are the fundamental elements of the algebraic function. We shall not in this paper make use of the notion of the Riemann surface of the function, but if we were to do so it would be found that to each such cycle there corresponds a unique place on the Riemann surface, and conversely. For this reason we may represent such a cycle by the symbol P, and if desirable, refer to it also as a place.

Consider now an arbitrary rational function $\eta(x,y)$ of the variables x and y. At each cycle P it has an expansion

$$\eta(x,y)=ct^{\nu}+c't^{\nu'}+\cdots \tag{3}$$

---

* See for example Picard, Traité d'Analyse, vol. 2 (1893), chapt. XIII; Appel and Goursat, Theorie des fonctions algébriques, chapt. IV; Hensel and Landsberg, Theorie der algebraischen Funktionen, pp. 39–52.

found by substituting in $\eta(x,y)$ the value $x=x_0+t^r$ (or $x=1/t^r$) and the series (2) for y. The numbers $\nu<\nu'<\cdots$ are positive or negative integers, and the smallest one ν is called the order of the function $\eta(x,y)$ at the cycle P. There are only a finite number of cycles at which ν is different from zero. These are called zeros of $\eta(x,y)$ when ν is positive and poles when ν is negative.

There are two expressions associated with every rational function $\eta(x,y)$ which are important for the developments of the following pages. These are the so-called *norm* and *trace* of η defined by the equations

$$N(\eta)=\eta(x,y_1)\cdots\eta(x,y_n),$$
$$T(\eta)=\eta(x,y_1)+\cdots+\eta(x,y_n),$$

where $y_1,\ldots,y_n$ are the n roots of the equation $f(x,y)=0$ corresponding to the value x. The values $\eta(x,y_i)$ of η corresponding to the different values y_i are called *conjugate values* of η. The norm and the trace are both representable as rational functions of x since they are symmetric in the roots $y_1,\ldots,y_n$ and therefore expressible rationally in terms of the coefficients of the powers of y in the equation $f=0$.

By means of the norm we can prove the well-known and useful theorem that *the sum of the orders of a rational function $\eta(x,y)$ is zero.* The r values of $\eta(x,y)$ corresponding to the r values of y defined by a cycle of the form (2) for a finite value $x=x_0$ are found by substituting in the expansion (3) the r values of the root $t=(x-x_0)^{\frac{1}{r}}$. Since the product of these r values of $\eta(x,y)$ occurs in the norm it follows that $N(\eta)$ has corresponding to each cycle P the factor $(x-x_0)^\nu$, and that its order at $x=x_0$ is the sum of the orders ν of $\eta(x,y)$ at the cycles for $x=x_0$. Similar remarks hold of course for the cycles for which $t=(1/x)^{\frac{1}{r}}$. We know that the sum of the orders of a rational function of x is zero, this being true in particular for $N(\eta)$, and the same result must therefore hold for the function $\eta(x,y)$.

The residue of a function $\eta(x,y)$ at a cycle (2) is defined to be the coefficient of $1/t$ in the expansion for the product $\eta\,dx/dt$, or, what is the same thing, it is the value of the integral

$$\frac{1}{2\pi i}\int_\Gamma \eta\frac{dx}{dt}dt$$

taken around a circle Γ in the t-plane with center at the origin $t=0$. Since when t describes this circle once the value $x=x_0+t^r$ describes r times a circle C about $x=x_0$ in the x-plane it follows that the residue is also expressible as an integral

$$\frac{1}{2\pi i}\int_C\left[\eta(x,y_1)+\cdots+\eta(x,y_r)\right]dx$$

where $y_1,\ldots,y_r$ are the r values of y corresponding to the cycle (2). Evidently the residue at $x=x_0$

$$\frac{1}{2\pi i}\int_C T(\eta)\,dx$$

of the rational functions of x designated by $T(\eta)$ is the sum of the residues of $\eta(x,y)$ at the cycles corresponding to $x=x_0$. Since the sum of the residues of a rational function of x is zero we now have the result that *the sum of the residues of a rational function $\eta(x,y)$ is also zero.*

If a rational function $\eta(x,y)$ has no pole it must be a constant. For the symmetric functions of the conjugate values $\eta(x,y_i)$ are then all rational in x and everywhere finite, and hence by a well known theorem are equal to constants. Consequently these values $\eta(x,y_i)$ are the roots of a polynomial with constant coefficients. It is easy to show on the Riemann surface of the algebraic function that these constants have all the same value, since there is always a path on the surface along which $\eta(x,y)$ varies continuously from the value $\eta(x,y_i)$ to the value $\eta(x,y_k)$.

A final remark concerning the expansions for the trace $T(\eta)$ will be helpful. The sum of the r values defined by the series (3) and corresponding to a cycle (2) has an expansion beginning with the term

$$(1+\omega^{\nu}+\omega^{2\nu}+\cdots+\omega^{(r-1)\nu})ct^{\nu},$$

where ω is a primitive r-th root of unity. This is true because the r values of the root t are exactly the values $\omega^k t$ $(k=0,1,\ldots,r-1)$. The value of the parenthesis is r when ν is an integral multiple of r, since then each term in it is unity, but zero otherwise since the sum of the roots of unity of every index r is zero. *If the exponent ν/r for a particular cycle at $x=x_0$ is smaller than the similar exponents of all the other cycles corresponding to $x=x_0$, then the first term in the expansion of $T(\eta)$ in powers of $(x-x_0)$ will have exactly the exponent ν/r when this quotient is an integer, and a larger exponent when it is a fraction.*

2. Bases for all rational functions. A set of functions $\eta_k(x,y)$ $(k=1,\ldots,n)$ is called a *basis for the totality of rational functions* $\eta(x,y)$ if the determinant $|\eta_k(x,y_i)|$ of their conjugate values is not identically zero.

THEOREM 1. *Every rational function $\eta(x,y)$ is expressible in terms of a basis in the form*

$$\eta=R_1(x)\eta_1+\cdots+R_n(x)\eta_n,$$

where the coefficients $R_k(x)$ are rational in x, and there is only one such expression for η.

To prove this we notice in the first place that a function $\zeta(x,y)$ for which the traces

$$T(\zeta\eta_k)=\zeta(x,y_1)\eta_k(x,y_1)+\cdots+\zeta(x,y_n)\eta_k(x,y_n) \qquad (k=1,\ldots,n)$$

of the products $\zeta\eta_k$ all vanish identically must itself be identically zero, since the determinant $|\eta_k(x,y_i)|$ does not vanish identically. Since the determinant $|T(\eta_i\eta_k)|$ $=|\eta_k(x,y_i)|^2$ is not identically zero the equations

$$T(\eta\eta_i)=R_1T(\eta_i\eta_1)+R_2T(\eta_i\eta_2)+\cdots+R_nT(\eta_i\eta_n) \qquad (i=1,\ldots,n)$$

determine the coefficients $R_1,\ldots,R_n$ uniquely as rational functions of x in such a way that the traces of the products $\zeta\eta_k$ for the function

$$\zeta=\eta-R_1\eta_1-\cdots-R_n\eta_n$$

all vanish identically. Hence ζ is identically zero and η is expressible uniquely as described in the theorem.

It is clear that the determinant $|T(\eta_i\eta_k)|=|\eta_k(x,y_i)|^2$ for a basis is symmetric in $y_1,\ldots,y_n$ and therefore rational in x. Since it is not identically zero, it can have only a finite number of poles and zeros. At all other values of x it is different from zero.

If $\zeta_1,\ldots,\zeta_n$ are expressible in terms of a basis in the form

$$\zeta_k=R_{k1}\eta_1+\cdots+R_{kn}\eta_n, \qquad (k=1,\ldots,n),$$

then the determinant $|\zeta_k(x,y_i)|$ is the product of $|R_{ik}|$ and $|\eta_k(x,y_i)|$ and a necessary and sufficient condition that $\zeta_1,\ldots,\zeta_n$ form a basis is that the determinant $|R_{ik}|$ of the rational functions $R_{ik}(x)$ be different from zero.

A special case of a basis is the set of functions $1,y,\ldots,y^{n-1}$. The determinant of the conjugates of these powers of y is the product of the differences of the roots y_i $(i=1,\ldots,n)$, as is well known, and it can not vanish identically since the roots of an irreducible polynomial $f(x,y)$ are distinct except at special values of x. All other bases are obtainable from this one by linear transformations such as are described in the last paragraph.

3. Divisors and their bases. If $P_1,\ldots,P_s$ are distinct cycles for an algebraic function and $\mu_1,\ldots,\mu_s$ a corresponding set of positive or negative integers, then the expression $Q=P_1^{\mu_1}\cdots P_s^{\mu_s}$ is called a *divisor* and the sum $q=\mu_1+\cdots+\mu_s$ is called the order of the divisor. A rational function $\eta(x,y)$ is a *multiple of the divisor* Q if its order at each cycle P_k is greater than or equal to μ_k, and if its orders at all other cycles of the algebraic function are greater than or equal to zero. The problem of determining the multiples of a divisor and their properties is a fundamental one for the theory of algebraic functions, as we shall see in the following pages. When it has been solved the determination of the Abelian integrals of various types associated with the algebraic function, and the proof of the important Riemann-Roch Theorem, are relatively simple matters.

As a preliminary to the determination of the multiples of a divisor Q we may study those rational functions $\eta(x,y)$ which have the properties of multiples except at the cycles corresponding to $x=\infty$ where no restriction whatever is now placed upon their behavior. A function of this sort is called a *multiple of Q except at infinity*, and the totality of such functions constitute the *ideal of Q* which may be denoted by the symbol $I(Q)$. In the determination of such an ideal only the places of Q which correspond to finite values $x=x_0$ have any effect, and we may without loss of generality suppose that Q contains only such places.

A *basis for the divisor Q* is a basis whose elements $\eta_1,\ldots,\eta_n$ are multiples of Q except at infinity, and which has the further property that the totality of such multiples is identical with the totality of functions $\eta(x,y)$ expressible in the form

$$\eta(x,y)=g_1(x)\eta_1+\cdots+g_n(x)\eta_n, \tag{4}$$

where the coefficients $g_k(x)$ are polynomials.

It is not a priori evident that there will be a basis with these properties for every divisor Q, and one of our first tasks will be to prove that such a basis exists. Before attempting the proof, however, it will be useful to deduce a characteristic property of such a basis. To do this let us consider a finite value x_0 where the algebraic function $y(x)$ has three cycles A, B, C providing, respectively, a, b, c values for $y(x)$, and at which the orders required by Q are λ, μ, ν. The methods to be used would be quite analogous if there were more or fewer than three cycles for $x = x_0$, but the notations would be more complicated.

If a basis $\eta_1, \dots, \eta_n$ has its elements all multiples of Q except at infinity then the expansions for η_k at the three places over x_0 will have the forms

$$\begin{aligned} &\text{at } A: \quad \eta_k = \alpha_{k0} t^{\lambda} + \cdots + \alpha_{ka-1} t^{\lambda+a-1} + \cdots, \\ &\text{at } B: \quad \eta_k = \beta_{k0} t^{\mu} + \cdots + \beta_{kb-1} t^{\mu+b-1} + \cdots, \\ &\text{at } C: \quad \eta_k = \gamma_{k0} t^{\nu} + \cdots + \gamma_{kc-1} t^{\nu+c-1} + \cdots. \end{aligned}$$

The determinant

$$\Delta = \begin{vmatrix} \alpha_{10} & \cdots & \alpha_{n0} \\ \cdots & \cdots & \cdots \\ \alpha_{1a-1} & \cdots & \alpha_{na-1} \\ \beta_{10} & \cdots & \beta_{n0} \\ \cdots & \cdots & \cdots \\ \beta_{1b-1} & \cdots & \beta_{nb-1} \\ \gamma_{10} & \cdots & \gamma_{n0} \\ \cdots & \cdots & \cdots \\ \gamma_{1c-1} & \cdots & \gamma_{nc-1} \end{vmatrix}$$

has numerous applications in the following pages and will be referred to always as the determinant Δ for the basis $\eta_1, \dots, \eta_n$ at the value $x = x_0$. It is clear that an analogous determinant can be constructed for every finite value x_0 no matter how many cycles the algebraic function $y(x)$ may have corresponding to it. We can now prove the following theorem:

THEOREM 2. *A necessary and sufficient condition that a set $\eta_1, \dots, \eta_n$ of multiples of Q except at $x = \infty$ be a basis for the divisor Q is that at every finite value x_0 their determinant Δ be different from zero.*

To prove the necessity of this condition suppose that the determinant Δ is equal to zero at a value x_0 for a set $\eta_1, \dots, \eta_n$ of multiples of Q except at infinity. Then there is a set of constants $C_1, \dots, C_n$ satisfying the linear equations whose coefficients are the rows of Δ, and the numerator of the function

$$r = \frac{C_1\eta_1 + \cdots + C_n\eta_n}{x - x_0}$$

has orders at least equal to $\lambda + a$, $\mu + b$, $\nu + c$ at the cycles A, B, C, while the denominator has orders a, b, c. Hence η is a multiple of Q except at infinity, not expressible in the form (4), and $\eta_1, \dots, \eta_n$ can not be a basis for Q.

The condition is also sufficient. For in the first place a set of functions $\eta_1,\ldots,\eta_n$ having the property of the theorem necessarily has its determinant $|\eta_k(x,y_i)|$ not identically zero, because at a value x_0 having n cycles distinct from those of Q the determinant $|\eta_k(x,y_i)|$ is exactly the determinant Δ and therefore different from zero by hypothesis. Every function $\eta(x,y)$ is therefore expressible in terms of $\eta_1,\ldots,\eta_n$ in the form

$$\eta=\frac{g_1(x)\eta_1+\cdots+g_n(x)\eta_n}{d(x)}$$

where $g_1(x),\ldots,g_n(x),d(x)$ are polynomials having no common factor. When a function η is a multiple of Q except at $x=\infty$ the denominator $d(x)$ must be a constant. For one may readily verify successively that in case $d(x)$ had a factor $(x-x_0)$ the two functions

$$\frac{g_1(x)\eta_1+\cdots+g_n(x)\eta_n}{x-x_0},\quad \frac{g_1(x_0)\eta_1+\cdots+g_n(x_0)\eta_n}{x-x_0}$$

would also be multiples of Q except at infinity. Since the determinant Δ is different from zero at $x=x_0$ the last one would surely have a lower order than that prescribed by Q at one at least of the places over $x=x_0$, which is a contradiction. Hence the denominator $d(x)$ has no factor $x-x_0$ and is constant. Every multiple $\eta(x,y)$ of Q except at infinity is therefore surely expressible in the form (4).

With the help of the last theorem we may proceed to the proof that there is a basis for every divisor Q. In the first place it is evident that a function $\eta(x,y)$ which is not a multiple of Q except at infinity can be made into one by multiplying it by a polynomial in x. For if $\eta(x,y)$ is multiplied by a sufficiently high power of $x-x_0$ the orders of the product at the cycles corresponding to x_0 may be made to exceed those required by Q. It is clear from this remark that a basis $\eta_1,\ldots,\eta_n$ can always be easily made over into one whose functions are all multiples of Q except at infinity.

If a basis $\eta_1,\ldots,\eta_n$ has its functions all multiples of Q except at infinity then the order of the determinant $|\eta_k(x,y_i)|$ at a value x_0 is surely greater than the sum of the orders required by Q at the cycles corresponding to x_0. For if at such a cycle Q requires the expansion of its multiples to begin with a term in $(x-x_0)^{\frac{\mu}{r}}$, then r rows of the determinant $|\eta_k(x,y_i)|$ will have a factor $(x-x_0)^{\frac{\mu}{r}}$ in each element and the determinant itself will have at least the factor $(x-x_0)^{\mu}$. The theorem which we wish to prove with the help of these remarks is now the following one:

THEOREM 3. *For every divisor Q there exists a basis $\eta_1,\ldots,\eta_n$ such that the totality of multiples of Q except at infinity is identical with the totality of functions expressible in the form*

$$\eta(x,y)=g_1(x)\eta_1+\cdots+g_n(x)\eta_n$$

where the coefficients $g_k(x)$ are polynomials in x.

To prove this, suppose that $\eta_1,\ldots,\eta_n$ is a basis of functions which are multiples of Q except at infinity. If it is not a basis for Q, there will be a value x_0 at which its determinant Δ vanishes. Let $C_1,\ldots,C_n$ be constants satisfying the linear equations whose coefficients are the rows of Δ, and suppose that C_k is one of them which is different from zero. Then the set of functions

$$\eta_1,\ldots,\eta_{k-1},\quad \frac{C_1\eta_1+\cdots+C_n\eta_n}{x-x_0},\quad \eta_{k+1},\ldots,\eta_n$$

is also a basis with elements multiples of Q except at infinity, and the determinant of its conjugates is that of the original basis multiplied by $C_k/(x-x_0)$. If the determinant Δ of the new basis still vanishes at x_0, this process may be repeated. It can be repeated a finite number of times only, however, before reaching a basis for which the determinant Δ is different from zero, since after each repetition the order of the determinant $\eta_k(x,y_i)$ at $x=x_0$ is decreased by unity, and we have seen above that for a basis of multiples of Q except at infinity the order of this determinant at $x=x_0$ has a minimum. If the determinant Δ has been made different from zero, as just described, at all the values x_0 corresponding to places in Q, and at all the places where the determinant of the conjugates of the basis originally vanished, then the basis will have $\Delta\neq 0$ at every x_0 and will be a basis for Q. We know that there are only a finite number of values x_0 at which alterations must be made since Q has only a finite number of factors and the determinant of conjugates only a finite number of zeros.

4. Multiples of a divisor. We have seen in §2 that a basis $\eta_1,\ldots,\eta_n$ for the totality of rational functions $\eta(x,y)$ can be transformed into an equivalent basis $\zeta_1,\ldots,\zeta_n$ by a linear transformation of the form

$$\zeta_k - g_{k1}\eta_1+\cdots+g_{kn}\eta_n \qquad (k=1,\ldots,n) \tag{5}$$

in which the coefficients g_{kl} are rational in x and have a determinant not identically zero. A similar relationship can be established for every pair of basis for a divisor Q, as indicated in the following theorem:

THEOREM 4. *If $\eta_1,\ldots,\eta_n$ is a basis for a divisor Q then a necessary and sufficient condition for $\zeta_1,\ldots,\zeta_n$ to be also such a basis is that $\zeta_1,\ldots,\zeta_n$ be expressible in the form* (5) *with coefficients $g_{kl}(x)$ polynomials in x and with a determinant $|g_{kl}|$ equal to a constant different from zero.*

We know that if $\zeta_1,\ldots,\zeta_n$ is to be a basis for Q its functions must be multiples of Q except at $x=\infty$ and hence uniquely expressible in terms of $\eta_1,\ldots,\eta_n$ with polynomial coefficients g_{kl}, since every such multiple is so expressible. Similarly $\eta_1,\ldots,\eta_n$ must be expressible in terms of $\zeta_1,\ldots,\zeta_n$ with polynomial coefficients h_{kl}. The determinants $|g_{kl}|$ and $|h_{kl}|$ are both polynomials in x and their product is unity. Hence both must be constants. Conversely one may easily see that a set of functions $\zeta_1,\ldots,\zeta_n$ related to $\eta_1,\ldots,\eta_n$ as described in the theorem will surely be a basis for Q, so that the theorem is completely established.

By means of the transformations described in the last theorem bases for Q can be found which have special properties of great assistance in the proofs which we shall discuss in this and the following sections. We may define the *column order* of a function $\eta(x,y)$ at $x=\infty$ as the minimum of the exponents in its n expansions at the cycles for $x=\infty$. If the column orders of the functions of a basis $\eta_1,\ldots,\eta_n$ for Q are the numbers $r_k(k=1,\ldots,n)$, then the conjugates of these functions will have at $x=\infty$ expansions of the form

$$\eta_k(x,y_i)=C_{ik}\left(\frac{1}{x}\right)^{r_k}+\cdots,\qquad (i,k=1,\ldots,n),\tag{6}$$

where one at least of each set $C_{1k},\ldots,C_{nk}$ is different from zero. A basis for Q is said to be *normal at* $x=\infty$ if the determinant $|C_{ik}|$ is different from zero. It is evident that for such a basis the order of the determinant $|\eta_k(x,y_i)|$ at $x=\infty$ is exactly $r_1+\cdots+r_n$.

THEOREM 5. *For every divisor Q there exists a basis $\eta_1,\ldots,\eta_n$ which is normal at $x=\infty$. When expressed in terms of such a basis a multiple*

$$\eta(x,y)=g_1(x)\eta_1+\cdots+g_n(x)\eta_n\tag{7}$$

of Q except at $x=\infty$ has its column order at $x=\infty$ the smallest of the numbers $r_k-\mu_k(k=1,\ldots,n)$ where r_k is the column order of η_k and μ_k the degree of the polynomial $g_k(x)$.

To establish these statements let us order the functions $\eta_1,\ldots,\eta_n$ of an arbitrarily selected basis for Q so that their column orders satisfy the inequalities $r_1\geqq r_2\geqq\cdots\geqq r_n$. If the determinant $|C_{ik}|$ of coefficients from the expansions (6) is equal to zero there will exist constants $C_1,\ldots,C_n$ not all zero satisfying the linear equations whose coefficients are the rows of this determinant. Let C_k be the last one which is different from zero. Then the basis $\eta_1,\ldots,\eta_{k-1},\eta_k',\eta_{k+1},\ldots,\eta_n$, with

$$\eta_k'=C_1x^{r_1-r_k}\eta_1+\cdots+C_{k-1}x^{r_{k-1}-r_k}\eta_{k-1}+C_k\eta_k,$$

is also a basis for Q and has the same column orders except that the order r_k' for η_k' is at least one greater than r_k. If the new basis is not normal at $x=\infty$ the process can be repeated. It can be repeated only a finite number of times before attaining a normal basis, however, since at each step the sum of the column orders of the basis is increased by at least unity, and this sum is at most equal to the order of the determinant of the conjugates of the basis, which is unchanged by the transformation.

Let μ be the smallest of the numbers $r_k-\mu_k$ so that $r_k-\mu_k\geqq\mu(k=1,\ldots,n)$. Then the coefficient $g_k(x)$ has degree $\mu_k\leqq r_k-\mu$, and has at $x=\infty$ an expansion of the form

$$g_k(x)=C_k\left(\frac{1}{x}\right)^{\mu-r_k}+\cdots.$$

At least one of the coefficients C_k is different from zero. The expansions of the function (7) at the cycles for $x=\infty$ have therefore the form

$$\eta(x,y_i)=\left(\frac{1}{x}\right)^{\mu}\sum_k C_{ik}C_k+\cdots\qquad (i=1,\ldots,n),$$

and one at least of the coefficients of $(1/x)^{\mu}$ is necessarily different from zero since the determinant $|C_{ik}|$ is not zero. The column order of η is therefore μ.

So far we have considered only "multiples of a divisor Q except at $x=\infty$," whose usefulness is of an auxiliary sort. The multiples which are of greater importance are those which have orders greater than or equal to the orders prescribed by Q at every cycle whatsoever of the algebraic function $y(x)$, including those at $x=\infty$. It is evident that some divisors will have no multiples, an example being a divisor Q which has positive but no negative exponents. A multiple $\eta(x,y)$ of such a divisor would necessarily have all of its orders greater than or equal to zero, and the sum of the orders of η could not be equal to zero as we know it must be. The following theorem describes the character and the number of the multiples of a divisor Q.

THEOREM 6. *If a divisor Q has a multiple, it has a set $\sigma_1(x,y),\ldots,\sigma_\nu(x,y)$ of linearly independent ones such that every multiple of Q is expressible in the form*

$$\sigma=c_1\sigma_1+\cdots+c_\nu\sigma_\nu$$

with constant coefficients. If Q has no cycles at infinity and if $\eta_1,\ldots,\eta_n$ is a basis for Q normal at $x=\infty$ with column orders

$$r_1 \geqq r_2 \geqq \cdots \geqq r_s \geqq 0 > r_{s+1} \geqq \cdots \geqq r_n, \tag{8}$$

then the number ν of such multiples is

$$\nu=(r_1+1)+\cdots+(r_s+1).$$

To prove the theorem we suppose that a transformation $x'=1/(x-x_0)$ has been applied so that none of the cycles of Q are at infinity. When $\eta_1,\ldots,\eta_n$ is a basis with the properties presupposed in the theorem the function

$$\eta=g_1(x)\eta_1+\cdots+g_n(x)\eta_n$$

will be a multiple of Q at all cycles, including those at $x=\infty$, if and only if its column order at $x=\infty$ is greater than or equal to zero. According to the last theorem this can never be so if all of the column orders r_k are negative, since then all the numbers $r_k-\mu_k$ are negative, and in this case the divisor Q has no multiples. When some of the column orders are positive, as indicated in the arrangement (8), the function η will have its column order greater than or equal to zero at $x=\infty$, according to Theorem 5, if and only if it is expressible in the form

$$\eta=g_1(x)\eta_1+\cdots+g_s(x)\eta_s \tag{9}$$

with $r_k-\mu_k \geqq 0$ for each coefficient $g_k(x)$ $(k=1,\ldots,s)$. It is clear from this remark that the degree μ_k of each $g_k(x)$ can be r_k but no greater. The functions

$$\eta_1, x\eta_1,\ldots,x^{r_1}\eta_1,\ldots,\eta_s, x\eta_s,\ldots,x^{r_s}\eta_s$$

therefore constitute a set of multiples of Q in terms of which all such multiples are expressible linearly with constant coefficients, and their number ν is that given in the theorem. They are linearly independent since no linear expression of the form

(9) can vanish identically when the determinant $|\eta_k(x,y_i)|$ is different from zero. This completes the proof of the theorem. We can infer in a similar manner the truth of the following useful corollary.

COROLLARY. *Let D be the divisor which is the product of the cycles at $x=\infty$, and let Q be a divisor having no cycles at $x=\infty$. If $\eta_1,\ldots,\eta_n$ is a basis for Q normal at $x=\infty$ and having there the column orders indicated in the arrangement* (8) *then the number of multiples of the divisor DQ is $\nu=r_1+\cdots+r_s$. Furthermore if*

$$r_1 \geqq r_2 \geqq \cdots \geqq r_t \geqq 2 > r_{t+1} \geqq \cdots \geqq r_n,$$

then the number of multiples of D^2Q is

$$\nu=(r_1-1)+\cdots+(r_t-1). \tag{10}$$

The proof of the first part of the corollary is like that of the theorem except that the degree of each $g_k(x)$ in the expression (9) can now not exceeed r_k-1 if we wish $\eta(x,y)$ to have a zero of order one at least at each cycle for $x=\infty$.

Similarly the multiples of D^2Q are the multiples of Q which have zeros of order two at least at the cycles for $x=\infty$. By an argument similar to the one just made it follows that these multiples are the functions

$$\eta=g_1(x)\eta_1+\cdots+g_t(x)\eta_t$$

for which each polynomial $g_k(x)$ $(k=1,\ldots,t)$ has degree at most equal to r_k-2. Hence their number is the number ν of the corollary.

It is not always easy to compute the number of linearly independent multiples of a divisor from the criteria given in the theorem above and its corollary, but we shall see that in a number of important cases this computation can be readily made with the help of the next theorem. Let the values of x corresponding to the cycles of a divisor $Q=P_1^{\mu_1}\cdots P_s^{\mu_s}$ be the finite values $x_1,\ldots,x_s$. Then the *ideal norm of the divisor Q* is defined to be the product

$$N(Q)=(x-x_1)^{\mu_1}\cdots(x-x_s)^{\mu_s}.$$

If a divisor has cycles at $x=\infty$ they are neglected in forming the ideal norm, but otherwise the definition is the same.

We have seen in § 1 that there are only a finite number of branch cycles for the algebraic function $y(x)$. The *divisor of the branch cycles* is defined to be the divisor $X=\prod P^{r-1}$, where P is a branch cycle and r the number of roots y_i of the equation $f(x,y)=0$ furnished by it, and the product is taken for all of the branch cycles. With the help of these notations the theorem which will now be proved is as follows:

THEOREM 7. *For every basis $\eta_1,\ldots,\eta_n$ of a divisor Q we have*

$$|\eta_k(x,y_i)|^2=cN(Q)^2N(X)$$

where c is a constant factor.

It is clear that the power of $x-x_0$ which occurs in the determinant $|\eta_k(x,y_i)|$ is the same for all bases of Q since for two equivalent bases the values of this determinant differ only by a constant factor (Theorem 4). Let us suppose therefore that the basis $\eta_1,\ldots,\eta_k$ has already been prepared by a linear transformation (5) with constant coefficients g_{kl} so that the determinant Δ for $\eta_1,\ldots,\eta_k$ at $x=x_0$ is the identity determinant. For the illustrative case used above, in which there are three cycles corresponding to $x=x_0$, the following table indicates the exponents of the lowest powers of $(x-x_0)$ in the expansions of the elements of the determinant $\eta_k(x,y_i)$ at the three cycles.

	η_1	η_2	$\cdots$	η_a	$\eta_{a+1}\cdots\eta_{a+b}$	$\eta_{a+b+1}\cdots\eta_n$
$\eta(x,y_1)$	$\frac{\lambda}{a}$	$\frac{\lambda+1}{a}$	$\cdots$	$\frac{\lambda+a-1}{a}$		
$\cdots$	$\cdots$	$\cdots$	$\cdots$	$\cdots$	$\geqq\frac{\lambda}{a}+1$	$\geqq\frac{\lambda}{a}+1$
$\eta(x,y_a)$	$\frac{\lambda}{a}$	$\frac{\lambda+1}{a}$	$\cdots$	$\frac{\lambda+a-1}{a}$		
$\eta(x,y_{a+1})$					$\frac{\mu}{b}\ \cdots\ \frac{\mu+b-1}{b}$	
$\cdots$			$\geqq\frac{\mu}{b}+1$		$\cdots\ \cdots\ \cdots$	$\geqq\frac{\mu}{b}+1$
$\eta(x,y_{a+b})$					$\frac{\mu}{b}\ \cdots\ \frac{\mu+b-1}{b}$	
$\eta(x,y_{a+b+1})$						$\frac{\nu}{c}\ \cdots\ \frac{\nu+c-1}{c}$
$\cdots$			$\geqq\frac{\nu}{c}+1$		$\geqq\frac{\nu}{c}+1$	$\cdots\ \cdots\ \cdots$
$\eta(x,y_n)$						$\frac{\nu}{c}\ \cdots\ \frac{\nu+c-1}{c}$

Out of each of the first a rows we may take the factor $(x-x_0)^{\frac{\lambda}{a}}$, and out of the $a-1$ columns following the first the factor $(x-x_0)$ raised to the power $1/a+\cdots+(a-1)/a$. Hence in all we have from these rows and columns the factor $(x-x_0)$ raised to the power $\lambda+(a-1)/2$. A similar process applied to the two remaining principal minors indicated in the diagram gives for the determinant $|\eta_k(x,y_i)|^2$ the factor $(x-x_0)$ raised to the power

$$2\lambda+2\mu+2\nu+(a-1)+(b-1)+(c-1),$$

which is exactly the power of $(x-x_0)$ occurring in the product $N(Q)^2N(X)$. The same method of proof applies when there are more or fewer than three cycles at $x=x_0$.

If we can show that, after the power of $(x-x_0)$ described above has been removed, the constant term in the expansion for $|\eta_k(x,y_i)|$ is different from zero, we shall have proved our theorem, for then the zeros and poles in the finite x-plane of the two rational functions $|\eta_k(x,y_i)|^2$ and $N(Q)^2N(X)$ are identical, and these two can differ only by a constant factor. When we set $x=x_0$ in the determinant Δ deprived of the factors $(x-x_0)$ as described, the three principal minors in the squares indicated in the diagram are the only ones which remain. The first one of these, for example, is

$$\begin{vmatrix} 1 & 1 & \cdots & 1 \\ \omega^{\lambda} & \omega^{\lambda+1} & \cdots & \omega^{\lambda+a-1} \\ \cdots & \cdots & \cdots & \cdots \\ \omega^{(a-1)\lambda} & \omega^{(a-1)(\lambda+1)} & \cdots & \omega^{(a-1)(\lambda+a-1)} \end{vmatrix}$$

where ω is a primitive a-th root of unity, and it is different from zero since no two of the roots $\omega^{\lambda+k}$ $(k=0,1,\ldots,a-1)$ are equal. A similar argument applies to the other minors, and the theorem is therefore proved.

As an immediate consequence of the preceding theorems we have

COROLLARY 1. *For every basis $\eta_1,\ldots,\eta_n$ of a divisor Q normal at $x=\infty$ the equation*

$$r_1+\cdots+r_n+q+\frac{w}{2}=0 \tag{11}$$

is true, where the integers r_k are the column orders of the basis at $x=\infty$, q is the order of the divisor Q, and w the order of the divisor X of the branch places.

To prove the corollary we note first that the order at $x=\infty$ of the determinant $|\eta_k(x,y_i)|^2$ for a basis normal at infinity is $2(r_1+\cdots+r_n)$, as we have seen above on page 12. According to the last theorem the sum of the orders of $|\eta_k(x,y_i)|^2$ at finite values of x is the same as that of the product $N(Q)^2N(X)$ which we know to be $2q+w$. The equation (11) then expresses the known fact that the sum of all the orders of the rational function $|\eta_k(x,y_i)|^2$ is zero.

COROLLARY 2. *The number*

$$p=\frac{w}{2}-n+1,$$

which is called the genus of the algebraic function defined by the equation $f(x,y)=0$, is always a positive integer or zero.

It is evident from the equation (11) that $w/2$ is an integer and hence that the number p in the last corollary is an integer. To prove that it is not negative consider a basis $\eta_1,\ldots,\eta_n$ normal at $x=\infty$ for the particular divisor $Q=1$. This basis must have all of its column orders $r_1,\ldots,r_n$ at $x=\infty$ zero or negative since no rational function can have positive orders at all cycles at $x=\infty$ and no negative ones elsewhere. One at least of the column orders must be zero since otherwise the column order of the function

$$\eta=g_1(x)\eta_1+\cdots+g_n(x)\eta_n$$

would always be negative (Theorem 5), and this is impossible since the function $\eta(x,y)=$constant is certainly a multiple of the divisor $Q=1$. On the other hand, two of the numbers $r_1,\ldots,r_n$ could not be zero since then both of the corresponding functions of the bases would have no singularities and would be constants, and they would not be linearly independent. Since now $r_1=0$ while all other integers $r_2,\ldots,r_n$ are negative, it follows with the help of equation (11), since the order of the divisor $Q=1$ is $q=0$, that

$$p=\frac{w}{2}-n+1=-r_1-\cdots-r_n-n+1\geqq 0.$$

5. Complementary bases. A basis $\zeta_1,\ldots,\zeta_n$ is said to be *complementary* to the basis $\eta_1,\ldots,\eta_n$ if the traces of the products of the functions composing the two bases satisfy the relations

$$T(\eta_k\zeta_k)=1,\quad T(\eta_k\zeta_l)=0\qquad(k\neq l).\tag{12}$$

We have seen in §2 that the coefficients R_k for the function

$$\zeta=R_1\eta_1+\cdots+R_n\eta_n$$

will be uniquely determined when the traces

$$T(\eta_i\zeta)=R_1T(\eta_i\eta_1)+\cdots+R_nT(\eta_i\eta_n)\qquad(i=1,\ldots,n)$$

are assigned. It follows readily that the functions ζ_k of a basis complementary to $\eta_1,\ldots,\eta_n$ are uniquely determined, and that the relation between the two bases is a reciprocal one.

THEOREM 8. *If a basis $\eta_1,\ldots,\eta_n$ is normal at $x=\infty$ with column orders $r_1,\ldots,r_n$, then its complementary basis $\zeta_1,\ldots,\zeta_n$ is also normal at $x=\infty$ and has the column orders $-r_1,\ldots,-r_n$.*

Let us denote the column orders which are to be determined for the basis $\zeta_1,\ldots,\zeta_n$, by $s_1,\ldots,s_n$, and let d_{ik} be the matrix of coefficients for this basis corresponding to the matrix C_{ik} in equation (6) for the original basis. Then at $x=\infty$ we have the expansion

$$T(\eta_i\zeta_k)=\left(\frac{1}{x}\right)^{r_i+s_k}\sum_{j=1}^{n}C_{ji}d_{jk}+\cdots.$$

These expansions must vanish identically when $i\neq k$ and be identically equal to 1 when $i=k$, on account of the relations (12). Since the determinant $|C_{ik}|$ is different from zero, and since the constants $d_{jk}(j=1,\ldots,n)$ for a fixed k are not all zero, it follows that the relations

$$\sum_{j=1}^{n}C_{ji}d_{jk}=0\quad(i\neq k),\qquad\sum_{j=1}^{n}C_{jk}d_{jk}=1,\quad r_k+s_k=0$$

must hold. We see that $s_k=-r_k$, and that the determinant $|d_{jk}|$ is different from zero since it is, except for interchange of rows and columns, the reciprocal of $|C_{ik}|$.

THEOREM 9. *If $\eta_1,\ldots,\eta_n$ is a basis for a divisor Q, then its complementary basis $\zeta_1,\ldots,\zeta_n$ is a basis for the divisor R defined by the equation $QRX=1$, in which X is the divisor of the branch places.*

To prove this consider again the illustrative case of a value $x=x_0$ having three cycles A, B, C providing, respectively, a, b, c roots of $f(x,y)=0$. The argument to be made would be quite similar if there were more or fewer than three places. At the three cycles A, B, C, respectively, let λ, μ, ν be the orders of Q and λ', μ', ν' the minima of the orders of the functions of the complementary basis $\zeta_1, \ldots, \zeta_n$. We suppose the notations for the cycles A, B, C chosen so that

$$(\lambda+\lambda')/a \leqq (\mu+\mu')/b \leqq (\nu+\nu')/c. \tag{13}$$

Consider now two functions

$$\eta = u_1\eta_1 + \cdots + u_n\eta_n, \qquad \zeta = v_1\zeta_1 + \cdots + v_n\zeta_n,$$

for which the coefficients u_i, v_i are constants. Since the determinant Δ on page 9 for the basis $\eta_1, \ldots, \eta_n$ is different from zero at the value x_0 it follows that when a number h of the set $0, 1, \ldots, a-1$ has been selected arbitrarily, the coefficients u_k can be determined so that the function η has at A, B, C, respectively, the expansions

$$\begin{aligned} r &= (x-x_0)^{\frac{\lambda+h}{a}} + \alpha(x-x_0)^{\frac{\lambda}{a}+1} + \cdots, \\ \eta &= \qquad\qquad\quad \beta(x-x_0)^{\frac{\mu}{b}+1} + \cdots, \\ \eta &= \qquad\qquad\quad \gamma(x-x_0)^{\frac{\nu}{c}+1} + \cdots. \end{aligned}$$

The coefficient v_k can then be in turn selected so that ζ has orders exactly λ', μ', ν' at A, B, C, respectively, and so that

$$T(\eta\zeta) = u_1v_1 + \cdots + u_nv_n \neq 0. \tag{14}$$

When the coefficients u_k, v_k have been chosen in this manner, the smallest exponent of $(x-x_0)$ in the expansions of the terms of $T(\eta\zeta)$ is seen with the help of the inequalities (13) to be $(\lambda+\lambda'+h)/a$. But the trace $T(\eta\zeta)$ is a constant different from zero, so that according to the remarks in the last paragraph of §1 we must have the exponent $(\lambda+\lambda'+h)/a$ less than zero when it is a fraction and equal to zero when it is an integer. Since there is one integer only in the set of numbers $(\lambda+\lambda'+h)/a$ for $h=0, 1, \ldots, a-1$ it follows that the largest number of this set must be zero, and therefore that the first of the relations

$$\lambda+\lambda'+a-1=0, \quad \mu+\mu'+b-1=0, \quad \nu+\nu'+c-1=0, \tag{15}$$

is true. A similar argument for the cycles B, C justifies the last two if at each step we use the relations (15) already found, and (13), and note that according to the last paragraph of §1 the smallest exponent in the terms of $T(\eta\zeta)$ must be $\leqq 0$. From the equations (15) we see therefore that the minimum orders λ', μ', ν' for the basis $\zeta_1, \ldots, \zeta_n$ at the cycles A, B, C for $x=x_0$ are exactly the orders $-\lambda-a+1, -\mu-b+1, -\nu-c+1$ prescribed by the divisor $R=1/QX$.

We can prove that at the value x_0 the determinant Δ on page 9 formed with respect to the divisor R for the basis $\zeta_1, \ldots, \zeta_n$ can not be zero. For if it had the value zero, constants v_k, not all zero, could be selected so that at A, B, C the function ζ has, respectively, the expansions

$$\zeta=\alpha'(x-x_0)^{\frac{\lambda'}{a}+1}+\cdots,$$
$$\zeta=\beta'(x-x_0)^{\frac{\mu'}{b}+1}+\cdots,$$
$$\zeta=\gamma'(x-x_0)^{\frac{\nu'}{c}+1}+\cdots.$$

Since for every choice of the constants u_k the orders of η at the cycles A, B, C are at least λ,μ,ν, the exponents of the terms in $T(\eta\zeta)$ would then all be greater than zero, on account of the relations (15). The trace (14) would be zero for every choice of the constants u_k, which is impossible.

Since the above reasoning applies at every finite value x_0 it follows that $\zeta_1,\ldots,\zeta_n$ is a basis for the divisor R, as stated in the theorem.

As a first application of the properties of the complementary basis we may establish formulas for the number of multiples of divisors of the form $D^2/(P_1\cdots P_\mu X)$ which will be of service in a later section. Let $\eta_1,\ldots,\eta_n$ be a basis for the divisor $Q=1/(P_1\cdots P_\mu X)$, normal at $x=\infty$ and with column orders

$$r_1\geqq r_2\geqq\cdots\geqq r_t\geqq 2>r_{t+1}\geqq\cdots\geqq r_n. \tag{16}$$

Since the order of Q is $q=-(\mu+w)$, the formula (11) of page 16 gives the relation

$$r_1+\cdots+r_n-(\mu+w)+\frac{w}{2}=0. \tag{17}$$

The basis $\zeta_1,\ldots,\zeta_n$ complementary to $\eta_1,\ldots,\eta_n$ is a basis for the divisor $R=1/XQ=P_1\cdots P_\mu$ and has at $x=\infty$ the column orders $-r_1,\ldots,-r_n$. Each of the functions ζ_k must have a negative column order at $x=\infty$ since each has zeros at the cycles $P_1,\ldots,P_\mu$ and must therefore have some poles at infinite places which are the only places where such poles are possible. The numbers r_k are therefore all positive, and those following r_t in the arrangement (16) are unity. Formula (17) with the equation (10) of page 14 therefore give

$$\nu=(r_1-1)+\cdots+(r_n-1)=\frac{w}{2}-n+\mu=p+\mu-1$$

as the number of multiples of $D^2/(P_1\cdots P_\mu X)$. When no cycles $P_1,\ldots,P_\mu$ are present, we have $R=1$ and it follows from the argument on page 16 that $r_n=0$, while the other column orders are positive. Hence we now have, from formulas (10) and (17) with $\mu=0$,

$$\nu=(r_1-1)+\cdots+(r_n-1)+1=\frac{w}{2}-n+1=p$$

as the number of multiples of D^2/X. This proves the following theorem:

THEOREM 10. *Let D be the divisor whose factors are the cycles at $x=\infty$, X the divisor of the branch cycles, and $P_1,\ldots,P_\mu$ arbitrarily chosen cycles. Then the number of linearly independent multiples of the divisor D^2/X is exactly $\nu=p$, and the number for $D^2/(P_1\cdots P_\mu X)$ ($\mu\geqq 1$) is*

$$\nu=p+\mu-1.$$

6. The invariant property of the genus number. If the poles and zeros of a rational function $\eta(x,y)$ are at the cycles $P_1,\ldots,P_s$ and have the orders $\mu_1,\ldots,\mu_s$, then the divisor $Q_\eta = P_1^{\mu_1},\ldots,P_s^{\mu_s}$ is called *the divisor of* η. Since the sum of the orders of a rational function is always zero, it follows that the order of the divisor Q_η must be zero. Conversely one might expect that there would be a rational function corresponding in this way to every divisor Q of order zero, but this is not always the case. It is true in the case of functions of a single variable x which may be regarded as rational functions of the algebraic function defined by the equation $y-x=0$. For we can easily construct out of factors of the type $|x-x_0|^\lambda$ a rational function of x with arbitrarily prescribed poles and zeros in the complex x-plane, provided only that the sum of the orders of these poles and zeros is zero. But for rational functions $\eta(x,y)$ of a more general algebraic function no such simple construction is possible, and it may be that for some divisors of order zero no corresponding rational function exists.

Some of the factors in Q_η have positive exponents and some negative. We may agree to denote the product of those with positive exponents by N_η, and may express the divisor in the form $Q_\eta = N_\eta / D_\eta$, where the meaning of D_η is evident. At every cycle P of a factor P^a of the denominator D_η the expansions (3) of page 5 for the algebraic function and its derivative have the forms

$$\eta = \frac{\beta}{t^a} + \cdots, \quad \frac{d\eta}{dt} = -a\beta \frac{t^{a-1}}{t^{2a}} + \cdots, \tag{18}$$

where β is a constant different from zero. At other cycles these expansions have the form

$$\eta = \alpha + \beta t^a + \cdots, \quad \frac{d\eta}{dt} = a\beta t^{a-1} + \cdots, \tag{19}$$

where β is again different from zero and α a constant which vanishes at the factors of N_η but not elsewhere. A cycle P of either type is called a *branch cycle for* $\eta(x,y)$ if $a \neq 1$. The *divisor of the branch cycles for* η is defined to be $X_\eta = \prod P^{a-1}$, where the product is taken for all the branch cycles P of η. We may for convenience denote the orders of X_η and D_η by w_η and n_η, respectively. For the function $\eta = x$, the notations which have just been introduced give the special cases X_x, D_x, w_x, n_x which have been denoted in the preceding pages by X, D, w, n.

THEOREM 11. *The divisor whose orders are identical with those of the derivative $d\eta/dt$ at the various cycles of an algebraic function is X_η / D_η^2. The order $w_\eta - 2n_\eta$ of this divisor is the same for all rational functions η.*

The first part of the theorem is evident after an examination of the exponents in the expressions (18) and (19) for $d\eta/dt$. To prove the second part we consider a second rational function $\xi(x,y)$ and notice that, by elimination of x and y from the three equations

$$\xi = \xi(x,y), \quad \eta = \eta(x,y), \quad f(x,y) = 0,$$

a relation $\varphi(\xi,\eta)=0$ can always be found in which $\varphi(\xi,\eta)$ is a polynomial. Since all of the expansions for ξ and η satisfy this equation identically it follows that

$$\frac{d\eta}{dt}=-\left(\frac{\partial\varphi}{\partial\xi}\Big/\frac{\partial\varphi}{\partial\eta}\right)\frac{d\xi}{dt},$$

which shows that at every cycle the order of $d\eta/dt$ is the sum of the order of a rational function of x,y and the order of $d\xi/dt$. The divisor of $d\eta/dt$ is therefore the product of the divisor of $d\xi/dt$ by that of a rational function, and the orders of the divisors of $d\eta/dt$ and $d\xi/dt$ are therefore always equal.

As a result of the invariance of the expression $w_\eta-2n_\eta$, we see that the genus of our algebraic function is also expressible in the form

$$p=\tfrac{1}{2}w_\eta-n_\eta+1,$$

since, according to Theorem 11, this number will have the same value no matter what rational function $\eta(x,y)$ is used in its computation. From Corollary 2 of page 16 we know that p is always a positive integer or zero and hence that $w_\eta/2$ is always an integer.

7. Construction of elementary integrals. Integrals of the form

$$I=\int\eta(x,y)dx,$$

where $\eta(x,y)$ is a rational function of x and y, are the so-called Abelian integrals associated with the algebraic equation $f(x,y)=0$. They play a most important and interesting role in the theory of the algebraic functions defined by such equations. If we write such an integral in the form

$$I=\int\eta(x,y)\frac{dx}{dt}dt$$

and substitute the expansions of η and dx/dt, it is evident that at each cycle the value of the integral will be expressible in terms of t by means of a series of the form

$$\frac{A_{-p}}{t^p}+\cdots+\frac{A_{-1}}{t}+A\log t+A_0+A_1t+\cdots. \tag{20}$$

If the series has no logarithmic term or terms with negative exponents, the cycle is called an ordinary cycle for the integral; if negative terms are present but no logarithmic term, it is a pole; if a logarithmic term is present, it is a logarithmic singularity. Integrals which have no singularities are called integrals of the first kind; those which have poles but no logarithmic singularities are of the second kind; while those with logarithmic singularities are of the third kind.

All integrals of the first kind may be thought of as elementary integrals, but an integral of the second kind is called an elementary integral only when it has no singularity except a single pole with an expansion of the form

$$\frac{1}{t^\mu}+A_0+A_1t+\cdots.$$

An integral of the third kind is an elementary integral if it has only two singularities at which its expansions are

$$\log t + A_0 + A_1 t + \cdots, \quad -\log t + B_0 + B_1 t + \cdots.$$

The coefficients A in the expansion (20) are the residues of the function $\eta(x,y)$ and we have found on page 7 that for every such function the sum of the residues is zero. We see then that the elementary integral of the third kind is as simple as one could hope to find, since every such integral must have at least two logarithmic singularities.

We shall see presently that every Abelian integral whatsoever is expressible as a sum of elementary integrals of the first, second, and third kinds multiplied by constant coefficients, but for this result to have significance we must be certain that elementary integrals of three kinds actually exist. For integrals I of the first kind it is evident that at every cycle the expansion for $\eta\, dx/dt$ must be without terms in negative powers of t, and the sum of the orders of η and dx/dt must therefore be positive or zero. This means that η is a multiple of the reciprocal of the divisor of dx/dt, and according to Theorem 11 of page 20 this reciprocal is D^2/X. From Theorem 10 of page 19 the number of linearly independent multiples of this divisor is p, and we have the following theorem.

THEOREM 12. *The number of linearly independent integrals of the first kind is exactly the genus p. If $I_1,\ldots,I_p$ are such integrals of the first kind then every other integral of the first kind is expressible uniquely with constant coefficients in the form*

$$I = c_1 I_1 + \cdots + c_p I_p + c.$$

For an elementary integral of the second kind with no singularity except a simple pole at a cycle P the product $\eta\, dx/dt$ can have no singularity except a pole of order two at P. It follows readily that the number of linearly independent functions η for which the product $\eta\, dx/dt$ has these properties is the same as the number of multiples of the divisor D^2/P^2X. According to Theorem 10 of page 19 this number is $\nu = p+1$. One at least of the multiples η must give the product $\eta\, dx/dt$ a negative order at P since otherwise all would be integrands of integrals of the first kind, and only p of these are linearly independent. For an η which provides a negative order the expansion at P must have the form

$$\eta \frac{dx}{dt} = \frac{A_{-2}}{t^2} + A_0 + A_1 t + \cdots, \quad (A_{-2} \neq 0), \tag{21}$$

the term in $1/t$ being absent since the sum of the residues of η is zero. The integrand $-\eta/A_{-2}$ gives the integral of the following theorem for the case when $\mu = 1$.

THEOREM 13. *For every cycle P there exists an elementary integral $J_\mu(P)$ of the second kind with an expansion at P of the form*

$$J_\mu(P) = \frac{1}{t^\mu} + A_0 + A_1 t + \cdots.$$

The integrals

$$J_\mu(P)+c_1I_1+\cdots+c_pI_p+c$$

are also of this type and there are no others.

The last statement is evident since the difference of two integrals $J_\mu(P)$ with the same μ is necessarily an integral of the first kind.

The proof for integers $\mu>1$ can be made by an induction. An integrand function η furnishing an integral $J_\mu(P)$ must give $\eta\,dx/dt$ no singularities except a pole of order $\mu+1$ at the cycle P. The number of linearly independent functions η providing only such a pole for $\eta\,dx/dt$ is the number $\nu=p+\mu$ (Theorem 10, page 19) of linearly independent multiples of the divisor $D^2/(P^{\mu+1}X)$. If Theorem 13 is true for all integers $\leqq \mu-1$, there can be only $p+\mu-1$ linearly independent η's giving $\eta\,dx/dt$ at most a pole of order μ at P, and among the $p+\mu$ giving poles of order $\mu+1$ at most there must be an η giving order exactly $\mu+1$. By subtracting suitable constant multiples of $J_1(P),\ldots,J_{\mu-1}(P)$ from the integral with η as its integrand, and multiplying finally by a constant if necessary, an integral $J_\mu(P)$ can be constructed with the properties of the theorem.

THEOREM 14. *For an arbitrary pair of distinct cycles P_1,P_2 there exists an elementary integral of the third kind $K(P_1,P_2)$ with expansions at P_1 and P_2, respectively, of the forms*

$$+\log t+A_0+A_1t+\cdots,\quad -\log t+B_0+B_1t\cdots.$$

The integrals

$$K(P_1,P_2)+c_1I_1+\cdots+c_pI_p+c$$

are also of this type and there are no others.

To construct such an integral $K(P_1,P_2)$ one must find an integrand function η which gives $\eta\,dx/dt$ no singularities except simple poles at P_1 and P_2, and which is therefore a multiple of the divisor D^2/P_1P_2X. According to Theorem 10 of page 19 there are $\nu=p+1$ linearly independent multiples of this divisor, one at least of which must give $\eta\,dx/dt$ a pole at one at least of the cycles P_1,P_2. But if η gives $\eta\,dx/dt$ a simple pole at P_1 with an expansion

$$\frac{A}{t}+A_0+A_1+\cdots,$$

it must also provide a simple pole at P_2 with an expansion

$$-\frac{A}{t}+B_0+B_1t+\cdots,$$

since the sum of the residues of η is zero. The integrand η/A furnishes the integral $K(P_1,P_2)$ of the theorem.

THEOREM 15. *Every Abelian integral I is expressible linearly in terms of elementary integrals with constant coefficients.*

Let L be a cycle at which the integral I has a logarithmic singularity and at which the residue of the integrand function η of I is denoted by A. We may select arbitrarily a cycle L_0 distinct from all the singularities L. The difference

$$I-\sum_{L} AK(L,L_0),$$

where the sum is taken for all the logarithmic singularities L, has no logarithmic singularities at the cycles L, and also none at L_0 since the sum of the residues A is zero. It may still have poles, however, at cycles P with expansions of the form

$$\frac{B_{-\mu}}{t^{\mu}}+\cdots+\frac{B_{-1}}{t}+B_0+B_1 t+\cdots.$$

The difference

$$I-\sum_{L} AK(L,L_0)-\sum_{P}\left[B_{-\mu}J_{\mu}(P)+\cdots+B_{-1}J_1(P)\right],$$

where the last sum is taken for all the poles P, is an integral of the first kind linearly expressible as indicated in Theorem 12 of page 22.

8. The Riemann-Roch Theorem. The theorem to be proved in this section is a famous one in the theory of algebraic functions which has many applications in both geometry and analysis. It is not the purpose of these pages to discuss these applications, but rather to present a proof of the theorem free from restrictive assumptions upon the singularities of the algebraic equation defined by the equation $f=0$, and based upon the theory of multiples of a divisor. For this purpose we need the following two lemmas.

LEMMA 1. *If Q is an arbitrarily selected divisor, and Q_{ξ} the divisor of a rational function $\xi(x,y)$, then the number of linearly independent multiples of Q is the same as that for $Q_{\xi}Q$.*

This is evident because, if σ is a multiple of Q, then the product $\xi\sigma$ is a multiple of $Q_{\xi}Q$ and conversely.

LEMMA 2. *If $P_1,\ldots,P_m$ are cycles selected arbitrarily among those of a divisor Q, there is always a rational function η such that the divisor $Q_{\eta}Q$ contains none of the places $P_1,\ldots,P_m$.*

We can always make a transformation $x'=1/(x-x_0)$ such that after the transformation none of the cycles $P_1,\ldots,P_m$ is at $x'=\infty$. Suppose that this has already been done so that none of them is at $x=\infty$, and let $\mu_1,\ldots,\mu_m$ be the exponents which they have in Q. The divisor $Q_1=P_1^{-\mu_1}\cdots P_m^{-\mu_m}$ has a basis $\eta_1,\ldots,\eta_n$ for which the determinant Δ of page 9 is different from zero at every finite value x_0. If we select the constant coefficients in the expression

$$\eta=c_1\eta_1+\cdots+c_n\eta_n$$

so that they do not satisfy the linear equations whose coefficients are the rows of the determinants Δ corresponding to the values x_0 for the cycles $P_1,\ldots,P_m$, then η

will have exactly the orders $-\mu_1,\ldots,-\mu_m$ at these cycles, and the divisor $Q_\eta Q$ will contain none of the cycles.

We are now ready to prove the following:

THEOREM 16. (Riemann-Roch.) *If Q and Q' are two divisors such that $QQ'=Q_\xi D^2/X$, where Q_ξ is the divisor of a rational function $\xi(x,y)$ and X/D^2 the divisor of dx/dt, then the numbers ν,ν' of linearly independent multiples of Q,Q' and the orders q,q' of these divisors satisfy the relations*

$$\nu=\nu'-q-p+1,\quad q+q'+2p-2=0,$$

which may also be written in the form

$$2\nu+q=2\nu'+q',\quad q+q'+2p-2=0.$$

Let us first of all make a transformation $x'=1/(x-x_0)$, if necessary, in order to remove the cycles of the divisor X from $x=\infty$. The relation between Q and Q' then becomes

$$QQ'=Q_\xi\frac{D^2}{D_{x'}^2}\frac{D_{x'}^2}{X}=Q_{\xi'}\frac{D_{x'}^2}{X_{x'}},\tag{22}$$

since $Q_\xi D^2/D_{x'}^2$ is readily seen to be the divisor of the rational function $\xi'=\xi/(x-x_0)^2$, and since the branch cycles in X are transformed into those of $X_{x'}$.

Suppose that this transformation has already been made so that none of the cycles of X is at $x=\infty$. We may write the divisor equation of the theorem in the form

$$Q_1Q_1'=\frac{1}{X},\quad Q_1=\frac{Q_\eta Q}{Q_\xi D},\quad Q_1'=\frac{Q'}{Q_\eta D},$$

and select the rational function η (Lemma 2) so that the divisor Q_1 contains no cycle at $x=\infty$. The divisor Q_1' then also has this property since X has no cycles at $x=\infty$. The numbers of linearly independent multiples of Q and Q' are now respectively equal (Lemma 1) to those numbers for DQ_1 and DQ_1'.

Let $\eta_1,\ldots,\eta_n$ be a basis for the divisor Q_1 with column orders arranged in the order (8) of page 13 so that the number of multiples of DQ_1 is given by the formula $\nu=r_1+\cdots+r_s$ of the corollary on page 14. The complementary basis $\zeta_1,\ldots,\zeta_n$ is a basis for the divisor Q_1' with column orders

$$-r_1\leqq -r_2\leqq\cdots\leqq -r_s\leqq 0<-r_{s+1}\leqq\cdots\leqq -r_n$$

(Theorem 8), and the number of multiples of DQ_1' (Corollary, p. 14) is

$$\nu'=-r_{s+1}-\cdots-r_n.$$

Hence by formula (11) of page 16 we have

$$\nu-\nu'=r_1+\cdots+r_n=-(q-n)-\frac{w}{2}=-q-p+1,\tag{23}$$

since the order of Q_1 is $q-n$ and $p=(w/2)-n+1$. Furthermore, the equation $QQ'=Q_\xi D^2/X$ shows that

$$q+q'=2n-w=2-2p.\tag{24}$$

The relations (23) and (24) are the first pair of the theorem which can easily be transformed into the second.

The theorem can be given the following slightly different form from which some of the properties of rational functions with prescribed poles can be more readily derived:

COROLLARY 1. *If Q and Q' are two divisors such that $QQ' = Q_\xi X / D^2$, then the orders q, q' of these divisors and the numbers μ, μ' of linearly independent multiples of $1/Q$ and $1/Q'$ satisfy the relations*

$$\mu = \mu' + q - p + 1, \quad q + q' = 2p - 2, \tag{25}$$

which may also be written in the form

$$2\mu - q = 2\mu' - q', \quad q + q' = 2p - 2.$$

It is evident that a divisor whose order is positive can have no multiple since the number of zeros of such a multiple would necessarily exceed the number of its poles, whereas, as we know, these numbers must be the same. If $q > 2p - 2$, then the second formula (25) shows that q' is negative and the number μ' of multiples of $1/Q'$ is zero. Consequently we have:

COROLLARY 2. *If Q is a divisor of order $q > 2p - 2$, then the number μ of multiples of $1/Q$ is exactly $\mu = q - p + 1$.*

If $P_1, \ldots, P_s$ are arbitrarily selected cycles of our algebraic function, and $\mu_1, \ldots, \mu_s$ an arbitrarily selected set of positive integers, this corollary gives exactly the number of linearly independent rational functions $\sigma_1, \ldots, \sigma_\mu$ which have no singularities except possibly poles of orders $\geqq -\mu_k$ at the cycles P_k, provided that

$$q = \mu_1 + \cdots + \mu_s > 2p - 2.$$

To prove this statement we have only to apply the Corollary 2 to the divisor $Q = P_1^{\mu_1}, \ldots, P_s^{\mu_s}$. When the sum q is less than or equal to $2p - 2$, we can examine the divisors $Q' = Q_\xi X / D^2 Q$. If the number μ' for one of them can be determined, then μ is again known from the first of equations (25).

2

THEOPHIL HENRY HILDEBRANDT

Professor T. H. Hildebrandt was born in Dover, Ohio, July 24, 1888. He entered the University of Illinois in 1902 at the age of 14, and completed his work for the bachelor's degree in 3 years. Being then only 17 years of age and too young to find a teaching position, he entered the University of Chicago in 1905 to do graduate work in mathematics. At this time the famed E. H. Moore was head of the department of mathematics at Chicago, and it was under Moore's direction that Hildebrandt studied. He completed his work for the Ph.D. in 1909, receiving the official award in 1910. In 1909 he went to the University of Michigan as an instructor in the engineering mathematics department. (At that time, there were two departments of mathematics at Michigan, one in the College of Literature, Science and the Arts, and one in the College of Engineering.) From 1935 until his retirement in 1957, Professor Hildebrandt was chairman of the (now combined) department of mathematics. During his tenure as chairman, the department at Michigan became nationally famous.

In addition to his work at Michigan, and occasional visiting professorships at other institutions, Professor Hildebrandt has been active in the affairs of the American Mathematical Society, being its Vice President in 1924-25, President 1945-46, and a member of the Board of Trustees for several years thereafter. He is also a member of the Mathematical Association of America, which he served as Vice President, 1936, and of the American Association for the Advancement of Science, of which he was a Vice President in 1935.

Professor Hildebrandt's field is Analysis. His paper, "The Borel Theorem and its Generalizations," published in the *Bulletin* of the American Mathematical Society in 1926, has been frequently cited both for its mathematical content and for the historical information that it contains regarding the development of covering theorems basic in both Analysis and Topology.

R. L. WILDER

THE BOREL THEOREM AND ITS GENERALIZATIONS*

T. H. HILDEBRANDT, University of Michigan

That any one should attempt to devote a paper to the subject of the Borel Theorem may at first glance seem a presumption. A brief investigation will however reveal the following facts. (a) The Borel Theorem is closely related to the fundamental postulates of linear order. (b) There are many extensions and analogs of the Borel Theorem, some of which are hidden away in papers on other subjects. (c) The Borel Theorem has held and still holds a central position in the development and analysis of general spaces.

The arrangement of topics in the paper is suggested by the previous paragraph. No claim is made for completeness with respect to the extensions and analogs of the Borel Theorem, due to the nature of the case. Nor do I claim any originality in the material presented. I hope that a systematic treatment of the Borel Theorem in general spaces will be suggestive and perhaps create further desirable interest and results in these spaces.

I. THE BOREL THEOREM AND ITS EXTENSIONS FOR THE LINEAR INTERVAL AND n-DIMENSIONAL SPACE

In order to give a simple statement of the Borel Theorem I shall use the phrase "a family $\mathfrak{F}$ of intervals I covers the point-set E" to mean that every x of E is interior to some interval I_x of $\mathfrak{F}$. Then the Borel Theorem in its simplest form may be stated as follows.

If the family $\mathfrak{F}$ of intervals I covers the closed interval (a,b) then a finite subfamily of $\mathfrak{F}$ covers (a,b).

1. Historical note.† As in the case of other important mathematical results, the conception of the Borel Theorem is an interesting chapter in mathematical history, to which the future will no doubt add contributions. So far the honor of having first stated an outright theorem on intervals belongs to Borel,‡ though Borel's formulation deals only with the reduction of a denumerable family of intervals to a finite one. The name Heine-Borel seems to be due to Schoenfliess,§ who noted the relationship of the Borel Theorem to Heine's proof of the uniform continuity of a function continuous on a closed interval, published in 1872.‖ That Heine was

* An address presented at the request of the program committee before the joint meeting of the American Mathematical Society and the American Association for the Advancement of Science, Section A, at Kansas City, December 30, 1925.

† Cf. ENZYKLOPÄDIE DER MATHEMATISCHEN WISSENSCHAFTEN, vol. II_3, pp. 882 et seq.

‡ Paris thesis, 1894, p. 43; ANNALES DE L'ECOLE NORMALE, (3), vol. 12 (1895), p. 51.

§ *Bericht über die Mengenlehre*, JAHRESBERICHT DER VEREINIGUNG, vol. 8 (1900), pp. 51 and 119. In a later edition (Schoenfliess-Hahn, *Entwickelungen*, vol. I (1913), p. 235), he reverts to the name Borel Theorem.

‖ Cf. JOURNAL FÜR MATHEMATIK, vol. 74 (1872), p. 188.

aware of the fact that an interval theorem lay hidden away in his proof seems rather doubtful. As a matter of fact priority on the uniform continuity seems to go back at least to G. Lejeune-Dirichlet, though he suffered the penalty of not publishing his result immediately. A proof of the theorem almost identical with that of Heine appears in an exposition of his lectures given in 1854 brought out in 1904 by G. Arendt.* Another result carrying within it the germs of the Borel Theorem is due to S. Pincherle,† who gave the following theorem in 1881.

If the positive-valued function $f(x)$ is bounded from zero in some neighborhood of each point of a closed interval, then there exists an e such that

$$f(x)>e>0$$

for all points of the interval.

He remarks that this theorem can be made the basis for proof of uniform continuity and of the uniform convergence of a series of functions, uniformly convergent at every point of a closed interval.

To whom shall go the honor of first having conceived of the possibility of extending the Borel theorem to the case where the given family of intervals is not necessarily denumerable is another debatable question. In a way it is true that Dirichlet, Heine, and Pincherle were dealing with this case. Closely related is also the following theorem for the plane, due to P. Cousin:‡

If to each point of a closed region there corresponds a circle of finite radius, then the region can be divided into a finite number of subregions such that each subregion is interior to a circle of the given set having its center in the subregion.

Lebesque, to whom this extension is usually credited, claims to have known the result in 1898,§ but first published it in 1904 in his *Leçons sur l'Intégration*. W. H. Young|| published a proof in 1902. As a matter of fact the statement and proof of the Borel Theorem given by Schoenfliess in his 1900 Bericht can easily be interpreted to be that of the extension in question.

In considering these divergent claims it seems simplest and most just to call the theorem the Borel Theorem, and in case a distinction is necessary indicate it as the denumerable-to-finite or any-to-finite form. Moreover the theorem which gives the reduction from any set of intervals to an equivalent denumerable family due to Lindelöf in a general space proves to be only another case of a more general Borel Theorem.

---

* *Vorlesungen über die Lehre von den bestimmten Integralen*, Braunschweig, 1904.

† MEMORIE DI BOLOGNA, (4), vol. 3 (1881), pp. 151ff.

‡ ACTA MATHEMATICA, vol. 19 (1895), p. 22; Fréchet: CONGRÈS SOCIÉTÉS SAVANTES, 1924, p. 68.

§ For his attitude on the priority question, see BULLETIN DES SCIENCES MATHÉMATIQUES, (2), vol. 31 (1907), pp. 132–4.

|| PROCEEDINGS OF THE LONDON SOCIETY, vol. 35 (1902), pp. 384–8.

2. Proofs of the Borel Theorem.*

(a) By Successive Subdivisions.† Of the proofs of the Borel Theorem, perhaps the simplest in form is the one which proceeds by successive subdivisions. It is an indirect proof. Assume that the theorem is not true for the interval (a,b). Then if we divide the interval into two or more equal parts, it will be not true for one of these parts. This process applies indefinitely, and we have a sequence of closed intervals, each containing the succeeding, with lengths converging to zero, giving a single point x of (a,b) common to the intervals. This point being interior to an interval I of the family $\mathfrak{F}$, I will contain the intervals of the sequence after a certain stage, thus yielding a contradiction.

We observe that in a way this proof connects the Borel Theorem with the Cantor Theorem: *An infinite sequence of closed sets of points, each containing the succeeding, have a common point.*

(b) By Use of the Weierstrass-Bolzano Theorem. Another method of applying a similar process is to note that if the theorem is not true for (a,b) then if we divide the interval into n equal parts, it will be not true for one of these parts. Let (a_n, b_n) be the interval of length $(1/n)(b-a)$ for which the theorem does not hold, and x_n a point belonging to this interval. Then by the Weierstrass-Bolzano theorem, since the sequence $\{x_n\}$ is bounded, there will exist a subsequence $\{x_{n_m}\}$ having as limit the point x_0 of the interval (a,b). If x_0 is interior to I of the family $\mathfrak{F}$, then, by the properties of limits of sequences, we find that one of the intervals (a_n, b_n) is interior to I.

This method of proof is not quite so simple nor so elegant as the first. It uses the Weierstrass-Bolzano theorem, which itself is usually derived by a process of successive subdivisions.

(c) Direct Proof by Subdivisions. The two preceding proofs on account of their indirect character do not give a method for actually selecting the intervals required by the theorem. We can however use the subdivision idea to make such a selection. Let d_x be the least upper bound of the values of d for which the interval $(x-d, x+d)$ is interior to some interval of $\mathfrak{F}$. Then the function d_x is bounded from zero on the closed interval (a,b), i.e., there exists a d such that for all x of (a,b) $d_x > d > 0$. This can be deduced from the fact that d_x is a continuous function on a closed interval, or connected with the two results (a) that there exists for each x a vicinity of x such that d_x is bounded from zero in this vicinity, and (b) that for any function defined on (a,b) there exists a point x such that the greatest lower bound of f for each vicinity of x is the same as the greatest lower bound of f on (a,b). The final step in the proof then is to divide (a,b) into intervals of length less

* Cf. Schoenfliess-Hahn, *Entwickelungen der Mengenlehre*, vol. 1, 1913, pp. 235 et seq.

† Cf. Borel (Baire), COMPTES RENDUS, vol. 140 (1905), p. 299; Capelli, NAPOLI RENDICONTI, (3), vol. 15 (1909), p. 151.

than d, each of which will be covered by some member of the family $\mathfrak{F}$ by the definition of d_x.

Observe that in this case we replace all the intervals to which a point is interior by a single member of another family, a family which is really involved in the proof of the uniform continuity theorem as given by Heine and by Dirichlet. The Pincherle result is also involved in the preceding proof, as well as the linear analog of Cousin's formulation of the Borel Theorem in the plane.* We might call attention to the following almost self-evident result involved in the preceding proof.

If every point of a bounded set E is the middle point of an interval of length $2d_x$, and d_x is bounded from zero, then all the points of E are interior to a finite number of these intervals.

(d) Denumerable-to-Finite.† When the given family is denumerable it is possible to select the finite family as follows. Suppose the intervals of $\mathfrak{F}$ are arranged in sequential order $I_1, I_2, \ldots, I_n, \ldots$. By step-by-step process, we obtain an equivalent series, retaining those intervals which contain at least one point not interior to the preceding intervals of the array. Let $I_1, I_{n_2}, \ldots, I_{n_m}, \ldots,$ be the members of the retained family. If this is finite, the theorem is proved. In the contrary case, we have an element x_m interior to I_{n_m} but not to the preceding intervals. By the use of the Weierstrass-Bolzano Theorem, this sequence contains a subsequence x_k' having a limit x_0 of (a,b). The contradiction arises from the fact that the interval I_{n_m} containing x_0 as an interior point will contain an infinite number of the x_k' as interior points, i.e., members of the sequence $\{x_m\}$ of index higher than n_m. We shall see later in the present paper that this method of proof is effective in general spaces.

(e) Borel's Proof. The Lebesgue Chain. The first proof of Borel also contains a scheme for selecting the finite subfamily, but it rests upon the properties of Cantor ordinal numbers, and the denumerability of the family, or of a family connected with the given family. Starting with the point a, there exists an interval I_1 containing a as interior point. If b_1 is the right-hand end-point of I_1, let I_2 contain b_1 as interior point. Continuing thus, we get a sequence of intervals $I_1, I_2, \ldots, I_n, \ldots$. If the point b has not been reached, we get a limiting point b_∞ of the left hand endpoints b_n and an interval I_∞ containing b_∞. By repeating the process, since the set of intervals is denumerable, we must eventually include the point b with an interval I_α where α is a transfinite ordinal of the second kind. We now extract from this well-ordered set of intervals a finite subset by noting that any interval I_α associated with a limit number α includes an infinite number of the end-points b_β

* Cf. also Baire, Annali di Matematica, (3), vol 3 (1899), pp. 13–15; Borel, Comptes Rendus, vol. 140 (1905), p. 299; Wirtinger, Wiener Berichte, vol. 108, IIa, pp. 1242–3.

† Cf. Young, W. H., Palermo Rendiconti, vol. 21 (1906), p. 127; Fréchet, Palermo Rendiconti, vol. 22 (1906), pp. 22–23.

of intervals preceding it. This enables us to select a decreasing array of ordinal numbers and such a decreasing set is finite.

This proof hinges on two facts, (a) that the system of intervals in denumerable, and (b) that a decreasing set of ordinal numbers selected from a well-ordered increasing set is finite.

While the denumerability of the family $\mathfrak{F}$ is postulated in this proof, the denumerability of the array $I_1, \ldots, I_\infty, \ldots$, leading to the point b can be deduced from the following fundamental theorem on intervals.

Any family of non-overlapping intervals is denumerable.

For the interval $(0,1)$, this follows from the fact that there exist at most n intervals of the set having a length greater than $1/n$, which gives a system for enumeration. Since the unbounded straight line can be broken up into a denumerable set of intervals of length unity, the extension of this result to any family of non-overlapping intervals is immediate.

In the preceding proof of the Borel Theorem the intervals $(b_\alpha, b_{\alpha+1})$ form a nonoverlapping family, which is therefore denumerable. Then the assumption that b would not be reached by a denumerable set of steps, would lead to a contradiction. In point of fact, there is contained implicitly in these considerations a theorem which has been called the Lebesgue Chain Theorem* which has been used by him in the discussion of lengths of curves. The theorem may be stated as follows.

LEBESGUE CHAIN THEOREM. *If the family $\mathfrak{F}$ of intervals I is such that to every point x of an interval (a,b) excepting perhaps b, there correspond intervals of the family having x as left-hand end-point, then there exists a finite or denumerable subfamily of these intervals without common points, containing each point of the interval, excepting possibly b, as an interior point or left-hand end-point.*

We shall give another proof of this theorem later on. Also we shall see that the idea of a Lebesgue chain and of applying a process of reduction via a decreasing set of ordinals underlies the proofs of other theorems similar to the Borel Theorem.

(f) Dedekind-Cut Proof.† Finally we call attention to a proof which depends on the Dedekind-Cut Axiom, i.e., is based on the properties of linear order. We observe that the point a is interior to an interval I of the family $\mathfrak{F}$. Let x' be the least upper bound of the points of (a,b) which can be reached with a finite number of intervals starting from a, i.e., x' is defined by a Dedekind cut. Now x' will belong to the interval (a,b) which is closed. Consequently there will be an interval $I_{x'}$ of the family $\mathfrak{F}$ to which x' is interior. It follows that x' is the point b.

An analysis of the proof shows that it rests upon the following principle.

---

* *Leçons sur l'Intégration*, p. 63.

† Lebesgue, *Leçons*, p. 105; O. Veblen, the BULLETIN OF THE AMERICAN MATHEMATICAL SOCIETY, vol. 10 (1904), pp. 436–9 (see also TRANSACTIONS OF THE AMERICAN MATHEMATICAL SOCIETY, vol. 6 (1905), p. 167, where it is pointed out that the Borel Theorem applies to any well-ordered set); F. Riesz, COMPTES RENDUS, vol. 140 (1905), pp. 244–6.

INDUCTION PRINCIPLE FOR LINEAR ORDER. *Suppose a statement S satisfies the following conditions relative to an interval* (a,b) (*which may be the infinite interval* $(-\infty, +\infty)$): (1) *there exists a point of the interval for which S is true*, (2) *if the statement S is true for all points preceding* x', *then there exists a point y beyond* x' *for which S is true. Under these conditions S holds for the entire interval* (a,b).

A. Khintchine* has pointed out that this Induction Principle is logically equivalent to the Dedekind-Cut Axiom.

That the Dedekind-Cut Axiom implies the Induction Principle is practically contained in the above proof of the Borel Theorem. On the other hand, assume the truth of the Induction Principle. Suppose we have divided the points of the closed interval (a,b) into two groups A and B, each containing at least one point, and such that every point of A is less than (precedes) every point of B, and that A and B contain all the points of (a,b). Let the statement S be "The point x is a member of A." Since the conclusion of the Induction Principle is not holding, it follows that either condition (1) or (2) is not holding. Now (1) is true, hence (2) must be false, i.e., there exists a point x' such that for every x preceding it the statement S is true, but S is not true for any point follwong x'. Since A and B contain all points of (a,b) it follows that the point x' belongs to one of these classes, i.e., is either the maximum of the class A, or the minimum of the class B.†

Along the same line is the observation that the Borel Theorem and Dedekind-Cut Axiom are logically equivalent; this remark is due to Veblen. The proof that the Borel Theorem has the Dedekind-Cut Axiom as a consequence, assumes that the interval is divided into the groups A and B, as conditioned in the preceding paragraph. With every point of the interval, we associate as far as possible an interval containing the point, and consisting only of points of group A or of group B. Either every point of (a,b) is interior to one of these intervals, or the contrary is true. In the first case, we have by the Borel Theorem a finite number of intervals, reaching from a to b, containing every point as an interior point. Since a belongs to A, the intervals overlap, and each interval contains only points of A or B, it follows that every point of (a,b) is a point of A, which is contrary to hypothesis. Hence there exists at least one point which is not interior to an interval containing only points of A, or of B, i.e., satisfies the conditions of the Cut Axiom. An analysis of this proof shows that the theorem is a close relative of the following theorem:

If a function is not invariant in sign throughout an interval (a,b), *then there exists a point of the interval in every vicinity of which the function is not invariant in sign.*‡

The Induction Principle also furnishes a method of proof for the Lebesgue Chain Theorem. Two have been given along similar lines. The first, due to J. Pal,§

---

* FUNDAMENTA MATHEMATICAE, vol. 4 (1922), pp. 164–6.

† To be a complete analog to mathematical induction, condition (2) should read: If S is true for all points preceding x', then it is true for x'. In this form, however, it does not seem to have the power which carries one to the end of the intervals.

‡ K. P. Williams, ANNALS OF MATHEMATICS, (2), vol. 17 (1915–6), pp. 72–3.

§ PALERMO RENDICONTI, vol. 33 (1912), pp. 352–3.

considers the case in which to every point of (a,b) there corresponds only a single interval having this point as left-hand endpoint. In that case, any chain beginning from a is necessarily unique. The second, due to G. C. Young,* applies to the general case, where the number of intervals associated with a given point is not specifed. Obviously in this case there may be many different chains leading from one point to another.

We say that the set $\mathfrak{G}$ of intervals I forms a chain from the point a to the point x in case $\mathfrak{G}$ is a set of non-overlapping intervals such that every point of (a,x), excepting perhaps x, is either an interior point or a left-hand end-point of one of the intervals of $\mathfrak{G}$. The point x may be interior to an interval, or an end-point of a chain, i.e., a right-hand end-point of an interval, or the limiting point of a sequence of intervals of the chain $\mathfrak{G}$.

In applying the Induction Principle assume that the point x is such that for every y less than x, there exists a chain from a to y. The case in which there is an end-point y of a chain from a to y, such that there exists an interval with y as left-hand end-point which contains x, and the case in which x is an end-point of a chain leading from a to x, are easily disposed of. If neither of these two cases are holding, let $e_1, e_2, e_3, \ldots, e_n \ldots$, be a monotonic decreasing sequence of positive numbers converging to zero. Then by hypothesis there exists a chain from a to x_1, where the distance of x_1 to x is less than e_1. If there exists a chain extending from x_1 beyond x, then by adding this chain to the chain from a to x_1, we extend beyond x. In the contrary case, let y_1 be the least upper bound of points reached by chains beginning at x_1. Then there exists a chain from x_1 to a point x_2 with $y_1 - x_2 < e_2$. By a repetition of this process, we get a sequence of points $x_1, x_2, x_3, \ldots$, having a limiting point x_0, and by the method of construction, by combining chains, we get a chain from a to x_0. Now x_0 is a left-hand end-point of an interval of the family $\mathfrak{F}$ whose length is greater than e_n for n sufficiently large. But this interval added to the chain from a to x_0 would give us a chain reaching beyond y_n for n sufficiently large, contrary to the definition of y_n. It follows that there is a chain from x_1 to x, and so beyond x.

A careful analysis of the proof shows that its basis is really the same as the proof of this theorem using the Cantor numbers, viz., the fact that there are at most a denumerable set of non-overlapping intervals.

3. Extensions of the Borel Theorem.

(a) To Closed Sets. An almost immediately obvious extension of the Borel Theorem is to replace the closed interval covered by the family $\mathfrak{F}$ by a closed set of points E. The methods of proof sketched above all apply excepting that in the case of (e) and (f) which involve order on a line, it may be necessary to take account of the intervals complementary to the closed set. An alternative method of procedure

* Bulletin des Sciences Mathématiques, (2), vol. 43 (1919), pp. 245–7.

is to take the smallest interval (a,b) containing the set E, and enlarge the family $\mathfrak{F}$ by the addition of the intervals belonging to the complement of E with respect to (a,b) and then apply the theorem for the interval.*

(b) Extension to *n*-Dimensional Space.† A further extension which is possible is to n-dimensional space. This was conceived almost as soon as the Borel Theorem, due to the consideration of functions of the complex variable.‡ We define the point *x interior to a set of points I of n-dimensional space* to mean that there exists a vicinity of x (i.e., an n-dimensional sphere or cube having x as center) containing only points of I. Then the Borel Theorem reads as follows:

If E is any closed set of n-dimensional space covered by a family $\mathfrak{F}$ of sets I (i.e., such that every point of E is interior to some member of the family) then a finite subfamily of $\mathfrak{F}$ covers E.

Some of the proofs given for the linear interval are immediately extensible to space. This is true of the methods (a) and (b) by successive subdivisions, and (c) in which the family $\mathfrak{F}$ is replaced by a family of spheres with the points of E as centers, also of (d) the denumerable case. (e) and (f) seem to use particularly linear order and consequently are effective mainly in proofs by induction, passing from the case of n dimensional to $(n+1)$-dimensional space. Lebesgue§ suggests an ingenious method of passing from the plane to the linear interval by using the Peano curve which maps the square on the linear interval.

4. Necessary Conditions. Lindelöf Theorem. The hypothesis of the Borel Theorem specifies as sufficient conditions that the set E be closed and bounded. These conditions are also necessary. If E is not closed then there exists a point x_0 limiting point of E not belonging to E. If we surround each point x of E by a sphere of radius one-half of the distance of x from x_0, then no finite subfamily of this family of spheres will cover E. The fact that E must be bounded is obvious.

In the same direction is the remark that the Weierstrass-Bolzano Theorem is a consequence of the Borel Theorem. For suppose an infinite set E_0 of points $x_1, \ldots, x_n, \ldots$, contained in a bounded closed region E. Then either for every point x of E there exists a sphere around x containing at most one point of E_0, or there exists a point x such that every sphere about x contains an infinity of points selected from E_0. In the first case, the spheres constitute a family $\mathfrak{F}$ for E to which the Borel Theorem applies. But a finite number of spheres of $\mathfrak{F}$ contain only a finite number of points of E_0. Hence E_0 has a limiting point in E.

---

* Cf. also W. H. Young, PROCEEDINGS OF THE LONDON SOCIETY, vol. 35 (1902), pp. 387–8.

† Cf. Schoenfliess-Hahn, *Entwickelungen*, vol. 1, 1913, pp. 239–241.

‡ Cf. Cousin above. The Cauchy-Goursat theorem is virtually based on a two-dimensional Borel Theorem. See TRANSACTIONS OF THE AMERICAN MATHEMATICAL SOCIETY, vol. 1 (1900), pp. 14–16.

§ *Leçons*, p. 119.

We note in passing that it is immaterial whether the family $\mathfrak{F}$ consist of open sets (containing only interior points) or arbitrary sets of points to make the Borel Theorem valid.

Returning to the linear case, the fact that the Borel Theorem does not hold for sets in general suggests the question what can be said about the set of points interior to any family of intervals. If by the term *two families of intervals are equivalent* we mean that they cover the same set of points, then we have on the one hand

Any family of intervals is equivalent to a family of non-overlapping intervals.

This is immediately evident. We need only take any point interior to some interval and proceed to the left and right until we meet a point which is not interior to any interval of the given family. In this way we define a group of non-overlapping sets which is denumerable.

As an analog to the Borel Theorem we have the following theorem, which is due to Lindelöf:*

LINDELÖF THEOREM. *In any family of intervals it is possible to find a denumerable sub-family having the same interior points.*

This theorem has been proved in different ways. Mention might be made of the following proofs:

(a) Using Density of Rational Points on a Line.† Let x be any point interior to some interval I of the family. Then there exists an interval R with rational end-points containing x, and entirely interior to I. This sets up a correspondence between a family of the given intervals and the family of intervals with rational end-points, which is denumerable.

We observe that this method of proof makes use of the density of the rational points on a line and their denumerability. An equivalent method of constructing the intervals with rational end-points, would be to use the rational points interior to the family of intervals, and make each of them the mid-point of an interval with rational end-points interior to some interval of the set. In this form, the proof of the corresponding theorem in n-space can be made.

(b) Via the Borel Theorem. Let (a,b) be one of the intervals of the equivalent family of non-overlapping intervals, and x_0 any point of (a,b). Let $x_0, x_{-1}, x_{-2}, \ldots,$ be a monotonic sequence of points approaching a, and $x_0, x_1, x_2, \ldots,$ be a monotonic sequence approaching b. Then the closed intervals (x_{-n}, x_n) can be covered by a finite number of intervals of the given family $\mathfrak{F}$. This gives us a method for selecting a denumerable set of intervals having the desired property relative to (a,b). The final result is a consequence of the properties of denumerability.

* COMPTES RENDUS, vol. 137 (1903), p. 697.

† Cf. W. H. Young, PALERMO RENDICONTI, vol. 21 (1906,) p. 125.

(c) Lindelöf Proof. Suppose the points covered by $\mathfrak{F}$ belong to a finite interval (a,b). For any x covered by $\mathfrak{F}$, let d_x be the maximum of the values of d for which $(x-d,x+d)$ is interior to one of the intervals of the family $\mathfrak{F}$, and consider the set E_n of points of the set E covered by $\mathfrak{F}$, for which $d_x > 1/n$. Then by the observation of §2(*c*), it follows that the set E_n can be covered by a finite number of intervals chosen from the family $\mathfrak{F}$. This gives a method for the enumeration of the subfamily. The extension to the unbounded interval results from the fact that it can be divided into a denumerable set of finite intervals.*

A similar method of procedure applies in n-space.

5. Strict Families of Intervals.† It is obvious that in general the selection of the finite subfamily of intervals in the Borel Theorem may be made in many different ways. Obviously too, some of the intervals of the finite family may be unnecessary in that all points interior to them are interior to other intervals of the family. We shall call a *strict* family of intervals, a family in which each interval is necessary, in the sense that it contains a point not interior to any of the other members of the family. We have then

Any finite family of intervals can be replaced by an equivalent strict family.

We note first that if three intervals have a common point they can be replaced by at most two of these intervals. Let the intervals be (a_1,b_1), (a_2,b_2), and (a_3,b_3), where since they contain a common interior point, the notation is chosen so that

$$a_1 \leqq a_2 \leqq a_3 \leqq b_1.$$

Then obviously if $b_2 < b_3$, we can dispense with (a_2,b_2) and if $b_2 > b_3$, then we can dispense with (a_3,b_3). By the use of this result the theorem stated is immediate. The process of deletion in any particular case may be tedious, especially if governed by other considerations, such as, for instance, the desire to make the sum of the lengths of the retained intervals a minimum.

More generally we have the following theorem:

If any family of intervals is such that every point interior to one of the intervals is interior to at most a finite number of intervals of the family, then we can select a strict subfamily, equivalent to the given family.

We note in the first place that the given family is necessarily denumerable. For to each point interior to an interval of the family there will correspond a vicinity which is common to a definite finite number n_x of intervals. By the Lindelöf Theorem these vicinities can be replaced by a denumerable set of the same vicinities, and since to each of the final vicinities there corresponds a finite number

* W. H. Young, Palermo Rendiconti, vol. 21 (1906), pp. 126–7, gives another proof utilizing somewhat similar procedure, which resulted in a seris of polemics with Schoenflies, the final shot being fired by the latter in Palermo Rendiconti, vol. 35 (1913), pp. 74–78. At best the method of proof, even as validated by Schoenflies, is not satisfactory.

† Cf. Denjoy, Journal de Mathématiques, (7), vol. 1 (1915), pp. 223–30.

n_x it follows that the original family of intervals is denumerable.

Let then the intervals be arranged in sequential array $I_1,\dots,I_n,\dots$. Let I_{n_1} be the first interval of the array covering a point not interior to any of the succeeding intervals. Such an interval will always exist. We determine I_{n_m} as the first interval of the sequence following $I_{n_{m-1}}$ containing a point not interior to $I_{n_1},\dots,I_{n_{m-1}}$, and a point not interior to any of the intervals following I_{n_m}. It is obvious that the resulting family is a strict family. It remains to show that every point x covered by the original family is covered by the subfamily. Since x is an interior point of a finite number of intervals of $\mathfrak{F}$ the indices of the sequence $I_1,\dots,I_n$ to which x is interior have a definite maximum N, i.e., there will be an index $n_m < N$ such that x is interior to I_{n_m}.

From the point of view of measure, a strict family of intervals has the property that the sum of the lengths of the intervals is less than double the sum of the lengths of the intervals covered. For consider any strict family of intervals. We note first of all that it is necessarily denumerable. For since each interval contains a point not interior to other intervals it contains a subinterval having no points in common with other intervals. These subintervals define a system of non-overlapping intervals which is denumerable, and consequently the original family, which is in one-to-one correspondence with it, is denumerable. Let x_n be a point interior only to I_n. Then the points x_n have as limiting points only points not interior to any interval of the family. From the contrary assumption would follow that some of the points x_n are interior to more than one interval of the given family. As a consequence the points x_n in each interval (a_k,b_k) of the equivalent family of non-overlapping intervals can be arranged in order so that between two points no further points of the sequence appear. Consequently no point of (a_k,b_k) appears in more than two intervals of the family.

Denjoy has given a condition under which the reduction of a family to an equivalent strict subfamily is possible. If a family $\mathfrak{F}$ of intervals is *upper semi-closed* in case every interval which is the limit of a sequence of intervals I_n chosen from the family is part of an interval of or belongs to $\mathfrak{F}$, then we have the following theorem:

From every upper semi-closed family $\mathfrak{F}$ of intervals it is possible to select an equivalent strict subfamily.

It will obviously be sufficient to show the possibility of selecting an equivalent subfamily $\mathfrak{F}_0$, such that every point covered by $\mathfrak{F}$ is interior to at most a finite number of intervals of $\mathfrak{F}_0$. Let (a,b) be an interval of the equivalent family $\mathfrak{G}$ of non-overlapping intervals. There are three possibilities:

(a) The points a and b are both end-points of intervals of $\mathfrak{F}$. Then, applying the Borel Theorem, we get a finite sub-family having every point of (a,b) excepting a and b as interior points.

(b) Both a and b are not end-points of intervals of $\mathfrak{F}$. Then by successive applications of the Borel Theorem, as in the proof of the Lindelöf Theorem (§3(b)), we construct an equivalent family which has the property that a or b is a limiting point of end-points of any infinite set of intervals selected from this family. If possible let x be an interior point of (a,b) interior to an infinite number of intervals of the resulting subfamily. Then a or b is a limiting point of end-points of these intervals. By the semiclosure of the family $\mathfrak{F}$ it follows that a or b is then an end-point of an interval reaching at least to x, contrary to the assumption. Hence every point interior to (a,b) is interior to at most a finite number of intervals of this subfamily.

(c) If only one of the end-points a or b is an end-point of an interval of $\mathfrak{F}$, then a process similar to case (b) utilizing only one end-point will apply.

As a corollary of this result we have the following theorem.*

If a family $\mathfrak{F}$ of intervals is such that for every e there are at most a finite number of intervals of $\mathfrak{F}$ of length greater than e, then there exists an equivalent strict subfamily of $\mathfrak{F}$.

6. Other Theorems on Reduction of Families of Intervals to Finite Subfamilies. We turn our attention briefly to a number of analogs of the Borel Theorem, most of which are due to W. H and G. C. Young.† A theorem which has the Borel Theorem as a consequence, but for which the converse has not yet been shown, has been called by the Youngs the Heine-Young theorem, because of its similarity to Heine's proof of the uniform continuity theorem.

HEINE-YOUNG THEOREM. *With every point x of a closed interval (a,b) there are associated two intervals, an R_x having x as left-hand end-point, and L_x having x as right-hand end-point. These intervals are connected by the condition that if x' is interior to the L_x for x, then $R_{x'}$ contains x as an interior point or an end-point. Then a finite number of the R intervals cover (a,b) without overlapping.*

The R intervals without the intervention of the L intervals are equivalent to a Lebesgue chain, and since the R_x is unique for every point there will be only one such chain. The presence of the L intervals insures the finiteness of the chain by preventing limiting points. For the L corresponding to a possible limiting point x would include an infinite number of x' whose $R_{x'}$ by hypothesis should reach up to or beyond x.

The finiteness of the R chain is also apparent from another point of view. The L intervals are equivalent to a chain. Now any L is covered by at most two R intervals. We can then as in the case of the Borel Theorem define a sequence of decreasing ordinals by beginning with R_a. For if a point x' is a limiting point of the L intervals, of the chain, the corresponding $R_{x'}$ will cover an infinte number of the L intervals, and have as right-hand end-point a point interior to an L or an end-point of an L or a point which is again a limiting point. In either of these cases we can proceed in the formation of our decreasing ordinal series, which is finite.

* Cf. R. L. Moore, PROCEEDINGS OF THE NATIONAL ACADEMY, vol. 10 (1924), 466–7.

† Cf. *Reduction of intervals*, PROCEEDINGS OF THE LONDON SOCIETY, (2), vol. 14 (1915), pp. 111–130.

While this theorem seems to be more general than the Borel Theorem, its usefulness is rather hampered by the peculiar way in which the two sets of intervals are interlaced.

Another theorem of the same type is due to Lusin*, which is stated by him as applying to non-dense perfect sets.

LUSIN THEOREM. *Let E be any non-dense perfect set, and $\mathfrak{G}$ the family of open intervals (a_k, b_k) complementary to E relative to an enclosing interval (a,b). Moreover, suppose that to every point x of E not an a_k or b_k there correspond intervals of a family $\mathfrak{F}$ to the right of x of the form (x, a_k), and all intervals to the left of the form (b_k, x) in a certain neighborhood of x; while to the points b_k, only intervals of first type correspond, and to points a_k only intervals of the second type; then a finite number of the intervals from $\mathfrak{F}$ and $\mathfrak{G}$ cover (a,b) or E without overlapping.*

The proof can be made along the lines of the Young Theorem; the right-hand intervals together with the intervals of $\mathfrak{G}$ can be formed into a chain, this being reduced to a finite set by the inclusion of the left-hand intervals. From this it is apparent that to prevent overlapping it is not sufficient to assume as Lusin does that only a single interval the right and left corresponds to each point of E.

W. H. and G. C. Young credit the following theorem for a closed interval to Lusin.

If to every point x of a closed interval (or closed point set) there correspond all the intervals to the left and right of the point in a given neighborhood of the point, then a finite number of these intervals suffice to cover the interval (or point set) without overlapping.†

With every x of (a,b) we associate an interval of a new family such that x is the middle point of its interval, and the interval consists of an L and an R of equal length selected from intervals associated with the point. A finite number of these intervals suffice to cover (a,b). We assume that the intervals retained give a strict covering for (a,b). Then if I_{n_1} contains a, let x_1 be its middle point. If $x_1 > a$, we reach x_1 by noting that (a, x_1) belongs to the given family. Let I_{n_2} overlap with I_{n_1} and x_2 be its middle point. Then $x_2 > x_1$. If x_2 belongs to I_{n_1} then (x_1, x_2) is an interval of the given family. If x_2 is not in I_{n_1} then we reach x_2 by taking the intervals (x_1, b_1) and (b_1, x_2), where b_1 is the right-hand endpoint of I_{n_1}. It is obvious that in this way we can construct a finite number of intervals as required.

A slightly more general theorem can be obtained by associating with every point all the intervals to the left in a certain vicinity, and only one or more to the right. In this form the proof given above via the Borel Theorem is not valid, and it is not clear whether the Borel Theorem can be used. It can be deduced by using the

* MOSCOW MATHEMATICAL SOCIETY TRANSACTIONS (SBORNIK), vol. 28 (1911–12), p. 270.

† This theorem can be made the basis of a proof of the theorem *If $f(x)$ has a Riemann integrable derivative $f'(x)$ on (a,b) then $f(b)-f(a) = \int_a^b f(x)dx$*, without using the mean-value theorem of the differential calculus, a desideratum in considerations in general functional space.

Lebesgue chain associated with the intervals to the right; or, following the Youngs, we get an equivalent method by associating with every point x of (a,b) as an R_x the smallest interval containing all the intervals of the given family having x as left-hand end-point and as L_x the given vicinity to the left. Then the Heine-Young Theorem applies; there exists a finite number of the R intervals reaching from a to b. Any of these R intervals can be replaced by at most two intervals of the original family.

In trying to avoid the infinite elements which pervade the Lebesgue Chain Theorem, the Youngs state the following lemma:

YOUNG'S LEMMA. *If to every point of the closed interval (a,b) there corresponds an interval of a family $\mathfrak{F}$ having this point as left-hand end-point, then for every e, there exists a finite subfamily $\mathfrak{F}_e$ of non-overlapping intervals, such that the sum of the complementary intervals is less than e.*

If we assume the Lebesgue Chain Theorem this result is immediate. The Youngs prove the theorem by an ingenious application of the Heine-Young Theorem. Choose n so that

$$2(b-a)<ne.$$

Then to every point x of (a,b) we make correspond as L_x an interval to the left of x whose length is $1/n$ of the least upper bound of the intervals of $\mathfrak{F}$ associated with x. If $(x,x+h_x)$ is any interval of $\mathfrak{F}$, then we add to $\mathfrak{F}$ all the intervals $(y,x+h_x)$ where

$$x-\frac{h_x}{n}\leqq y\leqq x.$$

Let R_x be the smallest interval containing all the intervals of this extended family which have x as left-hand end-point. Then it is apparent that the R and L intervals satisfy the conditions of the Heine-Young Theorem, so that a finite number of the R intervals extend from a to b without overlapping. The result desired follows from the fact that each R can be approximated up to $2/n$ of its length by an interval of the family $\mathfrak{F}$.

It is obvious that the Young Lemma will hold also for closed sets of points E, in which form it was originally stated.* A proof similar to this other proof given by Young has been made by Sierpinski† for the following more general theorem which utilizes the properties of upper measure.

If E is any bounded linear set, and $\mathfrak{F}$ a family of intervals such that to every x of E there corresponds an interval R_x of $\mathfrak{F}$, having x as left-hand end-point, then for every e it is possible to determine a finite subfamily $\mathfrak{F}_e$ of $\mathfrak{F}$, consisting of non-overlap-

---

* PROCEEDINGS OF THE LONDON SOCIETY, (2), vol. 9 (1911), pp. 325ff.

† Cf. FUNDAMENTA MATHEMATICAE, vol. 4 (1923), pp. 201–3.

ping intervals, and such that

$$\overline{m}(E-E\cdot F_e)<e.^*$$

The proof is as follows. Let E_n be the set of points of E for which there exists an R_x of length greater than $1/n$. Then the E_n form a monotonic increasing family of sets such that

$$E=\sum E_n,$$

and consequently

$$\lim_n \overline{m}E_n=\overline{m}E.$$

Choose n so that

$$\overline{m}E-\overline{m}E_n<\frac{e}{2}.$$

Let (a_1,b_1) be the smallest interval containing E_n, its length being l. Obviously if we use intervals of $\mathfrak{F}$ of length greater than $1/n$, then there will be at most nl possible non-overlapping intervals in (a_1,b_1). If between every two intervals we allow a space d such that

$$nld<\frac{e}{2},$$

then the points of E_n not covered will be of upper measure at most $e/2$. Since a_1 is a lower bound of points of E_n there will be a point x_1 of E_n in the interval (a_1,a_1+d) and an R_x of length greater than $1/n$. Let a_2 be the lower bound of the points of E_n to the right of $R_{x_1}=R_1$. Take x_2 so that x_2 belongs to (a_2,a_2+d) and E_n. Continuing in this manner we get a finite number of intervals $R_1,\ldots,R_m$. Let S be the set of points belonging to $R_1,\ldots,R_m$. Then we wish to show that

$$\overline{m}(E-ES)<e.$$

Now by the properties of upper measure we have †

$$\overline{m}E=\overline{m}(E-ES)+\overline{m}ES.$$

By the selection of E_n, we have

$$\overline{m}E-\overline{m}E_n<\frac{e}{2},$$

and by the selection of S,

$$\overline{m}E_n-\overline{m}E_nS<\frac{e}{2},$$

so that

$$\overline{m}E-\overline{m}E_nS<e.$$

But E contains E_n and so ES contains E_nS. Hence

$$\overline{m}(E-ES)=\overline{m}E-\overline{m}ES<\overline{m}E-\overline{m}E_nS<e.$$

---

* We follow the usual notations: $\overline{m}E$ or $\overline{\text{meas}}\,E$ for upper measure of E; $E\cdot F$ the set of points common to E and F; $E+F$ set of points belonging to either E or F; $E-F$ the sets of points of E not in F.

† Cf. Hausdorff, *Mengenlehre*, p. 415.

These theorems are of value in connection with the discussion of the distribution of infinite derivatives of a function of a single variable. The Young Lemma and the Sierpinski extension have been used by the Youngs* and by Rajchman and Saks† to obtain in a simple way forms of the theorem that any monotonic function and therefore any function of bounded variation has a finite derivative excepting at a set of points of measure zero.

Closely related to these theorems is a so-called "tile" theorem of the Youngs. If to a point correspond all the intervals in a certain vicinity of the point, then any such interval is called a *tile* and the point is called *the point of attachment*.

TILE THEOREM.‡ *Suppose a family of intervals* $\mathfrak{F}$ *such that to every point* x *of a linear bounded set* E, *there correspond all the intervals in a certain neighborhood of* x. *Then for every* e *and* d, *there exists a finite or denumerably infinite subfamily* $\mathfrak{F}_{de}$ *of* $\mathfrak{F}$, *such that*

(*a*) *every interval* I *of* $\mathfrak{F}_{de}$ *is of length less than* d,
(*b*) *the point associated with each interval of* $\mathfrak{F}_{de}$ *is interior only to that interval,*
(*c*) *every point of* E *is interior to some interval of* $\mathfrak{F}_{de}$, *and*

(*d*) $$\sum mI_n - \overline{m}E < e$$

where I_n *are the intervals of* $\mathfrak{F}_{de}$.

In case the set E is a closed interval, the set $\mathfrak{F}_{de}$ is finite.§ In this case the proof can be made via the Borel Theorem. We discard all intervals of $\mathfrak{F}$ of length greater than d, and then associate with each point x an interval of the remaining family associated with x, of which x is the middle point. Then replace this new family by a finite strict subfamily via the Borel Theorem. Let $x_1,\dots,x_n$ be the mid-points of the resulting intervals arranged in order. We can then select intervals from $\mathfrak{F}$ attached to the points $x_1,\dots,x_n$ in such a way that the overlap lies entirely between x_i, x_{i+1} and has for each i a length less than e/n.

In the case of any set E, we enclose E in a set of non-overlapping intervals J_n such that

$$\sum mJ_n - \overline{m}E < e.$$

Discard all the intervals of $\mathfrak{F}$ of length greater than d, and those which do not lie completely interior to some J_n. Associate with each point x of E an interval of the remaining intervals associated with x, having x as middle point, forming a family $\mathfrak{F}_0$. The family $\mathfrak{F}_0$ is equivalent to a denumerable family of non-overlapping intervals K_m lying interior to the J_n, which contains E, so that

$$\sum mK_m - \overline{m}E < e.$$

---

* PROCEEDINGS OF THE LONDON SOCIETY, (2), vol. 9 (1911), pp. 325–35.

† FUNDAMENTA MATHEMATICAE, vol. 4 (1923), pp. 204–13.

‡ PROCEEDINGS OF THE LONDON SOCIETY, (2), vol. 2 (1904), pp. 67–9; ibid., (2) vol. 14 (1915), pp. 122–126. The Youngs assume that E is measurable.

§ Also true if E is a closed set. Obviously this case is closely related to the Lusin-Young Theorem.

Now by following a process similar to that used in §5, and by discarding, at each stage, intervals not needed, we can reduce $\mathfrak{F}_0$ to an equivalent strict family $\mathfrak{F}$ of intervals. This strict family of intervals can be replaced by a family $\mathfrak{F}_{de}$ of intervals chosen from $\mathfrak{F}$, satisfying the conditions (b) and (d), by a line of reasoning similar to that used in connection with the case where E is a closed interval.

7. The Vitali Theorem. The tile theorem of the Youngs is very closely related to the Vitali Theorem, which plays a role in connection with measurable sets and derivatives of functions comparable to that of the Borel Theorem relative to closed sets and continuous functions. It was stated by Vitali* for the linear interval in a form equivalent to the following:

If $\mathfrak{F}$ is a family of intervals such that for every point x of the bounded measurable set E, there exists a set of intervals of $\mathfrak{F}$ containing x, whose lengths approach zero, then there exists a denumerable subfamily of $\mathfrak{F}$ consisting of non-overlapping intervals, the sum of whose lengths is greater than the measure of E.

This theorem has been extended in various ways, especially to higher dimensions, the chief extensions being due to Lebesgue† and Carathéodory.‡ Probably the simplest statement and proof of the fundamental extended form of the theorem has been given by Banach,§ viz.

VITALI THEOREM. *Let E be any bounded set of points in a space of n dimensions. Let $\mathfrak{F}$ be a family of closed sets of points I, such that to each point x of E, there corresponds a sequence I_n chosen from $\mathfrak{F}$ subject to two conditions:*

(1) *if $r_n(x)$ is the radius of the smallest sphere $C_n(x)$ of center x containing $I_n(x)$, then $\lim_n r_n(x)=0$, and*

(2) *there exists a quantity α independent of n and x such that*

$$\frac{\operatorname{meas} I_n(x)}{\operatorname{meas} C_n(x)} > \alpha.$$

Then for every e, there exists a finite or denumerable family of sets $I_n(x_n)=I_n'$ (1) without common points, and such that

$$\sum \operatorname{meas} I_n' < \overline{\operatorname{meas}}\, E + e \tag{2}$$

and

$$\overline{\operatorname{meas}}\left(E - \sum I_n' E\right) < e. \tag{3}$$

In this theorem it is immaterial whether the sets $I_n(x)$ contain x.|| Also it is possible to replace the spheres $C_n(x)$ by cubes having x as center, or rectangular

---

* ATTI DI TORINO, vol. 43 (1907), pp. 229–236.

† ANNALES DE L'ECOLE NORMALE, (3), vol. 27 (1910), pp. 391–5.

‡ *Vorlesungen über reelle Funktionen*, 1918, pp. 299–307.

§ *Sur le théorème de Vitali*, FUNDAMENTA MATHEMATICAE, vol. 5 (1924), pp. 130–6.

|| In Lebesgue's formulation (loc. cit., p. 391) the $I_n(x)$ contain x but then the spheres containing $I_n(x)$ need not have x as center.

parallelepipeds, the ratios of whose dimensions are bounded from infinity and zero. That the theorem is not true with an unconditioned set of rectangular parallelepipeds has been shown by Banach.*

We give the proof in two dimensions, the changes to be made for the n-dimensional case being obvious.

Since E is bounded, it is possible to find an open set U containing all the points of E and such that

$$\text{meas}\, U \leqq \overline{\text{meas}}\, E + e.$$

We then reduce the family $\mathfrak{F}$ to the family $\mathfrak{F}_1$, by retaining only those $I_n(x)$ which are contained in U, so that condition (2) of the conclusion is fulfilled if the selection can be made in accordance with condition (1).

Let k be any constant greater than unity. Then by selecting an

$$I_1' = I_{n_1}(x_1),$$

such that the radius $r_{n_1}(x_1)$ is sufficiently near to the least upper bound of the radii $r_n(x)$ we can make sure that

$$r_n(x) < k r_{n_1}(x_1) = k r(I_1'),$$

for every n and x. Similarly if $I_1', \ldots, I_{m-1}'$ have been determined, then by a similar method we can select the set

$$I_m' = I_{n_m}(x_m)$$

with properties (a) I_m' does not have any points in common with I_j' for $j \leqq m-1$, and (b)

$$r_n(x) < k r_{n_m}(x_m) = k r(I_m')$$

for all $I_n(x)$ which have no points in common with I_j' for $j \leqq m-1$. Since E and therefore also U is bounded, and consequently has finite upper measure, it follows that meas I_m' approaches zero with m and consequently $r(I_m')$ approaches zero with m, due to condition (2) of the hypothesis. Let C_m' be the circle of radius $(2k+1)r(I_m')$, center x_m. Then we show that for every m, all points of E belong to

$$\sum_1^{m-1} I_i' + \sum_m^{\infty} C_j'.$$

We observe first that the sets I_n' for $n \geqq m$ are chosen from the reduced family $\mathfrak{F}_m$ of $\mathfrak{F}$ with respect to the open set

$$U_m = U - \sum_1^{m-1} I_j'$$

as I_n', for $n \geqq 1$, was selected from the reduced family $\mathfrak{F}_1$ with respect to U. Let x_0 be any point of E. Then either x_0 belongs to some I_j' or it belongs to U_m for every m. Let $I_m(x_0)$ be the I of maximum radius contained in U_m. Then since $r(I_n)$

* Loc. cit., pp. 134–6.

approaches zero with n, there exists a value of n such that

$$r(I_m(x_0)) > kr(I'_n),$$

but

$$r(I_m(x_0)) \leqq kr(I'_j),$$

for $j = m, \ldots, n-1$. It follows that $I_m(x)$ must have parts in common with one of the sets $I'_m, \ldots, I'_{n-1}$, and hence is completely covered by the corresponding C'_j by the method of formation of the C'_j.

Now since

$$\sum_1^\infty \operatorname{meas} I'_j < \overline{\operatorname{meas}}\, E$$

it follows that $\Sigma_m^\infty \operatorname{meas} I'_j$ approaches zero with m. Hence since

$$\sum_m^\infty \operatorname{meas} I'_j > \alpha \sum_m^\infty \operatorname{meas} C'_j = \frac{\alpha}{(2k+1)^2} \sum_m^\infty \operatorname{meas} C'_j$$

it follows that $\Sigma_m^\infty \operatorname{meas} C'_j$ approaches zero with m, i.e., the set $\Sigma_1^\infty I'_j$ contains all points of E up to a set of measure zero.

The following generalization is possible:*

The Vitali Theorem is still valid in case the set E is any set, and the constant α of condition (2) *of the hypothesis is dependent on x, but independent of n.*

The case when E is any set can be reduced to the case when E is bounded, by consideration of the fact that n-dimensional space can be broken up into a denumerable set of compartments of finite magnitude. For the case where E is bounded and α is dependent on the point x, we consider the sets E_n of points of E which satisfy the conditions

$$\frac{1}{n+1} < \alpha(x) \leqq \frac{1}{n}$$

where $\alpha(x)$ is the greatest lower bound of the values of the fraction

$$\frac{\operatorname{meas} I_m(x)}{\operatorname{meas} C_m(x)}.$$

We can then apply the previous theorem to the sets E_n, successively. We determine $I'_1, \ldots, I'_n$ so that

$$\operatorname{meas}\left(E_1 - \sum_{j=1}^n E_1 I'_j\right) < \frac{e}{2},$$

* Lebesgue, loc. cit., p. 393; Carathéodory, loc. cit., p. 305.

then $I'_{n_1+1},\ldots,I'_{n_2}$ belonging to $U-\sum_{j=1}^{n_1} I'_j$, so that

$$\text{meas}\left(E_1+E_2-\sum_{j=1}^{n_2}(E_1+E_2)I'_j\right)<e\left(1-\frac{1}{2^2}\right),$$

and so on.

The Vitali Theorem gives a very elegant method for demonstrating the following generalization of the Lebesgue metric density theorem. For the statement of the theorem we define the upper metric density of a set E at a point x of E as

$$\lim_{r\to 0}\frac{\overline{\text{meas}}\ C_rE}{\text{meas}\, C_r}$$

where C_r is a sphere of radius r and center x. Then the theorem is*

The points of any set E at which the metric density is not unity form a set of zero measure.

For let E_0 be the set of points of E at which

$$\lim_{r\to 0}\frac{\overline{\text{meas}}\ C_rE}{\text{meas}\, C_r}$$

does not exist or is less than unity, and let E_k, k a positive integer, be the points of E for which there exists a sequence of circles with radii r_n converging to zero, such that

$$\overline{\text{meas}}\ C_{r_n}E<\left(1-\frac{1}{k}\right)\text{meas}\, C_{r_n}.$$

Then $E_0=\Sigma E_k$. The spheres C_{r_n} form a family $\mathfrak{F}$ as required by the Vitali theorem for the points of E_k. Hence for every e there exists a denumerable set of the C_{r_n}, C'_m satisfying the conditions

$$\overline{\text{meas}}\ E_k=\overline{\text{meas}}\ \sum_m C'_mE_k=\sum_m\ \overline{\text{meas}}\ C'_mE_k,$$

and

$$\sum_m \text{meas}\, C'_m<\overline{\text{meas}}\ E_k+e.$$

But since E_k is part of E, we have

$$\overline{\text{meas}}\ C'_mE_k<\overline{\text{meas}}\ C'_mE<\left(1-\frac{1}{k}\right)\text{meas}\, C'_m.$$

---

* See Lebesgue, loc. cit., p. 407; H. Blumberg, TRANSACTIONS OF THE AMERICAN MATHEMATICAL SOCIETY, vol. 24 (1923), pp. 122ff; and W. Sierpinski, FUNDAMENTA MATHEMATICAE, vol. 4 (1923), pp. 167–171, where other references are to be found.

Then

$$\overline{\text{meas}}\ E_k = \sum_m \overline{\text{meas}}\ C'_m E_k < \left(1 - \frac{1}{k}\right) \sum_m \text{meas}\, C'_m$$
$$< \left(1 - \frac{1}{k}\right)(\overline{\text{meas}}\ E_k + e),$$

from which

$$\overline{\text{meas}}\ E_k < (k-1)e,$$

i.e., E_k is of measure zero for every k, giving the same result for E_0.

The question naturally arises what it is possible to do in the matter of selecting the sets I_n so as to contain all points of E. H. Rademacher* has given the following which might be considered a generalization of the Young Tile Theorem:

If with every point x of E, there is associated a sequence of spheres, whose radii converge to zero, then for every e it is possible to find a denumerable set of these spheres E_n, such that

$$\sum \text{meas}\, C_n < \overline{\text{meas}}\ E + e$$

and every point of E is interior to at least one of the spheres.

This is a consequence of the Vitali Theorem and the result that under the hypothesis of the theorem, there exists a denumerable set of spheres covering E and a constant k (which depends upon the dimension of the space) such that

$$\sum \text{meas}\, C_n < k(\overline{\text{meas}}\ E + e).$$

Essentially the point is that under the given hypothesis it is possible for every e to cover a set of measure zero by a denumerable set of the given spheres, the sum of whose measures is less than the given e.

It is obvious that in the Rademacher theorem, the spheres can be replaced by cubes having the points x as center. To further extensions there are limitations. J. Splaya-Neumann† has shown that it is necessary that the points x be the centers of the spheres by giving an example of a plane closed set of measure zero, such that the sum of the measures of the covering circles is always greater than or equal to unity.‡

For the linear interval, J. C. Burkill § has given an exact covering theorem based on intervals, an extension of the Vitali Theorem. By observing that as a result of the metric density theorem the finite number of intervals of the Vitali Theorem can

* Monatshefte für Mathematik und Physik, vol. 27 (1916), pp. 189–190.

† Fundamenta Mathematicae, vol. 5 (1924), pp. 329–30.

‡ Cf. also K. Menger, Wiener Berichte, vol. 133 IIa (1924), pp. 425–7; and R. L. Moore, loc. cit., pp. 464–5, where examples based on intervals, rectangles, and squares are given, which are not centered relative to the points of association. Moore's example, however, is not with respect to a set of measure zero, as he claims.

§ Fundamenta Mathematicae, vol. 5 (1924), pp. 322–4.

be chosen so that each of the complementary intervals contains a point of the given set, and combining this with the Borel Theorem he obtains the following slightly complicated result:

If $\mathfrak{F}$ is a given family of intervals I satisfying the hypotheses of the Vitali Theorem relative to a closed set E, and $\mathfrak{G}$ a family of intervals J such that for every x of E all intervals in a certain vicinity of x belong to $\mathfrak{G}$, then there exists an interval K which for every e is completely covered without overlapping by a finite number of intervals from $\mathfrak{F}$ and $\mathfrak{G}$ fulfilling the conditions

$$\text{meas}(\Sigma I_n - (\Sigma I_n)EK) < e, \quad \textit{and} \quad \text{meas}(EK - (\Sigma I_n)EK) < e.$$

We note finally that in the Vitali Theorem, the sets I of the family $\mathfrak{F}$ may be replaced by measurable sets, but then the condition that the subfamily consist of non-overlapping sets must be dropped.*

II. The Borel Theorem in General Spaces

Probably no theorem of analysis has contributed more towards the analysis of general spaces than the Borel Theorem. The attempts to derive the theorem in increasingly general situations has led to interesting new properties and characterizations of spaces.

8. Metric Space. The first and simplest general space to which the Borel Theorem was extended is now generally called a metric space. The definition of the space and the proof of the theorem in this space were made by Fréchet in his Paris thesis.† A metric space $\mathfrak{D}$ consists of a set of general elements x. It is postulated that for every pair of elements x_1 and x_2 of the space there exists a positive real number $\delta(x_1,x_2)$ called distance and subject to the conditions

(1) $\delta(x_1,x_2)=\delta(x_2,x_1)$ for every x_1 and x_2,
(2) $\delta(x_1,x_2)=0$ if and only if x_1 and x_2 are identical,
(3) $\delta(x_1,x_2) \leqq \delta(x_1,x_3)+\delta(x_3,x_2)$ for every x_1, x_2 and x_3.

A sequence $\{x_n\}$ is said to have as *limit* the element x if $\lim_n \delta(x_n,x)=0$. A set E in the space $\mathfrak{D}$ is said to have x as *limiting element* if there exists a sequence of distinct elements $\{x_n\}$ extracted from E having x as limit. *Derived sets, closed sets,* and *perfect sets* are defined in the usual way, and the notation E' is used for the derived set of E. In such a space it seems natural to define a *sphere* center x_0 and radius r as the totality of points of $\mathfrak{D}$ satisfying the condition

$$\delta(x,x_0) \leqq r,$$

* See Lebesgue, loc. cit., p. 394; and Rademacher, loc. cit., pp. 191–2.

† *Sur quelques points du calcul fonctionnel*, Palermo Rendiconti vol. 22 (1906), pp. 1–72. Fréchet's results were stated with respect to what seemed to be a slightly more general situation, the equivalence with the above metric space being shown by E. W. Chittenden, Transactions of the American Mathematical Society, vol. 18 (1917), pp. 161–6.

also to speak of a sphere of radius e as a *vicinity* of its center. The notion x *interior to the set* E can be defined in the following equivalent ways:

(a) there exists a sphere having x as center all of whose points belong to E, or

(b) x belongs to E and is not a limiting element of any set consisting of points all of which do not belong to E. From the properties of δ it follows that all points x such that $\delta(x,x_0)<r$ are interior to the sphere center x_0 and radius r.

In a general metric space the Weierstrass-Bolzano Theorem is not a consequence of the boundedness of a set. Instead we have the property *compact*, a set E being compact, if every infinite subset has a limiting element. If every infinite subset of E has a limiting element in E, we shall call E *self-compact*,* which concept in a metric space is equivalent to compact and closed.

As in the linear case, we shall say that a family $\mathfrak{F}$ of sets I chosen from a space $\mathfrak{D}$ covers the set E if every point of E is interior to some set I of $\mathfrak{F}$. Then the Borel Theorem can be stated:

Any self-compact set E which is covered by a family $\mathfrak{F}$, *can be covered by a finite subfamily of* $\mathfrak{F}$.

It may occasionally be useful to refer to the property expressed in this theorem as the *Borel property*, i.e., E has the Borel property if from any family covering E, a finite subfamily covering E can be selected.

Fréchet proved the theorem first for the case in which the family is denumerable. The process is very much the same as that given in §2(d), for the linear case. The family is arranged in sequential order $I_1,\ldots,I_n,\ldots$, and by a step-by-step process replaced by an array $I_1,\ldots,I_{n_m},\ldots$, in which each set contains at least one element of E not interior to the preceding set. If this sequence is infinite, we get a sequence of distinct elements of $E, x_1,\ldots,x_m,\ldots$, where x_m is interior to I_{n_m} but not to any preceding set. By the self-compactness of E, this sequence has a limiting element x_0 of E. Now x_0 is interior to some set I_n and consequently I_n contains a sphere having x_0 as center, and hence an infinite number of the sequence x_m as interior points. This leads to a contradiction with the method of selecting I_{n_m} and x_m.

We call attention to the fact that the proof utilizes in particular two ideas, (a) the self-compactness of E, and (b) the fact that if x is interior to E, and a limiting element of E_1 then E contains as interior elements an infinite number of elements of E_1.

For the general Borel Theorem, Fréchet† originally postulated further properties of the space, viz., that the space $\mathfrak{D}$ is *separable*, i.e., can be considered as the derived set of a denumerable set of its elements. It was shown later that any compact set E of a space $\mathfrak{D}$ has the same property, i.e., for any compact set E, there exists a denumerable subset E_0 such that E is contained in E_0+E_0', thus

* Due to E. W. Chittenden, Bulletin of the American Mathematical Society, vol. 21 (1915), p. 18.

† Loc cit., pp. 25–27.

removing the restriction of separability for the space $\mathfrak{D}$.*

The proof of the Borel Theorem depends upon the following lemma:

LEMMA.† *If every point x of a compact set E is the center of a sphere of radius $r(x)$, and if there exists an e such that for all x of E*

$$r(x)>e>0,$$

then all points of E are interior to a finite number of these spheres.

Let $S(x_1)$ be the sphere about any point x_1 of E, and x_n any point not interior to $S(x_1), S(x_2), \ldots, S(x_{n-1})$. Then on account of the fact that $\delta(x_n, x_m)>e$ for every n and m and E is compact, the sequence x_n is finite.

With this lemma, the Borel Theorem in a metric space can be proved in two ways. Either (a) following the method of §2(c) on the linear interval, let $r(x)$ be the least upper bound of the radii of the spheres of center x interior to some set I of the given family. The self-compactness of the family yields immediately that the $r(x)$ have a positive lower bound, and the lemma then suggests a method for selecting the finite subfamily from $\mathfrak{F}$.

Or (b) from the lemma we conclude that for every n the points of a compact set E are interior to a finite number of spheres of radii $1/n$. For every n then we retain the spheres which are interior to some set I of $\mathfrak{F}$. We obtain in that way a denumerable family of spheres, covering E, and hence by the denumerable-to-finite Borel Theorem, we can select a finite subfamily of spheres, which in turn defines a finite subfamily of $\mathfrak{F}$ covering E.

This last proof contains practically the proof of the result that *any compact set E is separable*. For the centers of the spheres of radius $1/n$ having E as interior points will be a denumerable set E_0 having the property that E belongs to $E_0 + E_0'$.

Also there is present a special case of the Lindelöf Theorem in a metric space, viz.

If in a space $\mathfrak{D}$ E is any separable set covered by a family $\mathfrak{F}$ of sets I, then it is covered by a denumerable subfamily.

Obviously the reason why the Lindelöf theorem holds in linear or n-dimensional space is because these spaces are separable. The method of proof is entirely similar to these special instances. Let $E_0=[x_n]$ be a denumerable subset of E such that E belongs to E_0+E_0'. Consider the denumerable family of spheres, center x_n, and rational radii interior to some I of $\mathfrak{F}$. They will cover E and set up a one to one correspondence with a denumerable subfamily of $\mathfrak{F}$.

---

* See T. H. Hildebrandt, AMERICAN JOURNAL, vol. 34 (1912), pp. 278–281; W. Gross, WIENER BERICHTE, vol. 123 IIa (1914), pp. 809–812; Fréchet, BULLETIN DE LA SOCIÉTÉ DE FRANCE, vol. 45 (1917), pp. 1–8.

† See Hahn, *Reelle Funktionen*, 1921, pp. 89–93. See also Urysohn, FUNDAMENTA MATHEMATICAE, vol. 7 (1925), pp. 46–48, where the role of the axiom of choice in the proof of the Borel Theorem is emphasized.

The condition that E is self-compact is necessary for the validity of the Borel Theorem. For if E is not self-compact, then there exists a denumerable set of elements $[x_n]$ of E without a limiting element in E. To every point x of E there corresponds then a sphere containing at most one point of $[x_n]$. These spheres form a family covering E, but every finite subfamily contains only a finite number of points of $[x_n]$ and hence does not cover E.

Similarly the condition that E be separable is also necessary for the Lindelöf Theorem. Instead of proving this directly we relate this theorem and separability to a third property suggested in Lindelöf's paper. If we define an *element of condensation* of a set E, as a limiting point of E which remains a limiting point after the removal of any denumerable set from E, and a set as *self-condensed* if every non-denumerable set chosen from E has at least one element of condensation in E, then we have the equivalence of the following three properties in a metric space $\mathfrak{D}$:*

(A) *The set E is separable.*

(B) *The set E is self-condensed.*

(C) *The set E has the Lindelöf property*, i.e., if $\mathfrak{F}$ is any family of sets covering E, then a denumerable sub-family of $\mathfrak{F}$ covers E.

We have already shown that (A) implies (C). To show that (C) implies (B) follows the lines of the converse of the Borel Theorem above; the assumption of a non-denumerable set E_0 without an element of condensation in E yields via (C) a denumerable set of spheres each of which contains only a denumerable set of elements of E_0. To show finally that (B) implies (A), we show first that if E is condensed and if each point of E is the center of a sphere of radius greater than e, then all the points of E are interior to a denumerable subset of these spheres. Allowing e to take on successively the values $1/n$, we get a denumerable set of centers of spheres which is the desired denumerable subclass of E.

In so far as measure has not yet been effectively connected with a general metric space, it is not possible to give generalizations of the Vitali Theorem. However K. Menger† has given some results which are comparable to the Vitali Theorem.

We define the *diameter* of a set I as the diameter of the minimum sphere containing I. Then Menger is interested in the question: What properties of a set E in a metric space and what properties of a family $\mathfrak{F}$ of sets I such that every point of E is interior to a subfamily of sets I whose diameters converge to zero, are sufficient to make possible the selection of a finite or denumerable subfamily of $\mathfrak{F}$ of sets I whose diameters approach zero, and which covers E.

---

* See Gross, loc. cit., pp. 805–12; Fréchet, ANNALES DE L'ÉCOLE NORMALE, (3), vol. 38 (1921), pp. 349–356.

† *Einige Überdeckungssaetze der Punktmengenlehre*, WIENER BERICHTE, vol. 133IIa (1924), pp. 421–444.

If $d(I)$ is the diameter of I and $d(I,x)$ is the least upper bound of the diameters of spheres center x contained in $I(x)$, then Menger's principal result is that the selection desired is possible provided

(a) E is the sum of a denumerable set of compact sets;

(b) if $I(x)$ are the sets of $\mathfrak{F}$ associated with x, then for every x the condition that $d(I)$ approach zero is a consequence of the fact that $d(I,x)$ approaches zero. This latter condition is fulfilled in case there exists a positive-valued function $f(x)$ on E such that for each I associated with x

$$d(I)<f(x)\cdot d(I,x).$$

The proof is made first for a compact set E. If E_n is the subset of E for which there exists an I such that $d(I,x)>1/n$ then E_n is interior to a finite number of these sets I. As a consequence E can be covered by a finite or denumerable family of sets I_n from $\mathfrak{F}$, which if denumerable has the property that there exists a point x_n of I_n such that $d(I_n,x_n)$ approaches zero, so that $d(I_n)$ converges to zero as desired. The extension to the denumerable set of compact sets if obvious.

In addition to having contact with the Vitali Theorem, the results of Menger are also related to Young's Tile Theorem (§6) and R. L. Moore's reduction theorem (§5). The conditions of Menger are sufficient. No doubt further interesting results can be obtained by considering necessary conditions.

9. The Borel Theorem in a Space $\mathfrak{L}$ with Limit of Sequence Defined. Fréchet's thesis besides considering metric spaces also postulated a more fundamental space $\mathfrak{L}$, that in which limit of a sequence is defined. Limit is subject to three conditions: (1) the limit of a sequence is unique, (2) the limit of a sequence consisting of the same element repeated is this element, and (3) any subsequence of a sequence having a limit has the same limit. Obviously limiting element, interiority, and the other concepts can be defined in the main as suggested for metric spaces.

The first statement of a Borel Theorem in a space $\mathfrak{L}$ is due to E. R. Hedrick.* By analyzing the proof for the Borel Theorem in the denumerable-to-finite case, he observed that it could be effected provided the space $\mathfrak{L}$ had the following property† (called by Fréchet the Hedrick property):

(H) *If x is interior to a set E, then an infinite number of elements of any sequence having x as limit are interior to E.*

He found that this property was a consequence of the simpler property:

(S) *The derived class of any class is closed.*

---

* TRANSACTIONS OF THE AMERICAN MATHEMATICAL SOCIETY, vol. 12 (1911), pp. 285–7.

† Fréchet's statement (cf. ANNALES DE L'ÉCOLE NORMALE, (3), vol. 38 (1921), p. 348) of this property is: If x is interior to E and a limiting element of F, then x is a limiting point of a subset F_0 of F consisting entirely of elements interior to E. In an $\mathfrak{L}$ space this statement is equivalent to the one above.

For suppose x is interior to E and the limit of a sequence of distinct elements x_n, the result being obvious if the elements are not distinct. We show that there exists an n_0 such that for $n>n_0$, x_n is interior to E. The assumption of the contrary gives rise to a sub-sequence $\{x_{n_m}\}$ no point of which is interior to E, i.e., each member of the sub-sequence is the limiting element of a sequence of elements $\{x_{n_m k}\}$ not members of E. If we let E_0 be the class of elements $[x_{n_m k}]$, then E_0' contains the sequence $\{x_{n_m}\}$ and consequently the point x, by the property S. The element x would then be a limiting element of E_0 which would contradict the interiority condition of x to E.

It is now fairly obvious that the Borel Theorem is valid in the form

If the space $\mathfrak{L}$ *has the property* S *then any self-compact set E covered by a denumerable family* $\mathfrak{F}$ *of sets I, is covered by a finite sub-family of* $\mathfrak{F}$, i.e., any self-compact set E has the denumerable-to-finite Borel property.

The condition that E be self-compact is necessary in any space $\mathfrak{L}$.* For suppose $\{x_n\}$ is a sequence chosen from E not having a limiting element in E. Let I_m be the set obtained from the given space by deleting the elements x_n for $n>m$. Then obviously the set E is covered by the family $\mathfrak{F}$ of I_m, since no element of E is a limiting element of $\{x_n\}$, but no finite subfamily of $\mathfrak{F}$ contains all points of the sequence $\{x_n\}$.

Further the property S and so H is a necessary property in a space $\mathfrak{L}$ for the Borel Theorem in this form: if E is self-compact then E has the denumerable-to-finite Borel property.† Let if possible the set E be such that E' is not closed. Then there exists a sequence $\{x_n'\}$ of elements of E' with a limit x'' not belonging to E'. Let the elements $\{x_{mn}\}$ of E be such that x_n' is a limiting element of x_{mn} for $m=1,2,\ldots$. Consider the family $\mathfrak{F}$ consisting of (a) I_0, the set remaining after removing the elements x_{mn} from the fundamental space, and (b) I_n the set remaining after removing all members of the sequence $\{x_m'\}$ excepting x_n' from the fundamental space. Then the self-compact set E_0 consisting of the sequence $\{x_n'\}$ and x'' is covered by the family $\mathfrak{F}$ but by no finite subfamily of this family.

We note that *in a space* $\mathfrak{L}$ *there is then equivalence between these three properties*:

(S) *The derived class of every class E is closed.*

(H) *If x is interior to E, and is the limiting element of a set E_1 then an infinite subset of E_1 is interior to E.*

(B) *If E is self-compact, it has the denumerable-to-finite Borel property.*

The problem of determining conditions under which the any-to-finite Borel Theorem is valid in a space $\mathfrak{L}$ remained unsolved for some time. Obviously the methods of metric space could not be used. A method of attack is suggested by the

* Cf. E. W. Chittenden, *The converse of the Heine-Borel theorem in a Riesz domain*, BULLETIN OF THE AMERICAN MATHEMATICAL SOCIETY, vol. 21 (1915), pp. 179–183.

† Cf. Fréchet, BULLETIN DE LA SOCIÉTÉ DE FRANCE, vol. 45 (1917), pp. 1–8; Chittenden, BULLETIN OF THE AMERICAN MATHEMATICAL SOCIETY, vol. 25 (1918), pp. 60–66.

proof in the case just treated, viz., to utilize the theory of transfinite ordinals. The first solution of the problem was given by R. L. Moore.* He calls a *monotonic family of classes* G, a family such that for each pair G_1 and G_2 of the family, one contains the other. He then defines the concept called by Fréchet *perfectly compact*. The set E is perfectly compact in case every infinite monotonic family of sets chosen from E or the family of their derived sets has a common element. That a perfectly compact set is compact is obvious from a consideration of the monotonic family where G_m consists of the elements of the sequence $\{x_n\}$ for $n > m$.† The relationship of the property "perfectly compact" to the theorem "any monotonic sequence of closed compact sets has a common element" will appear later. Moore's result is as follows:

In a space $\mathfrak{L}$ *with the property* S, *a necessary and sufficient condition that E have the any-to-finite Borel property is that E be perfectly self-compact.*

Suppose there is given a family $\mathfrak{F}$ of sets I which covers E. Well-order this family, and then by a step-by-step process reduce the family so that each I contains as interior point at least one element of E not interior to any of the preceding sets in the array. Let the members of the resulting family be $I_1, I_2, \ldots, I_\alpha, \ldots$, and let x_α be interior to I_α but not to I_β for $\beta > \alpha$. Let E_α be the set of points x_β for $\beta > \alpha$. Then the family consisting of the E_α is a monotonic family from E, and since the E_α do not have a common point, either the family is finite or the derivatives E_α have a common point x'. In the latter case there exists a γ such that x' is interior to I_γ, and since x' is a limiting element of E_γ, by the property H it follows that I_γ contains interior points of the set of x_α for ordinals greater than γ, contrary to the definition of the x_α. Hence the number of members of the family of E_α is finite.

For the proof of the converse, let E be a set of the space $\mathfrak{L}$ having the any-to-finite Borel property. Then as shown above E is self-compact, i.e., compact and closed. Let G be any infinite monotonic family of sets E_α drawn from E. Then since E is closed E'_α will belong to E. Moreover the property S insures that the sets $\bar{E}_\alpha = E_\alpha + E'_\alpha$ are closed. Let the sets I_α of the family $\mathfrak{F}$ consist of the points remaining after deleting the elements of $\bar{E}_\alpha$ from the fundamental set. If the sets $\bar{E}_\alpha$ have no common element, it follows that E will be covered by the family $\mathfrak{F}$, but a finite subfamily of $\mathfrak{F}$ will not cover E since the sets $\bar{E}_\alpha$ contain elements for each α.

Another solution of this question of the Borel any-to-finite Theorem in a space $\mathfrak{L}$ was given in 1923 by Kuratowski and Sierpinski.‡ In so far, however, as their result is practically stated in a space in which vicinities are defined we shall postpone consideration of it to a later section of this paper.

* See Proceedings of the National Academy of Sciences, vol. 5 (1919), pp. 206–210.

† Fréchet (Annales de l'École Normale, (3), vol. 38, pp. 334–6) shows that in a metric space every compact set is also perfectly compact. This can also be deduced from the converse of the any-to-finite Borel Theorem in a metric space and in an $\mathfrak{L}$ space.

‡ *Le théorème de Borel-Lebesgue dans la théorie des ensembles abstraits*, Fundamenta Mathematicae, vol. 2 (1921), pp. 172–8.

10. Vicinity Spaces $\mathfrak{B}$. Hausdorff Form of the Borel Theorem. The first suggestions of sets or vicinities as the basis for consideration in a general space are to be found in the paper of Hedrick.* Another development based to some extent on the Riesz† postulates for limiting element was given by R. E. Root.‡ About the same time, Hausdorff in his book on Mengenlehre developed systematically the point set theory in a vicinity space. Later, in 1918, Fréchet§ gave an independent development of the same type of space showing in particular the relationship between the space characterized by the Riesz postulates and a space based upon vicinities. The Hausdorff postulates have come to be accepted as a satisfactory basis, and a space based on them is usually called a topologic space. The postulates are as follows:

I. To every point x there corresponds a family of sets $V(x)$, chosen from the given space, and containing x.

II. If $V_1(x)$ and $V_2(x)$ are vicinities of x, then there exists a common subvicinity $V_3(x)$.

III. For every pair of points x_1 and x_2, there exist vicinities $V_1(x_1)$ and $V_2(x_2)$ without common elements.

IV. If x_2 belongs to $V_1(x_1)$ then there exists a $V_2(x_2)$ contained in $V_1(x_1)$.

In a vicinity space *limiting element* of a class E can be defined either (a) x is a limiting element of E if every $V(x)$ contains at least one point other than x, or (b) x is a limiting element of E if every $V(x)$ contains an infinity of elements. In a $\mathfrak{B}$ space satisfying conditions I, II, and III these two definitions are equivalent.

More generally we can define x *is a limiting element of E of power* μ, if every vicinity of x contains a subset of power μ. Finally x is called a *complete limiting element of* E if every $V(x)$ contains a subset of E of the same power as E.

It is obvious that we can obtain an $\mathfrak{L}$ space in a $\mathfrak{B}$ space by assuming that $\lim_n x_n = x$ is defined "for every $V(x)$ there exists an n_0 such that if $n > n_0$ then x_n is contained in $V(x)$." But limiting element based on the sequence notion of this $\mathfrak{L}$ space need not agree with the limiting element of the given $\mathfrak{B}$ space, unless the $\mathfrak{B}$ space is subjected to additional conditions.‖ On the other hand given an $\mathfrak{L}$ space it is possible to define it as a $\mathfrak{B}$ space in which limiting elements are the same.#

The notion *interiority* can be defined in different ways, equivalent if we are in a Hausdorff $\mathfrak{B}$ space (i.e. subject to conditions I, II, III and IV)**: *x is interior to E* if x belongs to E and either (a) every set E_1 having x as a limiting element, contains at least one element other than x of E, or (b) every set E_1 having x as limiting

* Loc. cit., p. 289; Fréchet, TRANSACTIONS OF THE AMERICAN MATHEMATICAL SOCIETY, vol. 14 (1913), pp. 320–4, showed that the space postulated by Hedrick was a metric space.

† See ATTI DEL IV CONGRESSO INTERNAZIONALE (Roma) 1909, vol. 2, pp. 18–22.

‡ Cf. TRANSACTIONS OF THE AMERICAN MATHEMATICAL SOCIETY, vol. 15 (1914), pp. 51–70.

§ BULLETIN DES SCIENCES MATHÉMATIQUES, (2), vol. 42 (1918), pp. 138–156; called Fréchet I in the sequel. Fréchet considers a type of space that he has called "espace accessible," which is equivalent to a vicinity space subject to postulates similar to those of Hausdorff, IV and especially III being replaced by weaker ones.

‖ Cf. Root, loc. cit., pp. 67–71; Fréchet, I, p. 148.

Cf. Fréchet, I, pp. 140–148.

** Limiting element may be of power 2 or $\aleph_0$. Definition (c) is perhaps most satisfactory in so far as limiting element does not enter directly.

element contains an infinity of elements of E, or (c) there exists a $V(x)$ containing only elements of E.

Obviously x is interior to every $V(x)$.

An *open set* or *region* is a set containing only interior points. On account of condition IV every $V(x)$ is an open set. We note that the sum of two open sets is open, also the complementary set of a closed set is open.

Finally we note that on account of condition IV, a Hausdorff $\mathfrak{B}$ space has the property S.

The outstanding difference in Hausdorff's statement of the Borel Theorem from that stated by Fréchet is that the family $\mathfrak{F}$ of covering sets consists of open sets, which introduces an element of simplicity, since belonging to an open set is equivalent to being interior. The most elegant form of the Borel Theorem in a Hausdorff $\mathfrak{B}$ space has been given by P. Alexandroff and P. Urysohn,* by calling attention to the following equivalences:

THEOREM I. *The following properties of a set E in a Hausdorff space are equivalent*:

A_0. *E is self-compact.*

A_1. *Every denumerable subset of E has a complete limiting element in E.*

B. *If $\mathfrak{F}$ is a denumerably infinite monotonic family of closed sets F_n such that each F_nE contains at least one element, then the sets F_nE have a common element.*

C. *If E is covered by a denumerable family $\mathfrak{G}$ of open sets G, then E is covered by a finite subfamily of $\mathfrak{G}$.*

It is obvious that A_0 and A_1 are equivalent. The equivalence of B and C is a matter of taking complementaries, products and sums. Assume B, and let $\mathfrak{G}=[G_n]$. Then the sets $\Sigma_1^m G_n$ are open sets, and the sets $\mathfrak{B}-\Sigma_1^m G_n$ closed, as a matter of fact form a monotonic family. If the sets $E(\mathfrak{B}-\Sigma_1^m G_n)$ contain an element for every m, then by B they have a common element, i.e., the family $\mathfrak{G}$ does not cover E. Conversely, assume C and a monotonic family of closed sets $[F_n]$ such that each F_nE contains a point. Then the $\mathfrak{B}-F_n$ are open sets. If the F_nE do not have a common point, then the sets $\mathfrak{B}-F_n$ will contain all points of E. The finiteness of the equivalent set chosen from the $\mathfrak{B}-F_n$ leads to a contradiction.

The fact that A_1 implies B is a well-known result. The denumerable family F_nE leads to a set E_0, of points x_n where x_n belongs to F_nE. The complete limiting element of E_0 in E is also a limiting element for each F_nE, and hence in each member of the family.

To complete the equivalence we show that C implies A_0. If possible let E_0 be a denumerable set without a limiting element in E. Then $\bar{E}_0=E_0+E_0'$ will be closed and have no limiting elements in E. Then if x_n is any element of E_0, the sets

$$G_n=\mathfrak{D}-\bar{E}_0+x_n$$

will be open and will contain all points of E, but a finite subfamily of the G_n will not.

* Cf. MATHEMATISCHE ANNALEN, vol. 92 (1924), pp. 258–60.

More generally we have the following theorem:

THEOREM II. *The following properties of E in a Hausdorff $\mathfrak{B}$ space are equivalent:*

A. *Every infinite subset of E has a complete limiting element belonging to E.*

B. *If $\mathfrak{F}$ is any well-ordered monotonic decreasing family of closed sets $\mathfrak{F}$ such that for each α, $F_\alpha E$ contains an element, then the sets $F_\alpha E$ have a common element.*

C. *If E is contained in a family $\mathfrak{G}$ of open sets G, then it is contained in a finite subfamily.*

The proof of the equivalence of B and C follows the lines of the corresponding proof for Theorem I.

Assume A, and let $\mathfrak{G}$ be a family of open sets G.* Let μ be the power such that any subfamily of $\mathfrak{G}$ of power less than μ does not contain E, but there are subfamilies of power μ containing E. Let $\mathfrak{G}_0$ be a subfamily of $\mathfrak{G}$ of power μ containing E. Let Ω be the least ordinal of power μ. Then we can well-order $\mathfrak{G}_0$ in the form $G_1, G_2, \ldots, G_\alpha, \ldots$, such that $\alpha < \Omega$ and by possible deletion assume that each G_α contains at least one point x_α of E not in any preceding set. Consider the set $E_0 = [x_\alpha]$. By the property A if E_0 is infinite, E_0 has a complete limiting element x, belonging to E, and consequently to some member G_β of the family $\mathfrak{G}_0$. Consequently every vicinity of x and so also G_β will contain a subset of E_0 of power μ. Since the set of elements x_α for $\alpha < \beta$ is of power less than μ it follows that G_β contains points of E_0 of index greater than β, contrary to the method of choice of the x_α. Hence the set E_0 is finite.

To complete the equivalence we show that C implies A. For suppose E_0 is an infinite subset of E not having a complete limiting point in E. Then for every x of E there exists a vicinity $V(x)$ such that the power of $V(x)E_0$ is less than that of E_0. These vicinities being open sets, constitute a family of open sets containing E. But obviously a finite subfamily cannot contain all points of E_0.

Theorem I suggests an extension in which the word "denumerable" is replaced by "power less than or equal to μ," in the properties A_1, B and C, the proof being similar. Theorem II suggests the following extension:

THEOREM III. *The following properties of a class E in a Hausdorff $\mathfrak{B}$ space are equivalent:*

A. *Every infinite set of power μ has a complete limiting element in E.*

B. *If $\mathfrak{F}$ is a well-ordered monotonic decreasing family of closed sets F, of power $\geqq \mu$, such that for each α, $F_\alpha E$ contains at least one element, then the sets $F_\alpha E$ have a common element.*

C. *If E is contained in a family $\mathfrak{G}$ of open sets G, the power of $\mathfrak{G}$ being $\geqq \mu$, then E is covered by a subfamily of $\mathfrak{G}$ of power $\leqq \mu$.*

This theorem contains among others the Lindelöf Theorem as a special case. The proof is similar to that of Theorem II. It is interesting that in the general space

* This method of proof follows the lines suggested by Kuratowski and Sierpinski, loc. cit., pp. 174–5.

the Lindelöf Theorem and the Borel Theorem seem to join hands, a fact not to be foreseen by a consideration of n-dimensional space.

Alexandroff and Urysohn call a set E satisfying the conditions of Theorem II *bicompact*, because it is a meeting place of the generalization of Theorem I and of Theorem III. It is obvious however that property B is a special case of the *perfectly compact* property. Perhaps *completely compact* would be a better term.

It remains to consider to what extent the Hausdorff postulates on the $\mathfrak{V}$ space are needed. An analysis of the proofs shows that in any space in which closed and open sets are complementary the properties B and C are equivalent.*

In Theorem I, A_0 and A_1 are equivalent, and imply B and C if the $\mathfrak{V}$ space is subject only to conditions I, provided "limiting element" is the limiting element of power $\aleph_0$ (i.e. definition (b)). The converse that A_0 and A_1 follow from B and C requires condition IV, making $V(x)$ an open set, which condition is practically the property S: the derived set of a set is closed. If "limiting element" is of power 2, (i.e. definition (a)), then B and C follow from A_1, but A_0 and not A_1 follows from B and C under additional postulate IV.

Theorems II and III are true under a space satisfying postulates I and IV, the latter being required in the proof of the result "C implies A," for instance.

11. General Borel Theorem in a $\mathfrak{V}$ Space. We have pointed out that the Hausdorff statement of the Borel Theorem is based on families of open sets. It seems desirable to consider briefly what happens in case we are dealing with families of arbitrary sets, the deciding covering property being then interiority. At the same time, the attempt is to reduce the properties of the $\mathfrak{V}$ space to a minimum.

For most of this section, we shall assume considerations based on a $\mathfrak{V}$ space subject to condition I of Hausdorff. *Limiting element* is of power 2, i.e., x is a limiting element of E in case every vicinity of x contains a point of E other than x. Further x *is interior to* E if x belongs to E and every set having x as limiting element has a point other than x in common with E, or if E contains a vicinity of x. Compactness, derived sets, limiting points of power μ, complete limiting points are defined as above. We use finally a new concept: x is a *complete interior limiting point*† of E if every vicinity of x contains as *interior* points a set of the same power as E. The definition of interior limiting point of power μ is then obvious.

As in the previous section it seems proper to consider the following properties of a set E in this space as being connected with the denumerable-to-finite Borel Theorem:

A_0. E is self-compact.

A_1. Every denumerable subset of E has a complete limiting point in E.

A_2. Every denumerable subset of E has a complete interior limiting point in E.

B. If $\mathfrak{F}$ is a monotonic denumerable family of sets F chosen from E, either the sets F or their derived sets have a common element in E.

---

* Cf. Saks, FUNDAMENTA MATHEMATICAE, vol. 2 (1921), pp. 1–3.

† Chittenden (BULLETIN OF THE AMERICAN MATHEMATICAL SOCIETY, vol. 30 (1924), p. 556, referred to as Chittenden I in the sequel), calls such a point a hypernuclear point.

C. If E is covered by a denumerable family $\mathfrak{G}$ of sets G, then it is covered by a finite sub-family of $\mathfrak{G}$.

We obviously have that A_2 implies A_1 implies A_0.

The statement "A_2 implies C" is a form of the Borel Theorem whose proof is obvious. The converse C implies A_0 can be proved by assuming if possible E not self-compact. Then there exists a denumerable set E_0 of elements of E without a limiting element in E. Then for each point y_n of E_0 there exists a $V(y_n)$ containing only the point y_n of E, and for each point x of E not a point of E_0 a vicinity $V(x)$ containing no points of E_0. Then the sets

$$G_0 = \sum V(x), \qquad G_n = V_n(y_n)$$

cover E but a finite subfamily does not.

The proof of the fact that B follows from A_1 can be modelled after the more general result given below. The converse is obvious.

To obtain further results it seems necessary to add additional hypotheses on the fundamental space.

The assumption that $\mathfrak{B}$ satisfies the condition IV of Hausdorff, or has the property S, that derived classes are closed, gives a Hedrick property which can be stated as follows.

(H) *If x is interior to E and a limiting element of E_0 of power μ then E_0E contains a set of interior elements, whose power is μ.*

If derived classes are closed, then according to Fréchet* every vicinity $V(x)$ of x contains a subvicinity $V_0(x)$ all of whose elements are interior to $V(x)$. Now if x is interior to E, then there exists a vicinity $V(x)$ of x, which contains only points of E. The subvicinity $V_0(x)$ of V then defines a subset of E_0 of power μ all points of which are interior to E.

If $\mathfrak{B}$ has the property S and consequently H, then we can state the equivalence of B and C. The proof of this equivalence follows the lines of proof of Moore's form of the Borel Theorem in an $\mathfrak{L}$ space.†

The other possible equivalences seem to be linked up with the fact that in the general $\mathfrak{B}$ space with limiting element as defined, a finite class may have a limiting element. The conditions II and III of Hausdorff are sufficient to guarantee the contrary. Under these conditions it is possible to prove that C implies A_2 and A_0 implies A_1. The proof of the former of these statements follows the lines of the proof of the fact that C implies A_0 given above. The latter is obvious. We have as a consequence the following theorem.

THEOREM I. *If the $\mathfrak{B}$ space satisfies condition* IV *of Hausdorff, then properties* A_1, B *and* C *are equivalent. If the space is a Hausdorff $\mathfrak{B}$ space, then all the properties* A_0, A_1, A_2, B, *and* C *are equivalent.*

The extension of these results to the case where the word "denumerable" is replaced by "power less than or equal to μ" in the properties A_1, A_2, B, and C is obvious.

* Cf. Fréchet, I, p. 145.

† Chittenden (I, p. 519) has shown that in a general $\mathfrak{B}$ space the property S is not a consequence of the Borel Theorem in the form "A_0 implies C."

Theorem II of § 10 suggests the consideration of the following properties:

A_1'. Every subset of E has a complete limiting element in E.

A_2'. Every subset of E has a complete interior limiting element in E.

B′. E is perfectly self-compact, i.e., if $\mathfrak{F}$ is any monotonic family of sets F chosen from E then either the sets F or their derived sets have a common element in E.

C′. E has the Borel any-to-finite property, i.e., if E is covered by a family $\mathfrak{G}$ of sets G then E is covered by a finite subfamily of $\mathfrak{G}$.

Chittenden* has stated the following theorem.

THEOREM II. A_1 *is equivalent to* B′ *and* A_2' *is equivalent to* C′.

We prove first that A_1' implies B′. Suppose $\mathfrak{F}$ a monotonic family of sets F chosen from E. Let H be a set of elements of E such that every element of H is in some F and every F contains an element of H. Well-order H, x_α corresponding to the ordinal α. Then we determine a well-ordered subfamily of $\mathfrak{F}$ and a subset H_0 of H by the requirement that F_α contain no element of H with ordinal $\gamma < \beta_\alpha$ and x_{β_α} be the first element of H common to all F_γ for $\gamma \leqq \alpha$. Then for every F of $\mathfrak{F}$ there exists an α such that F contains F_α. For F contains an element x_γ of H, and for some α, x_γ will not belong to F_α, so that by the monotonic character of $\mathfrak{F}$, F contains F_α. From this it follows that if the derived sets F_α' have a common element, the same will be true of the sets F', i.e., if the theorem holds for a well-ordered monotonic family, it holds for any monotonic family.

Let $F_1, \ldots, F_\alpha, \ldots,$ be the elements of $\mathfrak{F}$ in order. Since every well-ordered set without final element is cofinal with a regular ordinal number,† it is obviously sufficient to prove the theorem for the case where every ordinal α precedes Ω the least transfinite ordinal corresponding to μ the power of $\mathfrak{F}$. Let x_α belong to F_α but not to $F_{\alpha+1}$. Then from the assumption concerning E it follows that the set H of elements x_α will have a complete limiting element x'. Every vicinity of x' contains a set of H of power μ, i.e., for every α an element x_β with $\beta > \alpha$, and consequently an element of F_α. Hence x' is common to the sets F_α'.

For the proof of the converse, that B′ implies A_1', we refer the reader to the paper by Sierpinski, which will appear in the next issue of the BULLETIN OF THE AMERICAN MATHEMATICAL SOCIETY.

The proof of the equivalence of A_2' and C′ is parallel to the corresponding equivalence in Theorem II of § 10.

Apparently to get complete equivalence it is necessary either to strengthen condition B′ or add further postulates on the fundamental space. Just what is necessary has not as yet been determined. A sufficient, but not necessary condition is that the space have the property S. Then the property H is valid, which added to

* I, pp. 514–8. Very recently, since this paper was in type, Professor Chittenden has called my attention to the fact that the proofs of the first part of this theorem contained in his paper are not correct. Correct proofs by Sierpinski, who first noted the error, will appear in the next issue of the BULLETIN OF THE AMERICAN MATHEMATICAL SOCIETY. The proof that is given here of the fact that A_1' implies B′ is a modification of Chittenden's proof, obtained before I was aware of an error in his work.

† Cf. Hausdorff, *Mengenlehre*, 1914, p. 132.

B′ gives C′ as in Moore's proof of the Borel Theorem in an $\mathfrak{L}$ space given in § 9. We thus have the following theorem.

THEOREM II′. *If the space* $\mathfrak{B}$ *has the property* S, *then properties* A_1', A_2', B′, *and* C′ *as applied to any set E of the space are equivalent.*

We note that since in a space $\mathfrak{B}$ with the property S, the set $\bar{E}=E+E'$ is closed, perfect self-compactness or condition B′ is equivalent to "If $\mathfrak{F}$ is any infinite monotonic family of closed sets such that each set $F \cdot E$ contains an element, then the FE have a common element," the relationship of which to condition B of Theorem II of § 10 is obvious.

Kuratowski and Sierpinski in a space $\mathfrak{L}$ define a $\mathfrak{B}$ space by the condition that any set V to which x is interior is a vicinity of x. Then their statement of the Borel Theorem is as follows:

In an $\mathfrak{L}$ *space having the property* S, *a necessary and sufficient condition that every self-compact set E have the Borel any-to-finite property is that every set of the space which is compact and whose derived set is compact have a complete interior limiting element.*

The relationship to Theorem II is apparent.

Finally it is obvious that Theorem II can be generalized as in § 10, giving a theorem corresponding to Theorem III, which contains the Lindelöf Theorem* as a special case, and might be labelled "The Borel-any-to-less-than-power μ Theorem."

In closing we cannot refrain from calling attention to a justification of the consideration of general spaces, in gathering under the same roof such apparently diverse results as the Borel and Lindelöf Theorems, and producing a result of greater scope.

APPENDIX TO "THE BOREL THEOREM AND ITS GENERALIZATIONS"

E. J. McSHANE, University of Virginia, Charlottesville

In the half-century since this paper appeared some words have changed meaning. The sets called "compact" in the paper are now called "sequentially compact"; the sets then called "bicompact" are now called "compact sets." The spaces called "topological spaces" are now Hausdorff spaces; topological spaces are now defined as having properties (I), (II) and (IV) of page 56. But besides these and a few other minor changes in terminology there have been two substantive changes, both concerned with Part II of the paper. Part II consists mainly of theorems resulting from striving after generality in the spaces while using only sequential convergence. These theorems were of interest in their day. None of the several mutually equivalent general theories of convergence had appeared in 1926, except the ancestor of them all, the Moore-Smith theory (1922); and this was not

* For an $\mathfrak{L}$ space, first given by Kuratowski and Sierpinski, loc. cit., pp. 176–8.

yet widely known, nor was it fully adequate for use with general topological spaces until J. L. Kelley devised his definition of "subnet." (Cf. "Partial orderings and Moore-Smith limits", in this volume.) Any of these general limit processes is far better suited to studies of compactness than is sequential convergence. For example, in J. L. Kelley's *General Topology* Moore-Smith convergence plays a fundamental role, while sequential compactness is mentioned only in two problems. The result of this shift to general limit processes is that most of the theorems of Part II have sunk into the oblivion that awaits almost all mathematical research.

The other great advance in the study of compactness is Tychonoff's theorem, which has been called "unquestionably the most useful theorem on compactness." The first statement and proof, by no means the most polished, was in a paper by A. Tychonoff in the *Mathematische Annalen* in 1930. Before we state and prove it, we need the ideas of product space and product topology.

Let T be a non-empty set, and for each t in T let Y be a non-empty set. Then the **cartesian product**

$$Y = \bigtimes_{t \in T} Y_t$$

is defined as usual to be the set of all functions $y : t \mapsto y_t$ on T such that for each t, y_t is in Y_t. For each such y, y_t is called the "**projection**" of y on Y_t, and is denoted by $P_t y$.

Suppose that each factor-space Y_t is a topological space. For each y in Y, the **basic neighborhoods** of y (in the product topology) are the cartesian products

$$V = \bigtimes_{t \in T} V_t$$

in which for each t in T, V_t is a neighborhood of y_t in Y_t, and for all but finitely many t, $V_t = Y_t$. A set G in Y is **open** if for each y in G there is a basic neighborhood of y that is contained in G.

We repeat four definitions from the appendix to the paper *Partial Orderings and Moore-Smith Limits* in this volume. A **direction** in a set D is a non-empty family $\mathfrak{A}$ of non-empty subsets of D such that if A_1 and A_2 are in $\mathfrak{A}$, there is a set A_3 in $\mathfrak{A}$ that is contained in both of them. A **directed function** is a pair $(f, \mathfrak{A})$ in which f is a function and $\mathfrak{A}$ is a direction in the domain of f; it is a **maximally directed function** if the direction $\mathfrak{A}$ is not a proper subset of any direction $\mathfrak{B}$ in the domain of f. If the values of f are in a topological space Z, $(f, \mathfrak{A})$ **converges** to a point z_0 of Z if to each neighborhood U of z_0 there corresponds a member A of $\mathfrak{A}$ such that for all x in A, $f(x)$ is in U.

The product topology is the topology of pointwise convergence, in the following sense.

THEOREM 1. *Let $(f, \mathfrak{A})$ be a directed function with values in the product space*

$$Y = \bigtimes_{t \in T} Y_t$$

of topological spaces Y_t, and let y^* be a point of Y. Then $(f, \mathcal{Q})$ converges to y^* in the product topology if and only if for each t in T, the directed function $(P_t f, \mathcal{Q})$ converges to y_t^* in the topology of Y_t.

Let $(f, \mathcal{Q})$ be a directed function with values in Y that converges to y^*. Let t be any point of T and U_t any neighborhood of y_t^* in Y_t. Define V to be

$$V = \underset{t' \in T}{\times} V_{t'},$$

in which $V_t = U_t$ and $V_{t'} = Y_{t'}$ for all $t' \neq t$. This is a basic neighborhood of y^* in the product topology, so there is a set A in $\mathcal{Q}$ such that for all x in A, $f(x)$ is in V. Then for such x, $P_t f(x)$ is in U_t, so $(P_t f, \mathcal{Q})$ converges to y_t^*.

Conversely, suppose that for all t in T, $(P_t f, \mathcal{Q})$ converges to y_t^*. Let

$$V = \underset{t \in T}{\times} V_t$$

be any basic neighborhood of y^* in the product topology. Then $V_t = Y_t$ except for t in a finite set $\{t(1), \ldots, t(k)\}$. For each $t(j)$ in that set, $(P_{t(j)} f, \mathcal{Q})$ converges to $y_{t(j)}^*$, so there exists a set $A(j)$ in $\mathcal{Q}$ such that for all x in $A(j)$, $P_{t(j)} f(x)$ is in $V_{t(j)}$. Since $\mathcal{Q}$ is a direction, there is a set A in $\mathcal{Q}$ that is contained in all the $A(j)$ $(j = 1, \ldots, k)$. If x is in A, it is in all the $A(j)$, so

$$P_t f(x) \in V_t$$

for all t in $\{t(1), \ldots, t(k)\}$. This relation also holds for all other t, since for them $V_t = Y_t$. So $f(x)$ is in V, and $(f, \mathcal{Q})$ converges to y^*.

We can now state and prove Tychonoff's theorem.

THEOREM 2 (TYCHONOFF). *If Y is the cartesian product of compact topological spaces, it is compact in the product topology.*

In the appendix to *Partial Orderings and Moore-Smith Convergence* we proved that a topological space Z is compact if and only if whenever $(f, \mathcal{Q})$ is a maximally directed function with values in Z, $(f, \mathcal{Q})$ converges to a point of Z. Let $(f, \mathcal{Q})$ be a maximally directed function with values in Y. Then for each t in T, $(P_t f, \mathcal{Q})$ is a maximally directed function in the compact space Y_t, so $(P_t f, \mathcal{Q})$ has a limit y_t^* in Y_t. By Theorem 1, $(f, \mathcal{Q})$ converges to $y^* : t \mapsto y_t^*$. So Y is compact.

In this proof we used a theorem about maximally directed functions whose proof depended on the Hausdorff maximal principle, which is equivalent to the axiom of choice. Tychonoff's theorem cannot be proved without the use of some equivalent of the axiom of choice, because Kelley has proved (Fund. Math., 37 (1950) 75–76) that Tychonoff's theorem implies the axiom of choice.

3

GODFREY HAROLD HARDY

Godfrey Harold Hardy was born on February 7, 1877, at Cranleigh, Surrey, England. At twelve years of age he passed his public examinations with distinction in mathematics at Cranleigh School and went to Winchester a year later on a scholarship. In 1896 he entered Trinity College at Cambridge where he was strongly influenced by A. E. H. Love who introduced him to analysis, thus beginning a career as a "real mathematician." He was elected a Prize Fellow in 1901, and in 1906 became a Lecturer in Mathematics, which position he held until 1919. In 1910 he was made a Royal Fellow and in 1914 was given the honorary title of Cayley Lecturer at the University of Cambridge. During this period he wrote *The Integration of Functions of a Single Variable* and his famous *A Course of Pure Mathematics*, first published in 1908. In 1912 he began a long and fruitful collaboration with J. E. Littlewood and together they produced over one hundred research papers. He brought the Indian mathematician Ramanujan to Cambridge in 1914 and they collaborated on many papers until the untimely death of Ramanujan in 1919.

Hardy was a disciple of Bertrand Russell and an admirer of Russell's colleague A. N. Whitehead. He transferred to New College at Oxford in 1919 where he lectured in geometry as well as analysis. In 1928-29 he was Visiting Professor at Princeton and the California Institute of Technology. While he was in this country he was invited to give the Josiah Willard Gibbs Lecture on "An Introduction to the Theory of Numbers," although he was unable to deliver it in person due to illness. He also lectured at Chicago on his analytic approach to Waring's problem.

In 1931 he returned to Cambridge where, until his retirement in 1949, he occupied the Chair of Pure Mathematics. He was an Honorary Fellow at New College and in addition received honorary degrees from Athens, Harvard, Manchester, Sofia, Birmingham, Edinburgh, Marburg, and Oslo. He was awarded the Royal Medal of the Royal Society in 1924, the Sylvester Medal in 1940 and the highest award, the Copley Medal at the time of his death in 1947.

During his career he wrote well over three hundred research papers, many in collaboration with other distinguished mathematicians. He wrote numerous books including *Orders of Infinity*, 1910, *The General Theory of Dirichlets Series*, with M. Riesz, 1915, *Fourier Series* with W. W. Rogosinski, 1934, *The Theory of Numbers* with E. M. Wright, 1938, a collection of lectures with Ramanujan, 1940, and at the time of his death had just finished the manuscript for a book on Divergent Series. In addition he edited the works of Ramanujan.

Hardy was strictly a research mathematician of the purest sort whose function was to "do something, to prove new theorems, to add to mathematics, and not to talk about what he or other mathematicians have done," [1]. To him "exposition, criticism, appreciation is the work of second-rate minds," [1]. Yet his student E. C. Titchmarsh reports that there could have been no more inspiring director of research of the work of others. "He was always at the head of a team of research workers whom he provided with an inexhaustible stock of ideas. He was an extremely kind-hearted man who could not bear any of his pupils to fail in their researches," [2]. As Norbert Wiener has stated, "Hardy was a staunch friend of all mathematicians and especially all young mathematicians," [3]. His award of the Chauvenet Prize attests to his ability at exposition and belies his apparent contempt for it. In fact, he was such a brilliant conversationalist that his seminars were known as "conversation classes." His book *A Mathematician's Apology* has been a source of inspiration and influence on many mathematicians who followed him.

References

1. G. H. Hardy, A Mathematician's Apology, Cambridge University Press, 1969.
2. E. C. Titchmarsh, Godfrey Harold Hardy, The Journal of the London Math. Soc., 25 (1950) 81-89.
3. Norbert Wiener, Godfrey Harold Hardy, Bull. A.M.S., 55 (1949) 72-77.

AN INTRODUCTION TO THE THEORY OF NUMBERS*

G. H. HARDY, Princeton University

PART I

1. Farey series. The theory of numbers has always occupied a peculiar position among the purely mathematical sciences. It has the reputation of great difficulty and mystery among many who should be competent to judge; I suppose that there is no mathematical theory of which so many well-qualified mathematicians are so much afraid. At the same time it is unique among mathematical theories in its appeal to the uninstructed imagination and in its fascination for the amateur. It would hardly be possible in any other subject to write books like Landau's *Vorlesungen* or Dickson's *History*, six great volumes of overwhelming erudition, better than the football reports for light breakfast table reading.

The excursions of amateur mathematicians into mathematics do not usually produce interesting results. I wish to draw your attention for a moment to one very singular exception. Mr. John Farey, Sen., who lived in the Napoleonic era, has a notice of twenty lines in the *Dictionary of National Biography*, where he is described as a geologist. He received as a boy "a good mathematical training." He was at one time agent to the Duke of Bedford, but afterwards came to London, where he acquired an extensive practice as a consulting surveyor, which led him to travel much about the country and "collect minerals and rocks." His principal work was a geological survey of Derbyshire, undertaken for the Board of Agriculture, but he also wrote papers in the *Philosophical Magazine*, on geology and on many other subjects, such as music, sound, comets, carriage wheels and decimal coinage. As a geologist, Farey is apparently forgotten, and, if that were all there were to say about him, I doubt that he would find his way into the *Dictionary of National Biography* today.

It is really very astonishing that Farey's official biographer should be so completely unaware of his subject's one real title to fame. For, in spite of the *Dictionary of National Biography*, Farey is immortal; his name stands prominently in Dickson's *History* and in the German encyclopedia of mathematics, and there is no number-theorist who has not heard of "Farey's series." Just once in his life Mr. Farey rose above mediocrity and made an original observation. He did not understand very well what he was doing, and he was too weak a mathematician to prove the quite simple theorem he had discovered. It is evident also that he did not consider his discovery, which is stated in a letter of about half a page, at all important; the editor of the *Philosophical Magazine* printed a very stupid criticism in the next volume, and Farey, usually a rather acrid controversialist, ignored it completely. He had obviously no idea that this casual letter was the one event of real importance in his life. We may be tempted to think that Farey was very lucky;

* The sixth Josiah Willard Gibbs Lecture, read at New York City, December 28, 1928, before a joint session of the American Mathematical Society and the American Association for the Advancement of Science.

but a man who has made an observation that has escaped Fermat and Euler deserves any luck that comes his way.*

Farey's observation was this. The *Farey series of order n* is the series, in order of magnitude, of the irreducible rational fractions between 0 and 1 whose denominators do not exceed n. Thus

$$\frac{0}{1}, \frac{1}{7}, \frac{1}{6}, \frac{1}{5}, \frac{1}{4}, \frac{2}{7}, \frac{1}{3}, \frac{2}{5}, \frac{3}{7}, \frac{1}{2}, \frac{4}{7}, \frac{3}{5}, \frac{2}{3}, \frac{5}{7}, \frac{3}{4}, \frac{4}{5}, \frac{5}{6}, \frac{6}{7}, \frac{1}{1},$$

is the Farey series of order 7. There are two simple theorems about Farey series; (i) if p/q and p'/q' are two consecutive terms, then

$$p'q-pq'=1,$$

and (ii) if $p/q, p'/q', p''/q''$ are three consecutive terms, then

$$\frac{p'}{q'}=\frac{p+p''}{q+q''}.$$

The second theorem (which is that actually stated by Farey) is an immediate consequence of the first, as we see by solving the equations

$$p'q-pq'=1, \quad p''q'-p'q''=1,$$

for p' and q'.

The theorems are not of absolutely first class imporatnce, but they are not trivial, and all of the many proofs have some feature of real interest. One of the simplest uses the language of elementary geometry. We consider the *lattice* or *Gitter* L in a plane formed by drawing parallels to the axes at unit distance from each other; the intersections, the points (x,y) with integral coordinates, are called the *points of the lattice*. It is obvious that the properties of the lattice are independent of the particular lattice point O selected as origin and symmetrical about any origin. The lattice is transformed into itself by the linear substitution

$$x'=\alpha x+\beta y, \quad y'=\gamma x+\delta y,$$

where α, β, γ, δ are integers and $\Delta=\alpha\delta-\beta\gamma=1$, since then there is a pair x,y which give any assigned integral values for x',y'.

The area of the parallelogram P based on the origin and two lattice points (x_1,y_1), and (x_2,y_2), not collinear with O, is

$$\delta=\pm(x_1y_2-x_2y_1).$$

We can construct a lattice L' (an oblique lattice) by producing and drawing parallels to the sides of P. A necessary and sufficient condition that L' should be equivalent to L, that is, that they should contain the same lattice points, is that $\delta=1$, that is, that δ should have its smallest possible value. It is clear that this is also a necessary and sufficient condition that *there should be no lattice point inside*

* It should be added that Farey's discovery had been anticipated 14 years before by C. Haros: see Dickson's *History*, vol. 1, p. 156. Cauchy happened to see Farey's note and attributed the theorem to him, and everyone else has followed Cauchy's example.

P, and it is easy to see that if there is such a point inside P, there is one inside, or on the boundary of, the triangular half of P nearer to O.

We may call the lattice point (q,p) which corresponds to a fraction p/q in its lowest terms a *visible* lattice point; there is no other lattice point which obscures the view of it from O. Let us consider all the visible lattice points which lie inside, or on the boundary of, the triangle bounded by the lines $y=0$, $x=n$, $y=x$. It is plain that these points correspond one by one to the fractions of the Farey series of order n. When the ray R from O to (q,p) rotates from the x-axis to the line $y=x$, it passes through each of these points in turn. If we take two consecutive positions of R, corresponding to the points (q,p), (q',p'), the parallelogram based on these two points contains no lattice point inside it, since otherwise there would be a lattice point inside its nearer triangle, and therefore a Farey fraction between p/q and p'/q'. It follows that

$$\delta=p'q-pq'=1,$$

which proves Farey's theorem.

2. Purpose of this lecture. So much then for Farey's discovery; it is a curious theorem, and its history is still more curious; but I have no doubt allowed myself to dwell upon it a little longer than its intrinsic importance deserves. My discussion of it will, however, help me to explain what I am trying to do in this lecture.

I shall imagine my audience to be made up entirely of men like Farey. I know that most of them are very much better mathematicians, but I shall not assume so; I shall assume only that they possess the common school knowledge of arithmetic and algebra. But I shall also assume that, like Farey, they are curious about the properties of integral numbers; one need after all be no Ramanujan for that.

Let us then imagine such a man playing about with numbers (as so many retired officers in England do) and puzzling himself about the curious properties which they seem to possess. What odd properties would strike him? What are the first questions he would ask? We must not try to be very systematic; if we do, we shall make no progress in an hour. We must aim merely at a rough preliminary survey of the ground. If in the course of our survey, we find the opportunity for any illuminating remark, we may delay to make it, as I have already delayed over Mr. Farey, even if it does not seem to fall in quite its proper logical place. Then, if time permits, we may return to examine a little more closely any important difficulties which our preliminary survey has revealed.

3. Congruences to a modulus. There is no doubt that the first general idea which we should have to explain is that of a *congruence*. Two numbers a and b are *congruent to modulus m* if they leave the same remainder when divided by m, that is, if m is a divisor of $a-b$. We write

$$a\equiv b \pmod{m}, \quad m|a-b.$$

It is obvious that congruences are of immense practical importance. Ordinary life is governed by them; railway time tables and lists of lectures are tables of congruences. The absolute values of numbers are comparatively unimportant; we want to know what time it is, not how many minutes have passed since the creation.

A great many problems both of arithmetic and of common life depend upon the solution of congruences involving an unknown x, such as

$$a_0x^n + a_1x^{n-1} + \cdots + a_n \equiv 0 \pmod{m}.$$

Such congruences may be classified like algebraical equations, as linear, quadratic,..., according to the value of n. Our first instinct in dealing with congruences is to follow up the analogy with algebra. In algebra a linear equation has one root, a quadratic two, and so on. We find at once that there are obvious and striking contrasts; even the linear congruence suggests a whole series of problems, and a full discussion of quadratic congruences involves quite an imposing body of general ideas.

Let us take the simplest case, the linear congruence, and suppose first that we are concerned only with one particular modulus, such as 7 or 24. We have then an example of a genuinely finite mathematics. Congruent numbers have exactly the same properties and cannot be distinguished, and our mathematics contains only *a finite number of things*. In such a mathematics any problem can be solved by enumeration; we can solve $2x \equiv 5 \pmod{7}$ by trying all possible values of x, and we find there is a unique solution, $x \equiv 6$. If we try to solve $2x \equiv 5 \pmod{24}$, we find that there is no solution; if I lecture every other day, I shall sooner or later lecture on Thursday, but if I lecture every other hour, I may never lecture at 5 P.M.

The difference is of course accounted for by the fact that 7 is *prime* and 24 is not. Here we encounter the notion of a prime, a number without factors, and all kinds of speculations suggest themselves. Can we tell, by any method short of trial of all possible divisors, whether any given number is prime or not? Are there formulas for primes? Are the primes infinite in number, and if so, what is the law of their distribution?

Again, it appears that all numbers are composed of primes, that primes are the ultimate material out of which the world of numbers is built up. We are bound to ask *how*; and here we meet our first big theorem, the "fundamental theorem of arithmetic," the theorem that factorization is unique. But we shall probably be wise to allow our enquirer to take this theorem for granted until he has acquired a little of the sophistication which comes with wider knowledge.

We may observe, however, before passing on, that the contrast between arithmetic and algebra becomes much more marked as soon as we consider congruences of higher degree. An equation of the fourth degree has, with appropriate conventions, just four roots. But

$$x^4 \equiv 1 \pmod{13}$$

has 4 roots, 1, 5, 8, and 12;

$$x^4 \equiv 1 \pmod{16}$$

has 8 roots, 1, 3, 5, 7, 9, 11, 13, and 15; and

$$x^4 \equiv 2 \pmod{16}$$

has none.

4. Regarding decimals. I pass to another subject that has an irresistible fascination for amateurs, the subject of *decimals*. Some decimals are finite and some recurring, but it is easy to write down decimals, such as

(a) 0.10100100010... (b) 0.11010001000...

which are neither. Here (a) the number of 0's increases by one at each stage, (b) the ranks of the 1's are 1,2,4,8,.... More amusing examples are

$$0.01101010001010\ldots \tag{c}$$

(in which the 1's have prime rank) and

$$0.23571113171923\ldots \tag{d}$$

(formed by writing down the prime numbers in order). The proof for (c) demands the knowledge that there is an infinity of primes, and that for (d) rather more.*

The answer to some of the obvious questions is immediate. A finite decimal represents a rational fraction $p/(2^\alpha 5^\beta)$, a pure recurring decimal a fraction p/q, where q is not divisible by 2 or 5, and a mixed recurring decimal a fraction in which q is divisible by 2 or 5 and also by some other number. The converses of these theorems are also true, but the proof demands a little genuinely arithmetical reasoning. I shall state the proof in the simplest case, since it depends upon the logical principle which is perhaps our most effective weapon in the elementary parts of the theory, where we are dealing with so simple a subject matter that our choice of arguments is naturally very restricted.

Suppose $p<q$ and q prime to 10. If we divide all powers 10^ν by q, there are only q possible remainders, and one at least must be repeated. It follows that there are a ν_1 and a $\nu_2>\nu_1$ such that

$$10^{\nu_2}\equiv 10^{\nu_1}, \quad 10^{\nu_1}(10^{\nu_2-\nu_1}-1)\equiv 0$$

to modulus q. It follows that, if we write $\nu_2-\nu_1=N$, we have $10^N\equiv 1$, so that $q|10^N-1$ and

$$\frac{p}{q}=\frac{P}{10^N-1}=P\cdot 10^{-N}+P\cdot 10^{-2N}+\cdots.$$

Since $P<10^N$, this is a pure recurring decimal with a period of at most N. The principles which we have used are (a) that *if there are more than q things of at most q kinds, there must be two of them of the same kind*; (b) *that if $10^\nu Q$ is divisible by q, and q is prime to* 10, *then Q is divisible by q*. In the second we are of course appealing to the "fundamental theorem." The first is the general logical principle to which I referred just now.

Let us take a slightly more complicated variant of this principle. *If there are two sets of objects*

$$a_1,a_2,\ldots,a_m, \quad b_1,b_2,\ldots,b_m,$$

no two of either set being the same; and if every b is equal to an a; then the b's are the a's arranged in a different order. We may apply this principle to obtain further

* See Pólya and Szegö's *Aufgaben aus der Analysis*, vol. 2, pp. 160, 383.

information about the period of our recurring decimal. I suppose now that q is prime. If q and a are given, and a is not a multiple of q, it is impossible that

$$ra \equiv sa \pmod q$$

unless $r \equiv s$. If (ra) is the remainder when ra is divided by q, the two sets

$$r,(ra) \qquad (r=1,2,\ldots,q-1)$$

satisfy the conditions of our principle and are therefore the same except in order. It follows that

$$(q-1)!a^{q-1} \equiv \Pi(ra) \equiv \Pi r = (q-1)! \pmod q,$$

and therefore that

$$a^{q-1} \equiv 1 \pmod q,$$

Fermat's Theorem. In the particular case in which we are interested, a is 10, and Fermat's Theorem shows that we may take $N=q-1$, so that the period of p/q cannot exceed $q-1$ figures. Observe that we have appealed to the fundamental theorem twice in the proof.

It is familiar to everyone that $\frac{1}{7}$ has 6 figures, the maximum number. We are bound to ask what other primes q possess this property; the values of q less than 50 are in fact 7, 17, 19, 23, 29, and 47, but here we begin to get into deeper water. I cannot stop to discuss this question now, but before passing on I must mention another familiar text-book theorem which I shall have to quote later. This is Wilson's Theorem, that

$$(q-1)!+1 \equiv 0 \pmod q$$

if and only if q is prime. Of the mass of proofs catalogued by Dickson, that of Dirichlet depends most directly on principles which we have used already. It is an immediate consequence of these principles that, if x is any one of the set $1,2,\ldots,q-1$, there is just one other, y, such that $xy \equiv 1 \pmod q$; we call y the *associate* of x. It is plain that 1 and $q-1$ are associated with themselves; and no other number can be, since $x_1^2 \equiv x_2^2$ implies $x_1 \equiv x_2$ or $x_1 \equiv q-x_2$. It follows that the numbers $2,3,\ldots,q-2$ are composed of $\frac{1}{2}(q-3)$ distinct pairs the product of each of which is congruent to 1. Hence

$$2\cdot 3 \cdots (q-2) \equiv 1^{(q-3)/2} = 1,$$
$$(q-1)! \equiv q-1 \equiv -1,$$

which is one half of Wilson's Theorem. The converse half is practically obvious, since $(q-1)!$ would be divisible by any factor of q.

5. Algebraic and transcendental numbers. The study of decimals leads directly to problems concerning *rationality and irrationality*. Our decimals such as 0.1010010001... must represent irrational numbers. What criteria are there for deciding whether a given number is rational or irrational? To ask this question is to go a little outside the theory of numbers proper, which is concerned first with integers, and then with rationals or irrationals of special forms, such as the form $a+b\sqrt{2}$, and not with irrationals as a whole or general criteria for irrationality. The problem is, however, one about which an amateur will certainly demand information.

The famous argument of Pythagoras shows that $\sqrt{2}$ is irrational; if a/b is in its lowest terms and $a^2=2b^2$, then a and b must both be even, a contradiction. It is obvious to us now that the Pythagorean argument extends at once to $\sqrt{3}, \sqrt{5}, \ldots, 2^{1/3}, \ldots$, and generally to $N^{1/m}$, where N is any number which is not a perfect mth power. There is a curious and very instructive historical puzzle connected with this argument. There is a passage in Plato's *Theaetetus*, discussed at length by Heath in his *History of Greek Mathematics*, about the attempt of Theodorus to generalize Pythagoras's proof. Theodorus, working some 50 years after Pythagoras, proved the irrationality of $\sqrt{N}$ for all values of N (except square values) up to 17 inclusive. Why, ask the historians did he stop? Why in any case should it have taken mathematicians like the Greeks 50 years to make so obvious an extension? Zeuthen in particular expended a great deal of ingenuity upon this question, but I think that the ingenuity was misplaced, and that the answer is obvious.

Theodorus *did not know the fundamental theorem of arithmetic*; there is something of a puzzle about the history of that theorem, but it cannot have been known to the Greeks before Euclid's time. The triviality of the generalization to us is due entirely to our knowledge of this theorem. Suppose, for example, we wish to prove that

$$a^2=60b^2,$$

where a and b are integers without common factor, is impossible. We argue that a^2 cannot be divisible by 3 unless a is divisible by 3; hence $a=3c$, $a^2=9c^2$, $3c^2=20b^2$, and a repetition of the argument shows that b also is divisible by 3. We can prove that $3|a^2$ implies $3|a$ *without the fundamental theorem*, by enumeration of possible cases, considering separately the cases in which $a\equiv 0, 1, 2 \pmod{3}$. If it were 17 instead of 3, the process would be a little tedious; and in any case such a classification of numbers would have been very novel in Theodorus's time. I am so far from being puzzled by the limitations of his work that I regard what he did as a very remarkable achievement.

There are very few types of numbers which present themselves at all naturally in analysis and which can be proved to be irrational. It is obvious that a number like $\log_{10}2$ is irrational, for a power of 2 cannot be a power of 10. The proof for e, from the exponential series, is quite easy, and that for e^2 not very much more difficult. That for π is decidedly more so, and when we come to numbers like e^3 and π^2, it ceases to be worth while to worry about elementary proofs; we may as well go the whole way and prove e and π are transcendental. The most famous constant in analysis, after e and π, is Euler's constant γ; and the proof of the irrationality of γ is one of the classical unsolved problems of mathematics. It has never been proved that $2^{\sqrt{2}}$, $3^{\sqrt{2}}$, and similar numbers are irrational; no plausible method for attacking such problems has even been suggested. I am inclined to think that the number which holds out the best hopes for new discovery is the number e^{π}, which presents itself so naturally in the formulas of elliptic functions.

I said just now that e and π were "transcendental." I must not stop to talk at length about this famous theorem of Lindemann,* which contains the final proof that the quadrature of the circle, in the classical sense, is impossible; but the *statement* of the theorem introduces a notion that we shall require, that of an *algebraic number*. An algebraic number is the root of an equation

$$a_0x^n + a_1x^{n-1} + \cdots + a_n = 0,$$

where the a's are integers. An *algebraic integer* is an algebraic number whose characteristic equation has unity for its leading coefficient. Thus $\sqrt{2}$ and $1+\sqrt{(-5)}$ are algebraic integers. A *transcendental* number is a number which is not algebraic; and Lindemann's Theorem is that *π is transcendental*. It is easy to show that all lengths which can be constructed by euclidean methods are algebraic, and indeed algebraic numbers of a quite special kind. It follows that the quadrature of the circle by any euclidean construction is impossible.

There is another direction in which we may be tempted to digress at this point, the theory of the approximation of irrationals by rationals, what is now called "diophantine approximation." There is just one theorem in this field that I shall mention, because it is connected so directly with what I have just been saying, and because it depends upon another of the stock arguments of number theory, the principle that *an integer numerically less than* 1 *is* 0. This is Liouville's theorem, that *there are transcendental numbers*. It is naturally much easier to prove this than to prove that a given number such as π is transcendental.

Liouville proves first that *it is impossible to approximate rationally to an algebraic number with more than a certain accuracy*. It is quite easy to see why. Suppose that ξ is an algebraic number defined by

$$f(\xi) = a_0\xi^n + a_1\xi^{n-1} + \cdots + a_n = 0.$$

We may suppose that the equation is irreducible, that is to say that $f(\xi)$ cannot be resolved into simpler algebraic factors of similar form; in this case we say that ξ is *of degree n*. We can obviously find a number M, depending only on ξ, such that

$$|f'(x)| < M$$

for x near ξ. Suppose now that p/q is a rational, near ξ. Then

$$f\left(\frac{p}{q}\right) = \frac{N}{q^n},$$

where N is an integer not zero. It follows from our general principle that $|N| \geqslant 1$

* See for example Hobson's *Trigonometry*, third edition, p. 305, or the same author's *Squaring the Circle*.

and

$$\left|f\left(\frac{p}{q}\right)\right| \geqq \frac{1}{q^n}.$$

But

$$f\left(\frac{p}{q}\right)=f\left(\frac{p}{q}\right)-f(\xi)=\left(\frac{p}{q}-\xi\right)f'(\eta),$$

where η lies between p/q and ξ. Hence, for all q,

$$\left|\frac{p}{q}-\xi\right|=\left|\frac{f(p/q)}{f'(\eta)}\right|>\frac{1}{Mq^n}.$$

It is impossible to approximate rationally to an algebraic number of degree n with an order of accuracy higher than q^{-n}.

On the other hand it is easy to write down numbers which have rational approximations of much higher accuracy than this; we have only to take a decimal of 0's and 1's in which the 1's are spaced out sufficiently widely. Thus

$$\xi=\frac{1}{10^{1!}}+\frac{1}{10^{2!}}+\frac{1}{10^{3!}}+\cdots=.11000100000\ldots$$

is approximated by its first k terms, that is, by a fraction

$$\frac{p}{q}=\frac{p}{10^{k!}}$$

with an error of order $10^{-(k+1)!}=q^{-k-1}$. Hence it is not an algebraic number of degree k and, since k is arbitrary, it must be transcendental. Obviously Liouville's argument enables us to *construct* transcendental numbers as freely as we please.

6. Arithmetic. Forms. The theory of irrationals starts from Pythagoras, and there is another great branch of the theory of numbers which also starts from him and about which I must now say something. This is the theory of *forms*.

Our interest in the theory of forms begins when we observe that there are Pythagorean triangles with integral sides; thus $3^2+4^2=5^2$. The first problem which suggests itself is that of determining all such triangles, and the solution given in substance by Diophantus, is easy. All the integral solutions of

$$x^2+y^2=z^2$$

are given by

$$x=\lambda(\xi^2-\eta^2),\quad y=2\lambda\xi\eta,\quad z=\lambda(\xi^2+\eta^2),$$

where the letters are integers and ξ and η are coprime and of opposite parity. This problem is trivial, but it suggests an infinity of others.

It is natural to begin by a generalization of the problem. Let us discard the hypothesis that the hypotenuse z is integral; then

$$n=z^2=x^2+y^2$$

is the sum of two squares, and we are led to ask what numbers n possess this property. This is the first and simplest problem in the theory of *quadratic forms*, and the answer to it shows that no such problem can be quite easy. Even linear forms are not quite trivial; the solution of $ax+by=n$ in integers is a quite interesting elementary problem. When we consider quadratic forms, we come up against difficulties of a different order.

The first theorem in the subject is another theorem of Fermat, that *$x^2+y^2=n$ is soluble when n is a prime $p=4m+1$* and, apart from trivial variations of the sign and order of x and y, uniquely. It is to be observed that the equation is plainly insoluble when n is $4m+3$, since any square is congruent to 0 or 1 to modulus 4. This theorem is one of the most famous in the theory of numbers, and very rightly so, since it was the first really difficult theorem in the subject proved by any mathematician. There is no really simple proof, and the most natural, that which depends on the Gaussian numbers $a+bi$, introduces a whole series of ideas of revolutionary importance.

The first stage of the proof consists in proving that *there is a number x such that*

$$x^2 \equiv -1 \pmod{p},$$

or $p|1+x^2$. Let us go back for a moment to the proof I sketched of Wilson's Theorem. Let us associate the numbers $x=1,2,\ldots,p-1$ in pairs x,y not, as then, so that $xy\equiv 1$, but so that

$$xy \equiv -1 \pmod{p}.$$

If any x is associated with itself, our proposition is established. If not, we have arranged the numbers from 1 to $p-1$ in $\frac{1}{2}(p-1)$ pairs of different numbers each satisfying the condition. Hence

$$(p-1)! \equiv \Pi xy \equiv (-1)^{(p-1)/2} = 1;$$

which is false, since, by Wilson's Theorem,

$$(p-1)! \equiv -1.$$

We thus obtain our proposition by reductio ad absurdum.

The second stage of the proof depends on much more novel ideas. We are concerned with the simplest case of an *algebraic field*. The field $K(i)$ is the aggregate of numbers

$$\xi = r+si = r+s\sqrt{(-1)},$$

where r and s are rational. This number satisfies the equation

$$\xi^2 - 2r\xi + r^2 + s^2 = 0,$$

and is an algebraic integer, in the sense I defined before, when $2r$ and r^2+s^2 are integers, that is, when r and s are integers. We may denote by $K^*(i)$ the aggregate of all the integers

$$\alpha = a+bi$$

of $K(i)$; a and b are ordinary integers. The numbers of $K^*(i)$ reproduce themselves by addition and multiplication, and we can define division in this field just as we define it in ordinary arithmetic. We can also define a *prime of* $K^*(i)$, and factorization of numbers into primes. There are four numbers, ± 1 and $\pm i$, which play a part in the new arithmetic similar to that of 1 and -1 in ordinary arithmetic. These are the "unities" or divisors of 1. If we define the *norm* of $\alpha = a + bi$ as

$$N(\alpha) = a^2 + b^2,$$

then the unities are characterized by the fact that their norm is 1. We do not count them as primes, just as, in the ordinary theory, we do not count 1 as a prime.

We now make an assumption, namely that *the analog of the fundamental theorem holds in the field* $K^*(i)$, that is to say that, apart from any trivial complications which may be introduced by the unities, *the factorization of a number of* $K^*(i)$ *into primes is unique*. This assumption is in fact correct. Returning now to the first stage of our proof, there is an x such that

$$p \mid 1 + x^2 = (1 + ix)(1 - ix).$$

It is obvious that p does not divide $1 + ix$ or $1 - ix$, so that *p divides the product of two numbers without dividing either of them.* Hence p cannot be a prime in $K^*(i)$. We may therefore write

$$p = \pi\lambda,$$

where $N(\pi) > 1$ and $N(\lambda) > 1$. But

$$N(\pi)N(\lambda) = N(p) = p^2,$$

so that $N(\pi)$ and $N(\lambda)$ must each be p. If we write

$$\pi = a + ib,$$

it follows that

$$p = N(\pi) = a^2 + b^2,$$

which is Fermat's theorem.

We may be tempted by our success to further efforts in the same direction. It is easy to satisfy ourselves, by considering particular cases, that *any prime $p = 20m + 1$ is of the form $a^2 + 5b^2$*: thus $61 = 4^2 + 5 \cdot 3^2$. Let us try to prove this theorem by a similar method. We must evidently consider now the field $K^*[\sqrt{(-5)}\,]$ formed of the algebraic integers of the form

$$\alpha = a + b\sqrt{(-5)}\ ;$$

it is easy to show that such a number is an algebraic integer if and only if a and b are ordinary integers. There is no difficulty in defining divisibility and primality in this field also.

The first step in our proof must plainly be to prove the existence of an x for which $p|1+5x^2$. This is not difficult, but it demands a little more knowledge of quadratic congruences than I can assume, and I must take it for granted.

We define the norm $N(\alpha)$ of a number of this field as a^2+5b^2. We then argue as before; we have

$$p|1+5x^2=\left(1+x\sqrt{(-5)}\right)\left(1-x\sqrt{(-5)}\right),$$

so that p divides a product without dividing either factor and is therefore not a prime. Hence, as before $p=\pi\lambda$, where $N(\pi)>1$ and $N(\lambda)>1$, and $N(\pi)$ and $N(\lambda)$ must each be p. It follows that

$$p=N(\pi)=a^2+5b^2,$$

the theorem we set out to prove.

At this point, however, there is a shock in store for us; we find that we can prove *too much*. The number

$$q=\left(2+\sqrt{(-5)}\right)\left(2-\sqrt{(-5)}\right)$$

is divisible by 3, while neither factor is so. Hence 3 is not a prime. Hence

$$3=\pi\lambda, \quad 9=N(\pi)N(\lambda),$$

and $N(\pi)$ and $N(\lambda)$ are each 3. It follows that

$$3=N(\pi)=a^2+5b^2.$$

Similarly we can prove that

$$7=a_2+5b_2;$$

and both of these theorems are obviously false.

There must therefore be a mistake somewhere in our argument, and if you examine it, and are prepared to believe that I have not been misleading you wilfully, you will see that there is only one step which can be questioned. In all three cases I concluded the argument by an appeal to the same theorem; *a number which divides the product of two numbers without dividing either of them cannot be prime*. This is true in ordinary arithmetic, because of the fundamental theorem; if 7 were a divisor of $15=3\cdot 5$, 15 would be factorable into primes in two distinct manners. It follows that *the analog of the fundamental theorem in the field* $K^*[\sqrt{(-5)}]$ *must be false*; and this is easily verified when once our suspicions have been excited; thus

$$2\cdot 3=\left(1+\sqrt{(-5)}\right)\left(1-\sqrt{(-5)}\right),$$

$$3\cdot 7=\left(1+2\sqrt{(-5)}\right)\left(1-2\sqrt{(-5)}\right),$$

and all of these numbers are prime in $K^*[\sqrt{(-5)}]$. The proof which I gave of the theorem concerning primes $20m+1$ was therefore fallacious, although the theorem is true. The proof of Fermat's theorem, on the other hand, was correct, since factorization *is* unique in $K^*(i)$.

7. Further problems. It is clear that we must go back to the beginning and study the theory of primes a little more closely; but before I do this I should like to call your attention to a series of further problems suggested by Fermat's theorem. We know now when a *prime* is the sum of two squares, and we have to consider the same problem for general n. Here in fact there are three different problems.

The first and most obvious problem is that of determining the necessary and sufficient conditions that n should be representable. This problem may be solved quite easily with the aid of the Gaussian numbers; n must be $2^{\alpha}M^2N$, where α is 0 or 1 and N contains prime factors of the form $4m+1$ only. We are then led naturally to the corresponding problem for other forms, first for the general binary quadratic form

$$ax^2+bxy+cy^2,$$

then for quadratic forms in a larger number of variables, such as

$$x^2+y^2+z^2, \quad x^2+y^2+z^2+t^2,$$

and then for forms of higher degree, such as x^3+y^3 and x^4+y^4. There is a highly developed theory of the general quadratic form; the most famous theorem is perhaps Lagrange's theorem, that *every number is the sum of four squares*. But as soon as we begin to consider cubic or higher forms we find ourselves on the boundary of knowledge. There is for example no criterion analogous to Fermat's by which we can decide whether a given number is the sum of two cubes.

The second problem about the form x^2+y^2 suggested by Fermat's theorem is that of determining the *number of representations*. This problem may be interpreted in two different ways. We may want an exact formula, in terms of the factors of n, and in this case the Gaussian theory again gives what we want; $r(n)$, the number of representations, is given by the formula

$$r(n)=4\{d_1(n)-d_3(n)\},$$

where $d_1(n)$ and $d_3(n)$ are the numbers of divisors of n of the forms $4m+1$ and $4m+3$ respectively. This is, however, not the most interesting interpretation of the problem. We may want, not a formula like this, but information concerning the *order of magnitude* of $r(n)$, whether $r(n)$ is generally large when n is large, whether numbers are usually representable freely or with difficulty. In this case our formula gives us very little help, and the solution of the problem requires quite different methods.

It is here that we come into contact for the first time with a new branch of the theory, the modern "analytic" theory. This theory has two special characteristics. The first is one of method; it uses, besides the methods of the classical theory, the methods of the modern theory of functions of a complex variable. The second is that it is concerned primarily with problems of order of magnitude and asymptotic distribution. The distinction is not a perfectly sharp one; there are "exact," "finite" theorems which have only been solved by "analytic" methods. For example, *every number greater than* $10^{10^{10}}$ *is expressible as the sum of* 8 *cubes*; this theorem includes no reference to "order of magnitude," and is a "finite" theorem in just the same sense as Fermat's theorem about the squares, but the only known proof is analytic. On the whole, however, it is the problems of asymptotic distribution which dominate the theory.

The answer given by the analytic theory to the special question which I raised is roughly as follows. The average value of $r(n)$ is π. It must be observed that representations which differ only trivially, that is, in the sign or order of x and y, are reckoned as distinct. If we allow for this, the average number of representations is rather less than a half; this is explained by the fact that, as we shall see, *most* numbers are not representable. On the other hand $r(n)$ tends to infinity with n with tolerable rapidity for numbers of appropriate forms, more rapidly for example than any power of $\log n$. The corresponding problems for cubes or higher powers present difficulties which are at present quite insuperable, and all that I can do is to mention a few curiosities. The smallest number representable by two cubes in two really distinct ways is

$$1729 = 1^3 + 12^3 = 9^3 + 10^3,$$

and the smallest representable in three ways is probably

$$175959000 = 70^3 + 560^3 = 198^3 + 552^3 = 315^3 + 525^3.$$

It can be proved that there exist numbers with as many different representations as we please. A. E. Western has carried out very heavy computations concerning representations by cubes; he has for example found 6 numbers, of which the smallest is 1,259,712, representable as the sum of *three* cubes in *six* different ways. The smallest number doubly representable by two fourth powers is probably

$$635318657 = 59^4 + 158^4 = 133^4 + 134^4;$$

there is, so far as I know, no known example of a number with three such representations, nor any proof that such a number exists.

The nature of the problems of the analytic theory becomes clearer when we consider the third problem suggested by Fermat's theorem. This is the problem of determining the *distribution* of the representable numbers. We want to know *how many numbers are representable*, or, to put it more precisely, how many numbers less than a large assigned number x are representable. If $Q(x)$ is the number of such numbers, what is the order of magnitude of $Q(x)$? Are nearly all numbers representable, or just a majority, or only a few? The answer is in fact that $Q(x)$ is approximately

$$\frac{Ax}{(\log x)^{1/2}},$$

where A is a constant; to put it roughly, quite a lot of numbers are representable, but strictly an infinitesimal proportion of the whole. This explains why the average number of representations turned out to be less than one.

This problem about $Q(x)$ is a very interesting one, but there is another of the same kind which is obviously still more interesting and much more fundamental. This is the problem of the distribution of the primes themselves; *how many primes are there less than x*? I shall say something about this problem in a moment; it is in any case time for us to return to the theory of primes, since all our enquiries have ended in questions about them, and it is obviously impossible to make serious progress until we know more both of their elementary properties and of the laws which govern their distribution.

PART II

8. The fundamental theorem. The *fundamental theorem* of arithmetic is the beginning of the theory of numbers, and it is plain that our first task must be to make this theorem secure.

There is another historical puzzle about the fundamental theorem. Who first stated the theorem, explicitly and generally? The natural answer is *Euclid*, since the *Elements* contain all the materials for the proof. Everything rests on Euclid's famous algorithm for the greatest common divisor. Given two numbers a, b, of which a is the greater, we form the table

$$a=bc+b_1,\quad b=b_1c_1+b_2,\quad b_1=b_2c_2+b_3,\ldots,$$

where $b_1, b_2, \ldots,$ are the remainders in the ordinary sense of elementary arithmetic. Since

$$b>b_1>b_2>\cdots,$$

b_n must sooner or later be zero. The last positive remainder δ has the properties implied by the words *greatest common divisor*, and it follows from the process by which δ is formed that any number which divides both a and b divides δ.

Let us note in passing that there is an analogous process in $K^*(i)$, but that the analogy fails in $K^*[\sqrt{(-5)}\,]$. In ordinary arithmetic, given a and b, we can find a number congruent to $a \bmod b$ and less than b. There is a similar theorem for the Gaussian numbers. Here there is no strict order of magnitude between different numbers, and we have to use the order of magnitude of their norms. Given α and β, there is a number, congruent to $\alpha \bmod \beta$, whose norm is less than that of β. There is no such theorem in $K^*[\sqrt{(-5)}\,]$, and the process analogous to Euclid's fails.

When the existence of δ is once established, the proof of the fundamental theorem is easy. We write

$$\delta=(a,b)$$

and we say that a is prime to b when $(a,b)=1$. The crucial lemma is that *if* $(a,b)=\lambda$ *and* $b|ac$, *then* $b|c$; in particular, *a prime cannot divide a product without dividing one or other of the factors*. This once granted, anybody can construct the proof of the fundamental theorem for himself; and you will remember that it was just this proposition which led to our troubles in $K^*[\sqrt{(-5)}\,]$.

The lemma itself may be proved as follows. We construct the euclidean algorithm for a and b, with the final remainder 1. If we multiply it throughout by c, we have the algorithm for ac and bc, and the final remainder is c. It follows that

$$(ac,bc)=c.$$

Since b divides ac, by hypothesis, and also bc, it divides c.

This is Euclid's own argument, and with it he had proved what is essential in the fundamental theorem. It is a very singular thing that he should then omit to state the magnificent theorem that he has proved. He is over the line and free, but apparently disdains the formality of touching down. I do not know of any formal statement of the theorem earlier than Gauss. The substance of the theorem, however, is in the *Elements*; it was plainly unknown, as I explained before, to the Greeks from 50 to 100 years before Euclid's time; and I see no particular reason for questioning the obvious view that it is Euclid's own.

As soon as we have proved the fundamental theorem our elementary knowledge falls into line. The theory of linear congruences, the theorems of Fermat and Wilson and all their consequences, the elementary theory of decimals and of the divisors of numbers, may be developed straightforwardly and without the introduction of essentially new ideas. I can now say something about the more modern side of the theory of primes.

9. Problems concerning primes. What are the most natural questions to ask about primes? I say deliberately the most *natural*; we must remember that a natural question does not always seem, on fuller reflection, to have been a *reasonable* one. It is natural to an engineer to ask us for a finite formula for

$$\int e^{-x^2}dx,$$

or for a solution of some simple looking differential equation in finite terms. If we fail to satisfy him, it is not because of our stupidity, but because the world does not happen to have been made that way.

So, if any one asks us (1) *to give a general formula for the nth prime* p_n, a formula in the sense in which

$$p_n=n^2,\quad p_n=n^2+1,\quad p_n=[e^n],$$

where $[x]$ denotes the integral part of x, would be a formula, I can only reply that it is not a reasonable question. It is, I will not say demonstrably impossible, but wildly improbable, that any such formula exists. The distribution of the primes is not like what it would have to be on any such hypothesis. I should make the same reply to a good many other questions which an amateur might be likely to ask, for example if he asked me (2) *to give a rule for finding the prime which immediately follows a given prime*. It would of course be perfectly reasonable that he should press me for the reasons why I gave so purely a negative a reply. On the other hand the problem (3) *to find the number of primes below a given limit* is, if interpreted properly, an entirely reasonable and a soluble problem. The problems (4) *to prove that there are infinitely many pairs of primes differing by* 2, and (5) *to prove that there are infinitely many primes of the form* n^2+1, are also entirely reasonable, and if (as is the case) we cannot solve them, it is quite reasonable to condemn our lack of ingenuity.

10. The distribution of primes. If we wish to classify these problems and to decide which of them are reasonable and which are not, the first essential is to understand broadly the present state of knowledge about the distribution, the distribution *in the large* or *asymptotic* distribution, of the primes. It is this theory which gives the solution of problem (3).

We denote by $\pi(x)$ the number of primes not exceeding x. The first step is to prove that (a) *the number of primes is infinite*; $\pi(x)$ *tends to infinity with* x. This is another of Euclid's great contributions to knowledge, and Euclid's proof is perhaps the classical example of proof by reductio ad absurdum. If the theorem is false, we may denote the primes by $2,3,5,\ldots,P$, and all numbers are divisible by one of these. On the other hand the number

$$(2\cdot 3\cdot 5\cdots P)+1$$

is obviously not divisible by any of $2,3,\ldots,P$, and this is a contradiction.

Another very interesting proof is due to Pólya.* It is easy to see that any two of the numbers

$$2+1, 2^2+1, 2^4+1, \ldots, u_n = 2^{2^n}+1$$

are prime to each other. For suppose that p is an odd prime and that $p|u_n, p|u_{n+k}$. Then also

$$p|2^{2^{n+k}}-1=u_{n+k}-2,$$

since

$$x^{2^k m}-1$$

is algebraically divisible by x^m+1, and therefore

$$p|u_{n+k}-(u_{n+k}-2)=2,$$

which is absurd. It follows that the number of primes less than u_n is at least n, and therefore that the number of primes is infinite. In fact the argument shows not merely that $\pi(x)\to\infty$ but that

$$\pi(x) > A \log\log x,$$

where A is a constant. Something in this direction, though a little less, can be proved by a refinement of Euclid's argument.

There is a third line of argument which is a little less elementary but may be made to prove a good deal more.* If $2,3,5,\ldots,P$ were the only primes, then every number would be of the form

$$2^a 3^b 5^c \cdots P^k.$$

---

* See Pólya and Szegö, loc. cit., pp. 133, 342.

* See Dickson's *History*, vol. 1, p. 414, where the proof is attributed to Auric.

If this number is less than x, then a fortiori 2^a is less than x, so that a is less than a constant multiple of $\log x$, and the same argument applies to $b,c,\ldots,k$. The number of possible choices of $a,b,\ldots,k$ is therefore less than a multiple of $(\log x)^{\pi}$, where π is the total number of primes. In other words the number of numbers less than x is less than

$$A(\log x)^{\pi},$$

where A is a constant, and this is impossible, since x tends to infinity more rapidly than any power of $\log x$. A refinement of the argument leads to the inequality

$$\pi(x) > A\frac{\log x}{\log\log x};$$

and the underlying principle may be stated roughly thus, that *if the number of primes were finite, there would not be enough numbers to go round.*

We are still a very long way from the ultimate truth. It is in fact possible to prove, and by comparatively elementary methods, that *the order of magnitude of* $\pi(x)$ *is* $x(\log x)^{-1}$. This theorem, conjectured by Legendre and Gauss, was first proved by Tchebycheff in 1848.

There are two much earlier theorems of Euler which point in this direction. The first is the theorem that (b) *the series*

$$\sum \frac{1}{p}$$

extended over all prime numbers p, is divergent. The proof of this theorem depends upon an identity, also due to Euler, upon which the whole of the modern theory of primes is founded. The identity is

$$\begin{aligned}1^{-s}+2^{-s}+3^{-s}+\cdots &= \sum n^{-s}\\ &= \frac{1}{(1-2^{-s})(1-3^{-s})(1-5^{-s})\cdots}\\ &= \prod\left(\frac{1}{1-p^{-s}}\right),\end{aligned}$$

and is valid for $s>1$; it is at bottom merely the analytical expression of the fundamental theorem, and its importance arises from the fact that it asserts the equivalence of two expressions of which one contains the primes explicitly while the other does not. From Euler's identity we deduce (b) roughly as follows: if Σp^{-1} were convergent, then

$$\prod\left(\frac{1}{1-p^{-1}}\right)$$

would be convergent, and therefore Σn^{-1} would be convergent, which is false. Of course the proof really needs a rather more careful statement.

Euler's second theorem is (c) *the quotient of $\pi(x)$ by x tends to zero*; or in symbols

$$\frac{\pi(x)}{x} \to 0,$$

or, as we write it now

$$\pi(x) = o(x).$$

The proportion of primes is ultimately infinitesimal, "*almost all*" *numbers are composite*. The theorem is a quite simple corollary of (b); roughly, if we remove from the numbers less than x all multiples of the primes $2, 3, \ldots, p$, other than these primes themselves, we are left something like

$$x\left(1-\frac{1}{2}\right)\left(1-\frac{1}{3}\right)\left(1-\frac{1}{5}\right)\cdots\left(1-\frac{1}{p}\right)$$

numbers. The product multiplying x tends to zero when $p \to \infty$, because of (b), and from this we can deduce Euler's second theorem.

It is rather curious that, although Euler's second theorem is a corollary of the first, the lessons which we learn from the two theorems concerning the distribution of the primes have exactly opposite tendencies. The second theorem tells us that the number of primes below a given limit is *not too great*, that the primes are in the end rather liberally spaced out; it is in fact exactly equivalent to the theorem that (d) *the nth prime* p_n *has an order of magnitude greater than* n, or

$$\frac{p_n}{n} \to \infty.$$

If on the other hand the order of magnitude of p_n were *much* greater than n, if it were for example n^2 or $n^{10/9}$ or $n(\log n)^2$, then the series $\sum p_n^{-1}$ would be *convergent*, which is just what Euler's first theorem denies. What we learn from the two theorems together is something like this. If, as we hope, the true order of magnitude of p_n can be measured by some simple function $\phi(n)$, then that function must be of order higher than n, but somewhere near the boundary of convergence of the series

$$\sum \frac{1}{\phi(n)}.$$

The most obvious function which satisfies these requirements is $n \log n$, and to say that p_n is of order $n \log n$ is the same thing as to say that $\pi(x)$ is of order $x(\log x)^{-1}$. This is just what is asserted by Tchebycheff's theorem.

11. Tchebycheff's theorem. The formal statement of Tchebycheff's theorem is (e) *the order of magnitude of* $\pi(x)$ *is* $x(\log x)^{-1}$; *there are constants A and B such that*

$$\frac{Ax}{\log x} < \pi(x) < \frac{Bx}{\log x}.$$

This theorem is precisely equivalent to (f) *the order of magnitude of* p_n *is* $n \log n$;

there are constants A and B such that

$$An\log n < p_n < Bn\log n.$$

The proofs of these theorems given by Tchebycheff have been simplified a good deal by Landau, and I can give you a sketch of one half of the proof which should enable you to understand without much difficulty the general character of the whole.

We begin by replacing $\pi(x)$ by another function. We can write $\pi(x)$ in the form

$$\pi(x) = \sum_{p \leqq x} 1;$$

count one for every prime up to x. A more convenient and really a more natural function is

$$\theta(x) = \sum_{p \leqq x} \log p,$$

the logarithm of the product of all primes up to x. This function seems at first sight a more complicated function, but it is easy enough to see why it is more convenient to work with. The most natural operation to perform on primes is *multiplication*, and this is the operation which we employ in forming $\theta(x)$. It is because it is natural to multiply primes and not to add or subtract them that problems like the problem of the prime pairs $(p, p+2)$, or Goldbach's problem of expressing numbers as sums of primes, turn out to be so terribly difficult.

Since $x/x^{1-\delta}$ tends to infinity, for any positive value of δ, we may expect that nearly all the primes which contribute to $\theta(x)$ will lie in the interval $(x^{1-\delta}, x)$, so that their logarithms lie between $(1-\delta)\log x$ and $\log x$. Hence we may expect $\theta(x)$ to be very much the same function as $\pi(x)\log x$, and in fact there is no difficulty in proving that

$$\theta(x) \sim \pi(x)\log x,$$

that is, that the ratio of the two functions tends to 1. It follows that the inequalities in (e) are equivalent to

$$Ax < \theta(x) < Bx.$$

I shall sketch the proof of the second inequality, which is rather the simpler.

Suppose that x is a power of 2, say 2^m. The primes between $x/2$ and x divide $x!$ but not $(x/2)!$, so that

$$\prod_{x/2 < p \leqq x} p \,\Big|\, \frac{x!}{(x/2)!(x/2)!}.$$

The expression on the right is *one term* in the binomial expansion of $(1+1)^x = 2^x$, and therefore

$$\prod_{x/2 < p \leqq x} p \leqq 2^x.$$

Replacing x by

$$x/2, x/4, x/8, \ldots$$

and multiplying the results, we find that

$$\prod_{p \leqq x} p \leqq 2^{x+x/2+x/4+\cdots} \leqq 2^{2x},$$

and

$$\theta(x) \leqq 2\log 2 \cdot x.$$

This proves the theorem when $x=2^m$. If

$$2^m < x < 2^{m+1}$$

we have

$$\theta(x) \leqq \theta(2^{m+1}) \leqq 4\log 2 \cdot 2^m < 4\log 2 \cdot x.$$

Hence we may take $B=4\log 2$. The proof of the second inequality is, as I said, not quite so simple, but does not involve essentially more difficult ideas. We have thus determined the order of magnitude of $\pi(x)$ and of p_n, and it is perhaps a little astonishing that a problem which sounds so abstruse should have so comparatively simple a solution.

12. The Prime Number theorem. Tchebycheff's solution of the problem is, however, one with which it is impossible to remain content for long, since the whole trend of our discussion has been to suggest that much more is true than we have proved. In fact Tchebycheff's work, fine as it is, is the record of a failure; it is what survives of an unsuccessful attempt to prove what is now called the *Prime Number Theorem.*

This is the theorem that (g) $\pi(x)$ *and* $x(\log x)^{-1}$ *are asymptotically equivalent; the ratio of the two functions tends to unity.* We express this by writing

$$\pi(x) \sim \frac{x}{\log x}.$$

The Prime Number Theorem is equivalent to

$$p_n \sim n\log n,$$

and we may express it very roughly by saying that *the odds are* $\log x$ *to* 1 *that a large number* x *is not prime.*

The Prime Number Theorem, the central theorem of the analytic theory of numbers, was proved independently by Hadamard and by de la Vallée-Poussin in 1896. The empirical evidence for its truth had for long been overwhelming, and I suppose that every number-theorist since Legendre had tried to prove it. The theorem differs from all those which I have discussed so far in that it is apparently impossible to prove it by properly elementary methods; there is no proof known which does not depend essentially on complex function theory. I do not mean to imply that there is any terrible difficulty in the proof; there are considerable difficulties of detail, but the fundamental ideas on which it depends are tolerably straightforward. They are, however, quite unlike any of those of which I have spoken, and I should require a whole lecture to explain them even to a strictly mathematical audience. Actually, a good deal more is known; it can be proved that $\pi(x)$ is approximated still more closely by the "logarithm-integral" of x,

$$\mathrm{Li}\, x = \int_2^x \frac{dt}{\log t},$$

that in fact

$$\pi(x) = \mathrm{Li}\, x + O\left\{ \frac{x}{(\log x)^k} \right\}$$

for every k, the error being of lower order than the quotient of x by *any* power of $\log x$; and it is probable, though not yet proved, that the order of the error does not very materially exceed that of $\sqrt{x}$.

13. Formulas for primes. I return now for a moment to a question which I discussed shortly before, the question whether it was reasonable to expect an "elementary formula" for the nth prime p_n. Let us imagine that my questioner was obstinate in his desire for such a formula; how could I refute his successive suggestions? If he suggested

$$p_n = n \log n,$$

I should have the obvious reply that $n \log n$ is not an integer. Suppose then that he modified his formula to

$$p_n = [n \log n].$$

I should reply that his formula did not agree with the known facts of the asymptotic theory. It agrees with $p_n \sim n \log n$, the first and most obvious deduction from the Prime Number Theorem itself; but the theory carries us much further; it enables us, for example, to show that

$$p_n = n \log n + n \log \log n + O(n),$$

which contradicts the formula. If, becoming more cautious, he asked me what ground I had for denying that p_n might be *some* elementary combination of

$$n, \log n, \log \log n, \ldots,$$

I should naturally find it harder to refute him, but I could advance three arguments which are enough in the aggregate to make up a tolerably convincing case. (i) Since $\mathrm{Li}\, x$ is a very good approximation to $\pi(x)$, the inverse function $\mathrm{Li}^{-1} n$ must be a very good approximation to p_n. Now it is demonstrable that neither the logarithm integral nor its inverse* is an elementary function. It is therefore very unlikely that there should be an elementary formula for p_n. (ii) If the "elementary formula" does

* This may be deduced from general theorems proved recently by J. F. Ritt.

not involve the symbol $[\cdots]$ of the "integral part," the function which it defines will generally not be integral for integral n. If it does, it loses all its simplicity and all its plausibility. (iii) An elementary function may be expected to behave with tolerable regularity at infinity, and so may all its *differences*. Now extremely little is known about the difference $p_{n+1}-p_n$ of two successive primes, but everything that is known, or seems probable from the evidence of the tables, suggests *extreme irregularity* in its behavior. The Prime Number Theorem shows that the *average* value of $p_{n+1}-p_n$ must be $\log n$, and tend to infinity with n. On the other hand there is overwhelming evidence that the smallest possible values of $p_{n+1}-p_n$, namely, $2,4,6,\ldots$, recur indefinitely. It seems practically certain, not merely that there are infinitely many prime pairs $(p,p+2)$ but that there are infinitely many triplets $(p,p+2,p+6)$, and so with any combination of successive primes that is arithmetically possible; such a combination as $(p,p+2,p+4)$ is naturally not possible, since one of these numbers must be divisible by 3. All this seems hopelessly inconsistent with the existence of such a formula as was suggested, and it is clear that speculation in this direction is a waste of time.

There are, however, questions which have a somewhat similar tendency and which cannot be dismissed so summarily. There is one, for example, mentioned in Carmichael's little book. The problem, as he states it, is "*to find a prime greater than a given prime*," which might be interpreted as meaning either "*to find an elementary function $\phi(n)$ such that $\phi(n)\to\infty$ and $\phi(n)$ is prime for every n, or for all n beyond a certain limit*" or as meaning "*to find an elementary function $\phi(p)$ such that $\phi(p)>p$ and $\phi(p)$ is prime whenever p is prime.*" With either interpretation, it is a reasonable challenge, and the problem has not been solved.

Let us take the first form of the problem, which is perhaps the more natural, and let us begin by demanding *less*, namely that $\phi(n)$ shall be prime only for *an infinity of values of n*. In this case the problem becomes trivial, since n is a solution, by Euclid's theorem. It is, however, very interesting to observe that even then n, and certain simple linear functions such as $4n-1$ and $6n-1$, are the *only* trivial solutions. Dirichlet proved that *any* linear function $an+b$ has the property required, provided only that b is prime to a, or in other words that *every arithmetical progression* (subject to the last reservation) *contains an infinity of primes*. This theorem is quite difficult, except in a few special cases such as those which I mentioned, and it exhausts our knowledge in this particular direction. No one has ever proved that any of the functions

$$n^2+1, \quad 2^n-1, \quad 2^n+1$$

is prime for an infinity of values of n. With functions of *two* variables we can progress a good deal farther; we know for example that every quadratic form $am^2+bmn+cn^2$ contains an infinity of primes, provided of course that a,b,c have no common factor and that $b^2\neq 4ac$, and we can study the law of their distribution.

To find a $\phi(n)$ prime for *every n* is naturally still more difficult. Here linear functions are obviously useless, and no solution of any kind is known. Fermat conjectured that

$$2^{2^n}+1,$$

is always prime, but Euler proved that this is false, since

$$2^{32}+1=4294967297=641\cdot 6700417.$$

So far as I know, no one else has ever advanced any other suggestion which is even plausible.

In view of the apparently insuperable difficulties of this problem, there is a certain interest in *negative* results. It is plain, first, that $an+b$ cannot be prime for all n, or all large n. More generally, no polynomial

$$f(n)=a_0n^k+a_1n^{k-1}+\cdots+a_k$$

can be prime for all or all large n; for if $f(m)=M$ then $f(rn+m)$ is divisible by M for all r. There are entertaining curiosities in this field; thus

$$n^2-n+41$$

is prime for the first 41, and

$$n^2-79n+1601$$

for the first 80 values of n. It is obvious that forms like

$$a^n-1,\quad a^n+1$$

cannot be prime for all large n, since, for example, $a^{3m}-1$ is divisible by a^m-1, and it is natural to suppose that the same is true for

$$P(n,2^n,3^n,4^n,\ldots,k^n),$$

where P is any polynomial with integral coefficients.*

14. The fundamental theorem in an algebraic field. I must not allow myself to succumb to the temptation of talking too long about the theory of the distribution of primes, which is after all only one chapter in arithmetic. There are other topics about which our imaginary enquirer will certainly demand more information, and of these I think one stands out; it is certain that he will want fuller explanations about the field $K^*[\sqrt{(-5)}\,]$ and the other algebraic fields in which the analog of the fundamental theorem fails. All ordinary arithmetic depends, it seems, upon the fundamental theorem; how then can there *be* an arithmetic in a field in which it is false? It would seem that the arithmetic of such a field can bear no real resemblance to ordinary arithmetic. I shall spend the rest of my time in an attempt to explain, in the very broadest outline, how order is restored.

I shall begin by quoting a remark of Hilbert which is trivial in itself but which shows us at once the direction in which we must look for a solution. Consider the numbers

$$1,5,9,13,17,21,\ldots,$$

---

* Morgan Ward of Pasadena has found a very simple proof of the theorem.

of the form $4m+1$. These numbers form a group for multiplication (though naturally not for addition), and we can define divisibility and primality in the group. The "primes" are the numbers

$$5, 9, 13, 17, 21, 29, 33, 37, 41, 49, \ldots,$$

which are greater than 1, of the form $4m+1$, and not decomposable into factors of this form. Thus 21, 57, 77, and 209 are "primes;" but

$$4389 = 21 \cdot 209 = 57 \cdot 77,$$

so that a number of the group may be resolved into "prime" factors in different ways.

In this case the solution of the mystery is obvious. The "*fundamental theorem*" fails *because of the absence from the group of the numbers* $4m+3$ *of ordinary arithmetic*. In fact

$$21 = 3 \cdot 7, \quad 57 = 3 \cdot 19, \quad 77 = 7 \cdot 11, \quad 209 = 11 \cdot 19$$

and

$$21 \cdot 209 = (3 \cdot 7)(11 \cdot 19) = (3 \cdot 19)(7 \cdot 11) = 57 \cdot 77.$$

We cannot give a proper account of the properties of the numbers $4m+1$ so long as we insist on excluding the numbers $4m+3$; *the numbers* $4m+1$ *do not form by themselves an adequate basis for arithmetic*. This observation has of course no intrinsic interest, since no reasonable person would expect that they would do so. It is trivial in itself, but it is not at all trivial in its suggestion, since it suggests that the troubles of the field $K^*[\sqrt{(-5)}\,]$ may be remedied *by considering the field as part of some larger field.*

This is in fact the solution found by Kummer. We consider the field $L[\sqrt{(-5)}\,]$ of numbers

$$\xi = \sqrt{\left(a + b\sqrt{(-5)}\,\right)},$$

where a and b are ordinary integers. This is only an approximate statement; we do not actually consider all such numbers, but only those satisfying certain further conditions; the greatest common divisor of a and b must be a square or five times a square, and a^2+5b^2 must be a square. The field L includes K^*. The numbers of L form a group for multiplication, and we can define divisibility and primality in the field. Finally, the analog of the fundamental theorem is valid; *factorization is unique in L.* The proof of this is quite simple, but requires a little attention to detail, and I must refer you for the details to Mordell's tract on *Fermat's Last Theorem.*

We can now give a simple account of the equations in $K^*[\sqrt{(-5)}\,]$ which puzzled us before. Consider for example the equation

$$3 \cdot 7 = \left(1 + 2\sqrt{(-5)}\,\right)\left(1 - 2\sqrt{(-5)}\,\right).$$

It is easily verified that

$$3^2=\left(2+\sqrt{(-5)}\right)\left(2-\sqrt{(-5)}\right),$$
$$7^2=\left(2+3\sqrt{(-5)}\right)\left(2-3\sqrt{(-5)}\right),$$
$$\left(1+2\sqrt{(-5)}\right)^2=-19+4\sqrt{(-5)}=-\left(2-\sqrt{(-5)}\right)\left(2+3\sqrt{(-5)}\right),$$
$$\left(1-2\sqrt{(-5)}\right)^2=-19-4\sqrt{(-5)}=-\left(2+\sqrt{(-5)}\right)\left(2-3\sqrt{(-5)}\right).$$

Hence, if we write

$$\alpha=\sqrt{\left(2+\sqrt{(-5)}\right)},\quad \alpha'=\sqrt{\left(2-\sqrt{(-5)}\right)},$$
$$\beta=\sqrt{\left(2+3\sqrt{(-5)}\right)},\quad \beta'=\sqrt{\left(2-3\sqrt{(-5)}\right)},$$

we have

$$3=\alpha\alpha', 7=\beta\beta', 1+2\sqrt{(-5)}=-\alpha'\beta, 1-2\sqrt{(-5)}=-\alpha\beta',$$
$$3\cdot 7=\alpha\alpha'\cdot\beta\beta'=\alpha'\beta\cdot\alpha\beta'=\left(1+2\sqrt{(-5)}\right)\left(1-2\sqrt{(-5)}\right);$$

and all of these equations are entirely natural. *In order to obtain a satisfactory theorem of factorization in K^*, we must conceive K^* as immersed in the larger field L.* The logic of the solution is exactly the same as that of the solution of the corresponding, but trivial, problem for the numbers $4m+1$.

On the other hand there is an obvious contrast between the two solutions. It is *natural* to think of the field "$4m+1$" as part of the field "m"; "m" is the more obvious and simpler field. It is not natural to think of K^* as part of L; K^* is a much simpler and more natural field than L, and we should like to do without the reference to the latter if we could. It will be very tiresome if, whenever we consider an algebraic field, we are to be compelled to construct some more elaborate field of which it is a part. We should prefer to tidy up the house without going out of doors.

We may look for a hint once more in the numbers $4m+1$. Some of these numbers are divisible by 7, a number outside the field; and these numbers stand in certain specific relations to one another inside the field. Could we give a rational account of these relations without explicit reference to the number 7? It is a very unnatural thing to try to do, since what is *important* about the numbers is precisely that they *are* divisible by 7, but we could do it; we could define the class

$$21, 49, 77, 105, \ldots,$$

of numbers $4m+1$ divisible by 7 in terms of the field $4m+1$ itself. For example, we could take the first two numbers 21 and 49, and say "the class in question is the class which begins with these two numbers and whose members recur at regular intervals in the field." It is of course an artificial definition, and it is impossible to conceal from ourselves what we are really doing.

It is often a very profitable exercise for a mathematician to force himself to solve some simple problem without the weapon obviously appropriate to the occasion, to throw away the key of the front door and insist on forcing himself in somehow through the window. The forced and unnatural solution of one problem will often turn out to contain the germ of a quite natural solution of another. So it proves in this case; it is natural to try to define the numbers of K^* divisible by ξ without going outside K^*; it is natural, and possible, and it gives us the key to what is, in the general case, the established method of constructing a satisfactory arithmetic.

It is obvious that, if α and β belong to K^*, and $\xi|\alpha$ and $\xi|\beta$, then $\xi|\lambda\alpha+\mu\beta$, where λ and μ are any numbers of K^*. The converse proposition is not true; it is not true that if I is any set of numbers of the field K^* which has the property "if α and β belong to I, then $\lambda\alpha+\mu\beta$ belongs to I, for every λ and μ of the field," then there exists a number ξ, belonging to K^* which divides every number of I. What *is* true is that every number of I is divisible by a ξ which belongs to L but not in general to K^*. The set I is identical with the set of numbers of K^* divisible by ξ. Such a set I, or the more general set based on any finite number of numbers $\alpha, \beta, \gamma, \ldots$, of K^*, is called an *ideal*, the numbers ξ, underlying K^* but not belonging to it, having been described by Kummer as "ideal numbers." In ordinary arithmetic ideals are simply the sets of numbers divisible by some special number such as 3, and there is nothing in particular to be gained by their introduction. In an algebraic field they are not, in general, the sets of numbers divisible by a number *of the field*, and their introduction is essential before arithmetic can get properly started. We can define multiplication and division of ideals, prime ideals, and so on, and when we have done this we find that the arithmetic of ideals has all the properties of ordinary multiplicative arithmetic. In particular, *every ideal can be resolved uniquely into prime ideals*; the fundamental theorem is true when stated in terms of ideals.

The proof of the fundamental theorem is not particularly difficult; Landau presents it, with all the preliminary definitions, in about a dozen pages of quite simple reasoning. But I would not commit the impertinence, even if I had the time, of assuming the airs of an expert in the algebraic theory of numbers, a subject which I admire only at a distance and in which I have never worked. It is ordinary rational arithmetic which attracts the ordinary man, and I have digressed outside it only because there is a good deal in it which it is impossible to appreciate properly without a little knowledge of the larger theory. It is impossible, for example, to appreciate Euclid's arithmetical achievements until we realize that there are arithmetics in which the most obvious analogs of his theorems are false.

15. Conclusion. Pedagogy. There are few things in the world for which I have less taste than I have for mathematical pedagogics, but I cannot resist the temptation of concluding with one pedagogic lesson. There was, and I fear still is, a popular English text book of algebra which I used at school and which contained a chapter on the theory of numbers. It might be expected that such a chapter would be among the most instructive in the book; we might suppose, for example, that Euclid's algorithm, with its elegance, its simplicity, and its far reaching consequences, would be an ideal text for the instruction of a bright young mathematician. In fact the algorithm was never mentioned; one was to find the highest common factor of 12091 and 14803, I suppose, by "trial"; and all that the authors had to say of the fundamental theorem was that "it is so evident that it may be regarded as a necessary law of thought." It is possible of course that all this may have been expunged from later editions. It is certain, however, that chapters on number theory in textbooks of algebra are usually quite intolerably bad, and it is conceivable that Oxford University may have been right in erasing the subject altogether from its more elementary examination schedule.

The elementary theory of numbers should be one of the very best subjects for early mathematical instruction. It demands very little previous knowledge; its subject matter is tangible and familiar; the processes of reasoning which it employs are simple, general and few; and it is unique among the mathematical sciences in its appeal to natural human curiosity. A month's intelligent instruction in the theory of numbers ought to be twice as instructive, twice as useful, and at least ten times as entertaining as the same amount of "calculus for engineers". It is after all only a minority of us who are going to spend our lives in engineering workshops, and there is no particular reason why most of us should feel any overpowering interest in machines; nor is it in the least likely that, on those occasions when machines are of real importance to us, we shall require the power of dealing with them by methods more elaborate than the simplest rule of thumb. It is not engineering mathematics that is wanted for the understanding of modern physics, and still less is it wanted by most of us for the ordinary needs of life; we do not actually drive cars by solving differential equations. There may be a case for subordinating mathematics to the linguistic and literary studies which are so much more obviously useful to ordinary men, but there is none for sacrificing a splendid subject to meet a quite imaginary need.

4

DUNHAM JACKSON

Dunham Jackson was born on July 24, 1888, in Bridgewater, Massachusetts. He entered Harvard in 1904 and received his Bachelor's degree in 1908 and his Master's degree in 1909. Thereafter he was granted a Harvard Sheldon scholarship for graduate study in Europe and worked under Professor Landau at Göttingen where he received his Ph.D. in 1911. He returned to Harvard as an instructor and later an assistant professor until 1916. He was a Captain in the Army working on mathematical ballistics under F. R. Moulton for the Ordinance Department during the years 1918 and 1919. He then went to the University of Minnesota where he remained until his death in 1946 at age 58, although his last six years were considerably hindered by a heart attack in 1940.

He was a member of the Council of the Society and its vice-president in 1921. He gave the Colloquium Lectures to the Society in 1925. A charter member of the Association and a member of its Board of Governors from 1923 to 1929; he served as Vice-President in 1924 and 1925 and as President in 1926. He was also a Vice-President of the American Association for the Advancement of Arts and Sciences (1927), a fellow of the American Academy of Arts and Sciences, and a member of the National Academy of Science.

He wrote approximately seventy-five research papers, mostly in the field of approximation theory by polynomials and trigonometric functions. His doctoral thesis contained important results on trigonometric sums. These results were useful in approximating periodic functions which led to later results on Fourier series and orthogonal polynomials, which in turn became powerful tools for the study of the convergence of Legendre series. He worked on degrees of convergence of Sturm-Liouville series, uniform summability, and approximation in the sense of mth powers. In addition, he contributed new results in boundary value problems, ordinary linear differential equations, functions of several complex variables, and sampling theory in statistics.

Professor Jackson was very devoted to teaching, particularly at the undergraduate level where he was recognized as a skillful teacher and an inspiring lecturer. He was very conscious of his responsibilities to his profession as an educator. William Hart describes him as devoted to "expositions of fundamental content, written at the level of readers satisfying only the minimum mathematical prerequisites, in order to make the material more accessible in its field of applications"; "He always preferred to have a substantial part of his teaching schedule at the undergraduate level, because he enjoyed personal contacts with young people and the opportunity to influence them scientifically at an early age." "Jackson was a man of high ideals, extremely unselfish, and very conscious of his responsibilities, not only in strictly personal affairs but also with respect to his profession as an educator and his duties as a citizen." [1]. His award of the Chauvenet Prize was well deserved and much of the material appeared in his prize paper in his famous Carus Monograph of *Fourier Series and Orthogonal Polynomials* in 1941. In addition to this book he also wrote *The Theory of Approximation*, a Colloquium publication of the American Mathematical Society (1930).

Reference

1. William L. Hart, Dunham Jackson, 1888–1946, Bull. A.M.S., 54 (1948) 847–860.

SERIES OF ORTHOGONAL POLYNOMIALS *

DUNHAM JACKSON, University of Minnesota

1. Introduction. The theory of systems of polynomials orthogonal over an interval with respect to a given weight function of more or less arbitrary character, and of the convergence of expansions in series of such polynomials, has undergone rapid development during recent years, and the essential facts of the theory are now well known.† In view of the increasing attention which has been given to the subject, however, and the diversity of possible applications, there may be room for a brief introductory account bringing out some of the more conspicuous facts with a degree of simplicity and comprehensiveness comparable with that found in existing elementary treatments of Fourier series. The purpose of this article is to give at least a part of such an exposition.

2. Fundamental formulas. For a considerable part of the formal theory, the range of the independent variable may be either finite or infinite in extent. In the present paper, however, the interval will be thought of as finite, and may without loss of generality be taken as that from -1 to $+1$. Let $\rho(x)$ be a function which is integrable over this interval (in the elementary sense or in the sense of Lebesgue),‡ nowhere negative, and actually positive at least for such a part of the interval that

$$\int_{-1}^{1} \rho(x)\,dx > 0.$$

Let $p_0(x), p_1(x), p_2(x), \ldots,$ be the corresponding sequence of normalized orthogonal polynomials, of degrees indicated by the subscripts respectively, satisfying the conditions that

$$\int_{-1}^{1} \rho(x)p_m(x)p_n(x)\,dx = 0 \qquad (m \neq n),$$

$$\int_{-1}^{1} \rho(x)p_n^2(x)\,dx = 1. \tag{1}$$

It is perhaps not necessary to reproduce here a proof that for any $\rho(x)$ of the

* Received November 16, 1932. Presented to the Mathematical Association of America at Los Angeles, August 30, 1932.

† Among outstanding contributions to the extensive literature, beyond the pioneer work of Tchebychef, the following may be mentioned: G. Darboux, *Mémoire sur l'approximation des fonctions de très-grands nombres, et sur une classe étendue de développements en série*, Journal de mathématiques pures et appliquées, (3), vol. 4 (1878), pp. 5–56, 377–416; a series of publications by W. Stekloff, one of which will be referred to specifically later; G. Szegö, *Über die Entwicklung einer willkürlichen Funktion nach den Polynomen eines Orthogonalsystems,* Mathematische Zeitschrift, vol. 12 (1922), pp. 61–94; G. Szegö, *Über den asymptotischen Ausdruck von Polynomen, die durch eine Orthogonalitätseigenschaft definiert sind,* Mathematische Annalen, vol. 86 (1922), pp. 114–139; S. Bernstein, *Sur les polynômes orthogonaux relatifs à un segment fini,* Journal de mathématiques pures et appliquées, (9), vol. 9 (1930), pp. 127–177, and vol. 10 (1931), pp. 219–286.

‡ For a formulation in terms of Stieltjes integrals see J. Shohat, *Stieltjes integrals in mathematical statistics,* Annals of Mathematical Statistics, vol. 1 (1930), pp. 73–94.

character specified such a sequence of polynomials exists.* For $\rho(x)\equiv 1$ the polynomials are those of Legendre, except for constant factors, and the convergence proof to be given below will be incidentally a proof of convergence of Legendre series. For $\rho(x)=(1-x^2)^{-1/2}$ the p's are essentially the "trigonometric polynomials" by which the cosine of n times an angle is expressed in terms of the angle itself: $p_n(x)=C_n\cos(n\arccos x)$. Both are particular cases of Jacobi polynomials, defined by the specification $\rho(x)=(x+1)^\alpha(1-x)^\beta$, $\alpha>-1$, $\beta>-1$. Other special cases will be referred to later.†

An "arbitrary" function $f(x)$ can be formally expanded in a series of the form

$$c_0p_0(x)+c_1p_1(x)+c_2p_2(x)+\cdots \tag{2}$$

with coefficients determined as in the case of Fourier series, on the basis of the property of orthogonality: if the series (2) is supposed uniformly convergent to the value $f(x)$, multiplication by $\rho(x)p_k(x)$ and integration from -1 to $+1$ with the use of (1) gives

$$c_k=\int_{-1}^{1}\rho(x)f(x)p_k(x)\,dx. \tag{3}$$

If $f(x)$ is any function such that $\rho(x)f(x)$ is integrable, coefficients c_k can be defined by (3) and used to form a series (2), and it may then be inquired whether this series does in fact converge and have $f(x)$ for its sum.

For any such function $f(x)$, the notation $s_n(x)$ will be used to represent the sum of the first $n+1$ terms of the series:

$$s_n(x)=\sum_{k=0}^{n}c_kp_k(x),$$

the coefficients c_k being given by (3). It follows at once from the definition that

$$s_n(x)=\int_{-1}^{1}\rho(t)f(t)K_n(x,t)\,dt, \tag{4}$$

where

$$K_n(x,t)=\sum_{k=0}^{n}p_k(x)p_k(t). \tag{5}$$

Since each $p_k(x)$ is actually of the degree indicated by its subscript, the identities expressing the p's as combinations of powers of x can be solved for the powers of x successively, and x^n, or any polynomial of the nth degree in x, can be expressed as a linear combination of $p_0(x),\ldots,p_n(x)$. If $q(x)$ is any polynomial

* The sequence may be constructed by what is commonly referred to as "Schmidt's process of orthogonalization"; see, for example, the writer's *Theory of Approximation*, American Mathematical Society Colloquium Publications, vol. XI (hereafter referred to as Colloquium), New York, 1930, pp. 89–90, 95.

† For the most important cases of orthogonal polynomials see Pólya and Szegö, *Aufgaben und Lehrsätze aus der Analysis*, vol. II, Berlin 1925, pp. 75–76, 91–94, 266–267, 287–295.

whatever of degree lower than the nth,

$$\int_{-1}^{1} \rho(x) p_n(x) q(x)\, dx = 0 \tag{6}$$

in consequence of (1), since $q(x)$ is a combination of p's with indices not greater than $n-1$.

3. The Christoffel-Darboux identity. On the basis of these fundamental properties of the p's, a formula for $K_n(x,t)$, to which the name of Christoffel is attached in the case of Legendre polynomials, can be obtained for general $\rho(x)$ by a method due to Darboux.*

Let a_k denote the coefficient of x^k in $p_k(x)$. The sign of a_k, which is not determined by (1), will be taken as positive. The product $xp_n(x)$, being a polynomial of degree $n+1$, can be linearly expressed in terms of $p_0(x),\ldots,p_{n+1}(x)$, and by comparison of the terms of highest degree can be written in the form

$$xp_n(x) = \frac{a_n}{a_{n+1}} p_{n+1}(x) + C_{nn} p_n(x) + C_{n,n-1} p_{n-1}(x) + \cdots + C_{n0} p_0(x), \tag{7}$$

where the C's are some constants. Let this identity be multiplied by $\rho(x) p_k(x)$ and integrated from -1 to $+1$, k being one of the numbers $0, 1, \ldots, n-1$. By (1) the result reduces to

$$\int_{-1}^{1} x\rho(x) p_n(x) p_k(x)\, dx = C_{nk}. \tag{8}$$

For $k \leqq n-2$, $xp_k(x)$ is a polynomial of degree not higher than $n-1$, which may be identified with $q(x)$ in (6), and it is seen that $C_{nk}=0$ for these values of k. For $k=n-1$, let n be replaced by $n-1$ in (7):

$$xp_{n-1}(x) = \frac{a_{n-1}}{a_n} p_n(x) + q_{n-1}(x),$$

where $q_{n-1}(x)$ is a polynomial of degree $n-1$ at most. Substitution of this in (8) gives

$$C_{n,n-1} = \frac{a_{n-1}}{a_n} \int_{-1}^{1} \rho(x) p_n^2(x)\, dx = \frac{a_{n-1}}{a_n}.$$

Thus the p's are found to be subject to the recursion formula

$$xp_n(x) = \frac{a_n}{a_{n+1}} p_{n+1}(x) + C_{nn} p_n(x) + \frac{a_{n-1}}{a_n} p_{n-1}(x), \tag{9}$$

the coefficient C_{nn} being left unspecified, as far as the present calculation is concerned.

---

* Loc. cit., p. 411. The notation is somewhat different, since Darboux takes the leading coefficient as unity.

With n replaced by k, (9) reads

$$xp_k(x)=\frac{a_k}{a_{k+1}}p_{k+1}(x)+C_{kk}p_k(x)+\frac{a_{k-1}}{a_k}p_{k-1}(x).$$

Let this be multiplied by $p_k(t)$, and the corresponding identity in t,

$$tp_k(t)=\frac{a_k}{a_{k+1}}p_{k+1}(t)+C_{kk}p_k(t)+\frac{a_{k-1}}{a_k}p_{k-1}(t),$$

by $p_k(x)$, and one subtracted from the other; the result is

$$(t-x)p_k(t)p_k(x)=\frac{a_k}{a_{k+1}}[p_{k+1}(t)p_k(x)-p_k(t)p_{k+1}(x)]$$
$$-\frac{a_{k-1}}{a_k}[p_k(t)p_{k-1}(x)-p_{k-1}(t)p_k(x)]. \qquad (10)$$

This holds for any value of $k \geqq 1$. For $k=0$, $p_0(x)\equiv a_0$, while $p_1(x)$ is of the form a_1x+b, so that

$$xp_0(x)=\left(\frac{a_0}{a_1}\right)p_1(x)-\left(\frac{a_0b}{a_1}\right),$$

and it follows immediately that

$$(t-x)p_0(t)p_0(x)=\frac{a_0}{a_1}[p_1(t)p_0(x)-p_0(t)p_1(x)]. \qquad (11)$$

If (10) is written for $k=1,2,\ldots,n$ successively, addition of (11) and the n identities (10) yields*

$$(t-x)[p_0(t)p_0(x)+\cdots+p_n(t)p_n(x)]=\frac{a_n}{a_{n+1}}[p_{n+1}(t)p_n(x)-p_n(t)p_{n+1}(x)],$$

$$K_n(x,t)=\frac{a_n}{a_{n+1}}\cdot\frac{p_{n+1}(t)p_n(x)-p_n(t)p_{n+1}(x)}{t-x}.$$

4. First convergence theorem. Since 1 is a polynomial of degree zero, it follows from (6) that

$$\int_{-1}^{1}\rho(t)p_k(t)\,dt=\int_{-1}^{1}\rho(t)p_k(t)\cdot 1\,dt=0$$

for $k \geqq 1$, while

$$p_0(x)\equiv a_0\equiv p_0(t), \qquad \int_{-1}^{1}\rho(t)p_0(x)p_0(t)\,dt=\int_{-1}^{1}\rho(t)p_0^2(t)\,dt=1.$$

---

* A more compact derivation of the formula for K_n, without intermediate use of the recursion formula, is given by Shohat, *On Stieltjes continued fractions*, American Journal of Mathematics, vol. 54 (1932), pp. 79–84.

Consequently, by direct substitution of the expression (5) for $K_n(x,t)$,

$$1 \equiv \int_{-1}^{1} \rho(t) K_n(x,t)\,dt,$$

a special case of (4) for $f(t) \equiv 1$. As $f(x)$ is a constant with respect to the variable of integration, the result of multiplying this identity by $f(x)$ may be written in the form

$$f(x) = \int_{-1} \rho(t) f(x) K_n(x,t)\,dt,$$

and subtraction of this from (4) gives

$$\begin{aligned} s_n(x) - f(x) &= \int_{-1}^{1} \rho(t)[f(t) - f(x)] K_n(x,t)\,dt \\ &= \frac{a_n}{a_{n+1}} \int_{-1}^{1} \rho(t)[f(t) - f(x)] \frac{p_{n+1}(t)p_n(x) - p_n(t)p_{n+1}(x)}{t-x}\,dt. \end{aligned} \tag{12}$$

The problem of convergence is to show that (12) approaches zero as n becomes infinite.

As a first step, it may be pointed out that* $a_n/a_{n+1} < 1$. Let a polynomial $\pi_{n-1}(x)$ be determined among all polynomials of degree $n-1$ by the requirement that the integral

$$\int_{-1}^{1} \rho(x)[x^n - \pi_{n-1}(x)]^2\,dx$$

shall be a minimum, and let γ_n be the corresponding minimum value of the integral. If $\pi_{n-1}(x)$ is expressed in terms of $p_0(x), \ldots, p_{n-1}(x)$, it is readily seen that a necessary condition for the minimum is that

$$\int_{-1}^{1} \rho(x)[x^n - \pi_{n-1}(x)] p_k(x)\,dx = 0$$

for $k = 0, 1, \ldots, n-1$, and hence that

$$\int_{-1}^{1} \rho(x)[x^n - \pi_{n-1}(x)] q(x)\,dx = 0$$

for every polynomial $q(x)$ of degree lower than the nth. Let this relation be multiplied by a_n and subtracted from (6); if the combination $p_n(x) - a_n[x^n - \pi_{n-1}(x)]$ is denoted by $r(x)$,

$$\int_{-1}^{1} \rho(x) r(x) q(x)\,dx = 0.$$

---

* Cf. J. Chokhatte [J. Shohat], *Sur le développement de l'intégrale... et sur les polynômes de Tchebycheff*, Rendiconti del Circolo Matematico di Palermo, vol. 47 (1923), pp. 25–46, pp. 31–33. A treatment of convergence can be based directly on the general least-square property, of which a special case is utilized here; see e.g., Colloquium, pp. 92–101.

But $r(x)$ is of degree lower than the nth, since the terms in x^n cancel; it is possible in particular to take $q(x)\equiv r(x)$, so that

$$\int_{-1}^{1} \rho(x)[r(x)]^2 dx = 0,$$

from which it follows that $r(x)\equiv 0$, $p_n(x)\equiv a_n[x^n - \pi_{n-1}(x)]$. Hence

$$\gamma_n = \frac{1}{a_n^2}\int_{-1}^{1} \rho(x) p_n^2(x)\,dx = \frac{1}{a_n^2}, \quad a_n = \gamma_n^{-1/2}.$$

Similarly, $a_{n+1} = \gamma_{n+1}^{-1/2}$, where γ_{n+1} is the minimum of

$$\int_{-1}^{1} \rho(x)[x^{n+1} - \pi_n(x)]^2 dx$$

as $\pi_n(x)$ ranges over all polynomials of the nth degree. But a particular polynomial of the nth degree is $x\pi_{n-1}(x)$, and hence

$$\gamma_{n+1} \leqq \int_{-1}^{1} \rho(x)[x^{n+1} - x\pi_{n-1}(x)]^2 dx = \int_{-1}^{1} x^2\rho(x)[x^n - \pi_{n-1}(x)]^2 dx.$$

As $x^2 < 1$ throughout the interior of the interval, the last integral is smaller than that defining γ_n, and $\gamma_{n+1} < \gamma_n$, $a_{n+1} > a_n$.

Let $f(x)$ be a function such that both $\rho(x)f(x)$ and $\rho(x)[f(x)]^2$ are integrable over $(-1,1)$. By use of the relations (1) and (3) it is seen that

$$\int_{-1}^{1} \rho(x)[f(x) - s_n(x)]^2 dx = \int_{-1}^{1} \rho(x)[f(x)]^2 dx - \sum_{k=0}^{n} c_k^2.$$

Since the left-hand member is positive or zero,

$$\sum_{k=0}^{n} c_k^2 \leqq \int_{-1}^{1} \rho(x)[f(x)]^2 dx.$$

As this holds for all values of n, while the integral on the right is independent of n, the series $\sum_0^\infty c_k^2$ is convergent, and

$$\lim_{k\to\infty} c_k = 0.$$

For the application to be made presently, this result is needed, not for the function $f(x)$ itself, but for another function related to it; for convenience of reference it may be restated in slightly different notation as follows:

LEMMA. *If $\varphi(t)$ is a function such that $\rho(t)\varphi(t)$ and $\rho(t)[\varphi(t)]^2$ are integrable over* $(-1,1)$,

$$\lim_{n\to\infty} \int_{-1}^{1} \rho(t)\varphi(t) p_n(t)\,dt = 0.$$

Now let the expression (12) for the remainder of the series be recalled. For a fixed value of x, let $\varphi(t)$ stand for the difference quotient $[f(t) - f(x)]/(t-x)$.

Then the remainder may be written in the form

$$s_n(x)-f(x)$$
$$=\frac{a_n}{a_{n+1}}\left[p_n(x)\int_{-1}^{1}\rho(t)\varphi(t)p_{n+1}(t)\,dt-p_{n+1}(x)\int_{-1}^{1}\rho(t)\varphi(t)p_n(t)\,dt\right]. \quad (13)$$

For simplicity in a first convergence theorem, to be reconsidered later in the interest of greater generality, let it be supposed that $f(t)$ is continuous except for a finite number of finite jumps, and that it has for $t=x$ a right-hand and a left-hand derivative, not necessarily equal to each other. Then $\varphi(t)$ is continuous except for a finite number of finite jumps throughout $(-1,1)$, and the integrals of $\rho\varphi$ and $\rho\varphi^2$ certainly exist. By the preceding paragraph, therefore, *the integrals in* (13) *approach zero for* $n\to\infty$. It has been shown that $a_n/a_{n+1}<1$. It will be seen presently that *in many important cases the polynomials* $p_n(x)$ *remain bounded as n increases, for a fixed value of x, in the interior of the interval at least,* and in fact are uniformly bounded throughout any closed interval interior to $(-1,1)$. In such cases the whole right-hand member of (13) approaches zero when x is an interior point (and at the ends of the interval also if the values of $p_n(\pm 1)$ are bounded). The present stage of progress may be summarized in

THEOREM I. *If the weight function* $\rho(x)$ *is such that* $p_n(x)$ *remains bounded as n increases, for a specified value of x in* $(-1,1)$, *and if* $f(t)$ *is continuous throughout the interval except for a finite number of finite jumps, the series* (2) *will converge to the value* $f(x)$ *if* $f(t)$ *has a right-hand and a left-hand derivative at the point* $t=x$.

In the case of convergence for $t=+1$, of course, only the left-hand derivative is involved, and similarly at the other end of the interval.

The trigonometric polynomials $\cos(n\arccos x)$, with or without the normalizing factor $(2/\pi)^{1/2}$, are uniformly bounded throughout the whole closed interval $(-1,1)$. The normalized Legendre polynomials $[(2n+1)/2]^{1/2}X_n(x)$ are uniformly bounded throughout any closed interval interior to $(-1,1)$, since $|X_n(x)|$ itself remains inferior to a constant multiple of $n^{-1/2}$ throughout such an interval;* and Theorem I thus applies directly to Legendre series. The fact that Jacobi polynomials in general possess the property in question is a consequence of an asymptotic formula due to Darboux,† and the requisite information for broad classes of weight functions is obtainable from the memoirs of Szegö and Bernstein. While it would be far outside the scope of the present "elementary" treatment to reproduce the arguments of these authors, it is easy to give an immediate extension of the range of applicability of Theorem I somewhat beyond the fundamental cases of cosine series and Legendre series.

Let $q_0(x),q_1(x),q_2(x),\ldots,$ be the normalized orthogonal polynomials corresponding to a weight function $\rho_1(x)$, and let $p_0(x),p_1(x),p_2(x),\ldots,$ be those corresponding to the weight function $\rho(x)=\Pi(x)\rho_1(x)$, where $\Pi(x)$ is a polynomial

* See e.g. L. Fejér, *Abschätzungen für die Legendreschen und verwandte Polynome*, Mathematische Zeitschrift, vol. 24 (1926), pp. 285–298; Colloquium, pp. 27–28.

† Cf. Szegö, Mathematische Zeitschrift, vol. 12, loc. cit., p. 89.

which is non-negative for $-1 \leqq x \leqq 1$. Let $\rho_1(x)$ be such that the polynomials $q_k(x)$ are uniformly bounded over a point set E contained in the interval $(-1, 1)$, say $|q_k(x)| \leqq H$ for all points in E and for all values of k. For the present general formulation, the set E may consist of one point or several points or an infinity of points, and may or may not include one or both of the points ± 1. Let it be supposed that $\Pi(x)$ remains different from zero and has a positive lower bound, say $\Pi(x) \geqq g > 0$, for all values of x in E. The purpose is to show that the property of boundedness in E carries over from the q's to the p's.

Let the degree of $\Pi(x)$ be m. Since $\Pi(x)p_n(x)$ is a polynomial of degree $n+m$, it can be written in the form*

$$\Pi(x)p_n(x) = c_1 q_1(x) + c_2 q_2(x) + \cdots + c_{n+m} q_{n+m}(x);$$

the c's of course vary with n, but it is not necessary to complicate the notation for the sake of making this explicitly apparent. *In each case* $c_k = 0$ *for* $k < n$, since

$$c_k = \int_{-1}^{1} \rho_1(x) \cdot \Pi(x) p_n(x) \cdot q_k(x)\, dx,$$

and $p_n(x)$ is orthogonal to every polynomial of lower degree with respect to the weight function $\Pi(x)\rho_1(x)$. Consequently

$$\Pi(x)p_n(x) = c_n q_n(x) + \cdots + c_{n+m} q_{n+m}(x),$$

the number of non-vanishing coefficients on the right being at most $m+1$, *and so remaining finite as n increases*. If x belongs to the set E,

$$|\Pi(x)p_n(x)| \leqq H(|c_n| + \cdots + |c_{n+m}|).$$

Let G be the maximum of $\Pi(x)$ for $-1 \leqq x \leqq 1$. For x in E,

$$|p_n(x)| \leqq (H/g)(|c_n| + \cdots + |c_{n+m}|).$$

For any value of k from n to $n+m$, obviously,

$$|c_k|^2 = c_k^2 \leqq c_n^2 + \cdots + c_{n+m}^2, \quad |c_k| \leqq \left(c_n^2 + \cdots + c_{n+m}^2\right)^{1/2},$$

and†

$$|c_n| + \cdots + |c_{n+m}| \leqq (m+1)\left(c_n^2 + \cdots + c_{n+m}^2\right)^{1/2}.$$

* Cf. Szegö, *Über die Entwickelung einer analytischen Funktion nach den Polynomen eines Orthogonalsystems*, Mathematische Annalen, vol. 82 (1921), pp. 188–212, footnote on pp. 190–191; Bernstein, Journal de mathématiques, vol. 9, loc. cit., pp. 138–139; J. A. Shohat, *On the development of continuous functions in series of Tchebycheff polynomials*, Transactions of the American Mathematical Society, vol. 27 (1925), pp. 537–550, p. 542.

† By Schwarz's inequality,

$$\begin{aligned}(|c_n| + |c_{n+1}| + \cdots + |c_{n+m}|)^2 &= (|c_n| \cdot 1 + |c_{n+1}| \cdot 1 + \cdots + |c_{n+m}| \cdot 1)^2 \\ &\leqq (c_n^2 + c_{n+1}^2 + \cdots + c_{n+m}^2)(1^2 + 1^2 + \cdots + 1^2),\end{aligned}$$

$$|c_n| + \cdots + |c_{n+m}| \leqq (m+1)^{1/2}\left(c_n^2 + \cdots + c_{n+m}^2\right)^{1/2}.$$

But the cruder inequality in the text serves the present purpose equally well.

But inasmuch as the q's are orthogonal and normalized for the weight function $\rho_1(x)$, while the p's have the same property with respect to the weight function $\Pi(x)\rho_1(x)$, and $\Pi(x) \leqq G$,

$$c_n^2+\cdots+c_{n+m}^2=\int_{-1}^{1}\rho_1(x)(c_nq_n+\cdots+c_{n+m}q_{n+m})^2\,dx$$
$$=\int_{-1}^{1}\rho_1(x)[\Pi(x)p_n(x)]^2\,dx$$
$$\leqq G\int_{-1}^{1}\Pi(x)\rho_1(x)[p_n(x)]^2\,dx=G.$$

It follows that $|c_n|+\cdots+|c_{n+m}| \leqq (m+1)G^{1/2}$, and

$$|p_n(x)| \leqq (m+1)HG^{1/2}/g$$

for x in E, the right-hand member being independent of n.

Without the restriction that $\Pi(x)$ shall not vanish in E, the same reasoning shows that

$$|p_n(x)| \leqq (m+1)HG^{1/2}/\Pi(x)$$

at any point of E where $\Pi(x)\neq 0$; the right-hand member, though now involving x, is still independent of n.

Thus Theorem I is applicable whenever $\rho(x)$ satisfies the conditions of the last paragraphs. The property of boundedness in the interior of the interval for the normalized sequence can be carried over, for example, from the Legendre polynomials $X_n(x)$ to their qth derivatives* $X_n^{(q)}(x)$, which are orthogonal with weight function $(1-x^2)^q$, so that a convergence theorem is obtained for developments in terms of the "associated Legendre functions" $(1-x^2)^{q/2}X_n^{(q)}(x)$; or from the polynomials $X_n(x)$ to the polynomials†

$$S_n(x)=\sum_0^n\frac{1}{2}(2k+1)X_k(x)$$

corresponding to the weight function $1-x$; or from the "trigonometric polynomials" $\cos(n\arccos x)$, with weight function $(1-x^2)^{-1/2}$, to the polynomials† $\sin(n+1)\theta/\sin\theta$, $\theta=\arccos x$, with weight function $(1-x^2)^{1/2}$.

5. Generalization of the convergence theorem. At the expense of resorting to somewhat less elementary considerations, the generality of Theorem I can be considerably increased by lightening the unnecessarily severe restrictions imposed on the difference quotient $\varphi(t)$ in the application of the Lemma, and still further by re-examining the hypothesis to which the function $\varphi(t)$ was subjected in the Lemma itself. It was supposed in the Lemma that $\rho(t)[\varphi(t)]^2$ is integrable from -1

* See e.g. Courant-Hilbert, *Methoden der Mathematischen Physik*, vol. I (second edition), Berlin 1931, p. 282.

† See Pólya and Szegö, *Aufgaben und Lehrsätze*, loc. cit.

to 1. *Under the assumption that the polynomials $p_n(t)$ are uniformly bounded throughout an interval* $-1+\eta \leqq t \leqq 1-\eta$, let it be demanded of $\varphi(t)$ merely that it be defined throughout $(-1,1)$, except possibly for a set of measure zero, that $\rho(t)\varphi(t)$ be summable in the sense of Lebesgue over the whole interval $(-1,1)$, and that $\rho(t)[\varphi(t)]^2$ be summable from -1 to $-1+\eta$ and from $1-\eta$ to 1. It is understood that $\rho(t)$ itself is summable from -1 to 1. *The conclusion of the Lemma is still valid.*

For any positive N, let E_N be the point set consisting of the intervals $(-1,-1+\eta)$ and $(1-\eta,1)$, together with those points of $(-1+\eta,1-\eta)$ where $|\varphi(t)| \leqq N$, and let CE_N be the complementary set, made up of the points of $(-1+\eta,1-\eta)$ where $|\varphi(t)|>N$. Let $\varphi_N(t)$ be defined as equal to $\varphi(t)$ on E_N, and equal to 0 on CE_N. As this function is bounded from $-1+\eta$ to $1-\eta$, $\rho(t)[\varphi_N(t)]^2$ is summable from -1 to 1. Let $|p_n(t)| \leqq H$ for $-1+\eta \leqq t \leqq 1-\eta$. Let ε be any positive quantity. From the fact that $\varphi(t)$ is defined almost everywhere it follows that the measure of CE_N approaches zero as N increases without limit. Hence, by the summability of $\rho(t)\varphi(t)$,

$$\lim_{N\to\infty}\int_{CE_N}\rho(t)|\varphi(t)|\,dt=0.$$

Let N be taken so large that the value of the last integral is less than $\varepsilon/(2H)$, and then held fast. It is thus assured that

$$\left|\int_{CE_N}\rho(t)\varphi(t)p_n(t)\,dt\right| \leqq H\int_{CE_N}\rho(t)|\varphi(t)|\,dt<\frac{\varepsilon}{2}$$

for all values of n. In the resolution

$$\begin{aligned}\int_{-1}^{1}\rho(t)\varphi(t)p_n(t)\,dt&=\int_{E_N}\rho(t)\varphi(t)p_n(t)\,dt+\int_{CE_N}\rho(t)\varphi(t)p_n(t)\,dt\\&=\int_{E_N}\rho(t)\varphi_N(t)p_n(t)\,dt+\int_{CE_N}\rho(t)\varphi(t)p_n(t)\,dt\\&=\int_{-1}^{1}\rho(t)\varphi_N(t)p_n(t)\,dt+\int_{CE_N}\rho(t)\varphi(t)p_n(t)\,dt,\end{aligned}$$

the next to the last integral approaches zero as n becomes infinite, while N is held fast, because $\rho(t)[\varphi_N(t)]^2$ is summable. For n sufficiently large, therefore, this integral also is inferior to $\varepsilon/2$ in absolute value, and

$$\left|\int_{-1}^{1}\rho(t)\varphi(t)p_n(t)\,dt\right|<\varepsilon,$$

which is the substance of the conclusion to be proved.

In the corresponding adaptation of the proof of Theorem I for convergence at a specified interior point of the interval, the properties in question carry over directly from $f(t)$ to $\varphi(t)$, except in the neighborhood of the point $t=x$. On the assumption that the polynomials $p_n(x)$ are uniformly bounded throughout any closed interval interior to $(-1, 1)$, *it is sufficient for convergence of* $s_n(x)$ *toward* $f(x)$ *that* $\rho(t)[f(t)]^2$ *be summable over an interval extending to the right from* -1 *and over an interval extending to the left from* $+1$, *that* $\rho(t)f(t)$ *be summable throughout* $(-1, 1)$, *and that for the difference quotient* $\varphi(t)=[f(t)-f(x)]/(t-x)$ *there exist an interval about the point* x *over which* $\rho(t)\varphi(t)$ *is summable.* The last condition can be further interpreted as is customary in the theory of Fourier series.*

Another important fact can be brought out in this connection. The requirement as to summability of $\rho(t)\varphi(t)$ will certainly be met if $f(t)$ vanishes for $t=x$ and throughout an interval containing the point x in its interior, and if the other conditions are fulfilled $s_n(x)$ will converge to the value zero. Let this observation be applied to the series for the difference $g(t)$ of two functions $f_1(t)$ and $f_2(t)$, both satisfying the hypotheses of summability previously imposed on $f(t)$, and identical throughout an interval surrounding the point x. The partial sum $s_n(x)$ of the series for $g(x)$ is the difference of the partial sums for $f_1(x)$ and $f_2(x)$ separately. Since $s_n(x)$ approaches zero, it follows that if the series for $f_1(x)$ converges the series for $f_2(x)$ will converge to the same value; such difference as there may be between the functions outside the interval of identity about the point x has no effect on the convergence at that point. Under the hypothesis still that the polynomials $p_n(x)$ are bounded as before in the interior of $(-1, 1)$, it is seen that *beyond the requirement that* $\rho(t)f(t)$ *be summable from* -1 *to* 1 *and that* $\rho(t)[f(t)]^2$ *be summable near the ends of the interval, the convergence of the series at a specified interior point depends only on the behavior of* $f(t)$ *in the neighborhood of the point in question.*

6. More general weight function; second convergence theorem. While the proof of uniform boundedness of the polynomials $p_n(x)$ in the interior of the interval has been given only under restrictive hypotheses as to the nature of the weight function, it is possible to assign an upper bound to their order of magnitude under very general conditions.†

Let $\rho(x)$ be any function which is summable over $(-1, 1)$ and has a positive lower bound throughout the interval, $\rho(x) \geqq v>0$. Let $p_n(x)$ be any polynomial of the nth degree such that

$$\int_{-1}^{1} \rho(x)[p_n(x)]^2 dx=1;$$

the argument is not restricted for the moment to the particular polynomial which belongs to the normalized orthogonal sequence. Since $1 \leqq \rho(x)/v$,

$$\int_{-1}^{1}[p_n(x)]^2 dx \leqq (1/v)\int_{-1}^{1} \rho(x)[p_n(x)]^2 dx=1/v.$$

* Cf. e.g. H. Lebesgue, *Leçons sur les séries trigonométriques*, Paris, 1906, Chapter III.

† For this fact and its application to the theory of convergence of the series, cf. e.g. W. Stekloff, *Sur les problèmes de représentation des fonctions à l'aide des polynômes, etc.*, Proceedings of the International Mathematical Congress held in Toronto, August 11–16, 1924, vol. I, pp. 631–640; Shohat, Transactions, vol. 27, loc. cit.

Let $Q(x)$ denote the polynomial

$$Q(x)=\int_{-1}^{x}[p_n(x)]^2\,dx,$$

of degree $2n+1$. Inasmuch as the integrand is positive throughout the interval, $0 \leqq Q(x) \leqq Q(1) \leqq 1/v$ for $-1 \leqq x \leqq 1$. Hence, by the theorems of Markoff and Bernstein* (absolute value signs being superfluous as long as the quantities written down are positive), the polynomial $Q'(x)=[p_n(x)]^2$ has the upper bound $(2n+1)^2/v$ for $-1 \leqq x \leqq 1$, and the upper bound $(2n+1)/[v(1-x^2)^{1/2}]$ for $-1<x<1$. It follows that† $|p_n(x)|$ *has an upper bound of the order of magnitude of* n *for* $-1 \leqq x \leqq 1$, *and an upper bound of the order of* $n^{1/2}$ *throughout any closed interval interior to* $(-1,1)$.

The proof just given depends essentially on the hypothesis that the weight function stays away from zero. Let this requirement now be relaxed, and let it be supposed instead that $[\rho(x)]^{-1}$ as well as $\rho(x)$ is summable over $(-1,1)$. By Schwarz's inequality,

$$\begin{aligned}\left[\int_{-1}^{1}|p_n(x)|\,dx\right]^2&=\left[\int_{-1}^{1}[\rho(x)]^{-1/2}[\rho(x)]^{1/2}|p_n(x)|\,dx\right]^2\\&\leqq\int_{-1}^{1}[\rho(x)]^{-1}dx\int_{-1}^{1}\rho(x)[p_n(x)]^2dx\\&=\int_{-1}^{1}[\rho(x)]^{-1}dx.\end{aligned}$$

If the square root of the value of the last integral is denoted by I,

$$\left|\int_{-1}^{x}p_n(x)\,dx\right| \leqq \int_{-1}^{x}|p_n(x)|\,dx \leqq \int_{-1}^{1}|p_n(x)|\,dx \leqq I,$$

and as the first integral in this chain of inequalities is a polynomial of degree $n+1$, with $p_n(x)$ for its derivative, $|p_n(x)|$ has an upper bound of the order of n^2 for $-1 \leqq x \leqq 1$, and an upper bound of the order of n throughout any closed interior interval.

It would be possible to pursue this line of inquiry further, but the results already obtained will be allowed to suffice for the present.

---

* The theorems in question may be stated as follows: If $P_n(x)$ is a polynomial of the nth degree such that $|P_n(x)| \leqq L$ for $-1 \leqq x \leqq 1$, then $|P'_n(x)| \leqq n^2L$ throughout the same interval, and $|P'_n(x)| \leqq nL/(1-x^2)^{1/2}$ throughout the interior of the interval. See e.g. Marcel Riesz, *Eine trigonometrische Interpolationsformel und einige Ungleichungen für Polynome*, Jahresbericht der Deutschen Mathematiker-Vereinigung, vol. 23 (1914), pp. 354–368; Colloquium, pp. 80–82, 92.

† For the method of proof cf. J. Shohat, *On the polynomial and trigonometric approximation of measurable bounded functions on a finite interval*, Mathematische Annalen, vol. 102 (1930), pp. 157–175, pp. 165–166.

To make use of these facts in a proof of convergence, it is important not only to know that the general coefficient in the expansion of an arbitrary function approaches zero, but also to be able to specify conditions under which it approaches zero with a preassigned degree of rapidity.

Let $f(x)$ be an arbitrary function in the interval $(-1,1)$, subject to conditions presently to be imposed, and let c_n be the coefficient of $p_n(x)$ in its expansion. There is occasion now to use the least-square property of the development, according to which the partial sum $s_n(x)$ of the series is distinguished among all polynomials of the nth degree by the fact that

$$\gamma_n=\int_{-1}^{1}\rho(x)[f(x)-s_n(x)]^2\,dx$$

is a minimum.* It has been noted that

$$\gamma_n=\int_{-1}^{1}\rho(x)[f(x)]^2\,dx-\sum_{k=0}^{n}c_k^2.$$

Comparison of this with the corresponding formula for γ_{n-1} shows that $c_n^2=\gamma_{n-1}-\gamma_n$. As $\gamma_n\geqq 0$, this implies that $c_n^2\leqq\gamma_{n-1}, |c_n|\leqq\gamma_{n-1}^{1/2}$. Information can be obtained with regard to the magnitude of γ_n, in view of the least-square property, from general theorems on approximation by means of polynomials. If $P_n(x)$ is a polynomial of the nth degree such that $|f(x)-P_n(x)|\leqq\varepsilon_n$ for $-1\leqq x\leqq 1$,

$$\gamma_n\leqq\int_{-1}^{1}\rho(x)[f(x)-P_n(x)]^2\,dx\leqq\varepsilon_n^2\int_{-1}^{1}\rho(x)\,dx.$$

If approximating polynomials $P_n(x)$ are constructed for the various values of n successively, with ε_n as an upper bound for the error in each case, $|c_n|\leqq\mu\varepsilon_{n-1}$, where μ is independent of n. So theorems on the order of magnitude attainable for ε_n can be translated at once into theorems on the magnitude of c_n.

In particular, it is possible to make $\lim_{n\to\infty} n^{1/2}\varepsilon_n=0$, if $f(x)$ is such that $\lim_{\delta\to 0}\omega(\delta)/\delta^{1/2}=0$, where $\omega(\delta)$ is the maximum of $|f(x_2)-f(x_1)|$ for $|x_2-x_1|\leqq\delta$. This condition will be more than satisfied if $f(x)$ satisfies the Lipschitz condition $|f(x_2)-f(x_1)|\leqq\lambda|x_2-x_1|$. If $f(x)$ has a continuous derivative, $n\varepsilon_n$ can be made to approach zero, and if $f''(x)$ is continuous throughout the interval, $n^2\varepsilon_n$ can be made to approach zero.† If follows that $n^{1/2}c_n\to 0$ or $nc_n\to 0$ or $n^2c_n\to 0$, in the various cases respectively.

For application in (13) as a means to a convergence proof, hypotheses corresponding to those just described are to be imposed on the difference quotient $\varphi(t)=[f(t)-f(x)]/(t-x)$. It is not difficult to show, by appropriate use of the mean value theorem, that $\varphi(t)$ (defined with the value $f'(x)$ for $t=x$) will have a

* See e.g. Colloquium, pp. 77–80, 95.

† See Colloquium, Chapter I, Theorem VI and Corollary of Theorem VIII.

continuous derivative* if $f''(t)$ is defined throughout an interval containing x in its interior (or having x as an end point, if $x=\pm 1$) and continuous for $t=x$, and $f'(t)$ continuous throughout $(-1,1)$; and $\varphi(t)$ will have a continuous second derivative if $f''(t)$ and $f'(t)$ in the statement just formulated are replaced by $f'''(t)$ and $f''(t)$ respectively. For simplicity, the case of the factor $n^{1/2}$ may be covered by supposing here also that $f(t)$ satisfies the conditions under which $\varphi(t)$ has a continuous derivative.† From (13) it is possible then to read off

THEOREM II. *If the weight function $\rho(x)$ is summable and has a positive lower bound, and if $f(t)$ has a continuous derivative throughout $(-1,1)$, the series (2) will converge to the value $f(x)$ at any point x of the closed interval in the neighborhood of which $f(t)$ has a second derivative, continuous for $t=x$; if $\rho(x)$ is any function such that $\rho(x)$ and $[\rho(x)]^{-1}$ are summable, the same conditions on $f(t)$ will be sufficient for convergence at an interior point of the interval, and the series will converge to the right value for $x=\pm 1$ also if $f''(t)$ is continuous throughout the interval and $f'''(t)$ defined near and continuous at the point in question.*

---

* Since

$$f(t)-f(x)=(t-x)f'(x)+\frac{1}{2}(t-x)^2 f''(\xi),$$

it is seen that

$$\varphi(t)=f'(x)+\frac{1}{2}(t-x)f''(\xi),$$

and

$$[\varphi(t)-\varphi(x)]/(t-x)=\frac{1}{2}f''(\xi),$$

which approaches $\frac{1}{2}f''(x)$ as t approaches x, so that $\varphi(t)$ has a derivative equal to $\frac{1}{2}f''(x)$ for $t=x$. For $t\neq x$, $\varphi(t)$ has the continuous derivative

$$\varphi'(t)=\frac{(t-x)f'(t)-[f(t)-f(x)]}{(t-x)^2}.$$

By an alternative application of the mean value theorem,

$$f(x)=f(t)+(x-t)f'(t)+\tfrac{1}{2}(x-t)^2 f''(\eta),$$

so that the numerator in the expression for $\varphi'(t)$ reduces to $\frac{1}{2}(t-x)^2 f''(\eta)$; hence

$$\varphi'(t)=\tfrac{1}{2}f''(\eta),$$

which approaches $\frac{1}{2}f''(x)$ when t approaches x, and $\varphi'(t)$ is continuous for $t=x$ as well as elsewhere. Similar reasoning shows the existence and continuity of $\varphi''(t)$ when $f''(t)$ and $f'''(t)$ satisfy the appropriate conditions.

† With the use of properties of the "modulus of continuity" $\omega(\delta)$ (see, for example, C. de la Vallée Poussin, *Leçons sur l'approximation des fonctions d'une variable réelle*, Paris, 1919, pp. 7–8; E. L. Mickelson, *On the approximate representation of a function of two variables*, Transactions of the American Mathematical Society, vol. 33 (1931), pp. 759–781, pp. 761–762) it can be seen that if $f'(t)$ has a modulus of continuity $\omega(\delta)$, $\varphi(t)$ will have a modulus of continuity not exceeding $2\omega(\delta)$. Since $\varphi(t)=f'(\xi)$, with a

It would not be profitable to attempt refinement of the conditions for convergence by further elaboration of details, as more general results than those expressed in the first part of the theorem can be obtained by reasoning along other lines.* The interest lies in the extension of the method of proof of Theorem I to more general weight functions, and in the intermediate results which make the extension possible. It is to be noted also that the latter part of Theorem II gives convergence, within the hypotheses stated, even at points where the weight function vanishes.

7. Uniform convergence. The discussion that has been outlined above is admittedly incomplete in important respects. It has not given any information as to convergence at a point of discontinuity—though in Theorem I the function may have discontinuities at points other than the one where convergence is being investigated—and it has not said anything about uniformity of convergence. For the former question, as well as for a thoroughgoing treatment of the latter, reference must be made to other presentations. Certain simple facts with regard to uniform convergence, however, are immediately obtainable from the results already noted.

---

value of ξ between x and t, it is recognized that if $|t-x| \leqq \delta$,

$$|\varphi(t)-\varphi(x)| = |f'(\xi)-f'(x)| \leqq \omega(|\xi-x|) \leqq \omega(\delta).$$

If t_1 and t_2 are distinct from x and include x between them,

$$|\varphi(t_2)-\varphi(t_1)| \leqq |\varphi(t_2)-\varphi(x)|+|\varphi(t_1)-\varphi(x)| \leqq \omega(|t_2-x|)+\omega(|t_1-x|) \leqq 2\omega(|t_2-t_1|).$$

If t_1 and t_2 are on the same side of x, and $|t_2-t_1| \leqq \delta$,

$$\begin{aligned}\varphi(t_2)-\varphi(t_1) &= (t_2-t_1)\varphi'(\xi) \qquad (\xi \text{ between } t_1 \text{ and } t_2)\\ &= (t_2-t_1)\frac{(\xi-x)f'(\xi)-[f(\xi)-f(x)]}{(\xi-x)^2} \qquad \text{(see preceding footnote)}\\ &= (t_2-t_1)\frac{(\xi-x)f'(\xi)-(\xi-x)f'(\eta)}{(\xi-x)^2} \qquad (\eta \text{ between } x \text{ and } \xi)\\ &= \frac{t_2-t_1}{\xi-x}[f'(\xi)-f'(\eta)],\end{aligned}$$

$$|\varphi(t_2)-\varphi(t_1)| \leqq \left|\frac{t_2-t_1}{\xi-x}\right|\omega(|\xi-\eta|) \leqq \left|\frac{t_2-t_1}{\xi-x}\right|\omega(|\xi-x|);$$

if $|\xi-x| \geqq \delta$,

$$\frac{\omega(|\xi-x|)}{|\xi-x|} \leqq 2\frac{\omega(\delta)}{\delta}, \quad |\varphi(t_2)-\varphi(t_1)| \leqq 2\omega(\delta),$$

while if $|\xi-x|<\delta$ it must be that $|t_1-x|<2\delta, |t_2-x|<2\delta$,

$$|\varphi(t_2)-\varphi(t_1)| = |f'(\xi_2)-f'(\xi_1)| \leqq \omega(|\xi_2-\xi_1|) \leqq \omega(2\delta) \leqq 2\omega(\delta),$$

the points ξ_1 and ξ_2 being on the same side of x and distant from x by less than 2δ.

For the convergence proof it is sufficient that a suitable condition with regard to modulus of continuity be satisfied by $f'(t)$ in the neighborhood of $t=x$, and by $f(t)$ itself throughout the rest of the interval.

* See e. g. Colloquium, pp. 92–101.

If the series for a continuous function $f(x)$ converges uniformly throughout the closed interval, it must converge to the value $f(x)$, under very general hypotheses as to the weight function $\rho(x)$, by an argument which is well known and, in its essentials, frequently encountered in connection with still more general problems. The proof may be given as follows.

Being uniformly convergent, the series represents some function $g(x)$ which is continuous for $-1 \leqq x \leqq 1$. Multiplication of the series by $\rho(x)p_k(x)$ and integration over the interval, with due regard for the properties of the orthogonal polynomials, gives

$$\int_{-1}^{1} \rho(x)\, g(x)\, p_k(x)\, dx = c_k.$$

If the difference $f(x)-g(x)$ is denoted by $\psi(x)$, comparison of the last equation with the formula (3) defining c_k shows that

$$\int_{-1}^{1} \rho(x)\psi(x)\, p_k(x)\, dx = 0$$

for all values of k, and consequently, as any polynomial can be expressed linearly in terms of the p's,

$$\int_{-1}^{1} \rho(x)\psi(x) q(x)\, dx = 0$$

for any polynomial $q(x)$ whatever. By Weierstrass's theorem, the continuous function $\psi(x)$ can be uniformly approximated by a polynomial throughout the interval with any assigned degree of accuracy, or, as an equivalent statement, can be represented by a uniformly convergent series of polynomials. (For, if $q_1, q_2, q_3, \ldots,$ are a succession of approximating polynomials, q_n can be regarded as the sum of the first n terms of the series $q_1+(q_2-q_1)+(q_3-q_2)+\cdots$.) If this series is multiplied by $\rho(x)\psi(x)$ and integrated term by term from -1 to 1, the result of the integration is 0 for each term, and

$$\int_{-1}^{1} \rho(x)[\psi(x)]^2\, dx = 0.$$

Let it be assumed that $\rho(x)$, *if it vanishes at all, vanishes only at a set of points of measure zero*, as must necessarily be the case, for example, if $[\rho(x)]^{-1}$ is summable; or, more generally, that it is positive over a set of positive measure in every subinterval of $(-1, 1)$. Then, if $\psi(x)$ were not identically zero, $\rho(x)[\psi(x)]^2$ would be positive over a set of positive measure, and the value of the integral could not be zero. So it must be that $\psi(x) \equiv 0$, and $g(x) \equiv f(x)$. It is obvious that the theorem can not be proved without some restriction on the vanishing of $\rho(x)$; if $\rho(x)$ were to be identically zero throughout a subinterval of $(-1, 1)$, the values of $f(x)$ in the subinterval would have no influence on the determination of the coefficients, and there would be an infinity of different functions having the same formal expansion.

If $\rho(x)$ has a positive lower bound, let $f(x)$ be assumed to have a continuous second derivative satisfying a Lipschitz condition of order α,

$$|f''(x_2)-f''(x_1)| \leqq \lambda |x_2-x_1|^{\alpha}, \qquad \alpha > 0.$$

Then* it is possible to construct approximating polynomials so that the upper bound of error, denoted above by ϵ_n, does not exceed a constant multiple of $1/n^{2+\alpha}$. It follows that $|c_n|$ has an upper bound of the same order, while $|p_n(x)|$ has an upper bound of the order n, so that $|c_n p_n(x)|$ is uniformly less than a constant multiple of $1/n^{1+\alpha}$ for $-1 \leqq x \leqq 1$, and the series is uniformly convergent throughout the interval. If the restriction on the weight function is merely that ρ and $1/\rho$ be summable, uniform convergence will be assured if $f(x)$ has a third derivative satisfying a Lipschitz condition of positive order α.

The reasoning of the last few paragraphs has not presupposed the conclusions of Theorems I and II, or required the use of the Christoffel-Darboux formula, on which the proofs of those theorems depended. On the basis of what is known as to the magnitude of the polynomials $p_n(x)$ at interior points, uniform convergence throughout any closed interval interior to $(-1,1)$ can be deduced with less severe restrictions on $f(x)$, if reference is then made to the theorems for assurance that the series converges to the right value.

8. The problem of convergence with respect to the representation of statistical frequencies. One other point is worth noting in this rapid summary. It has been proposed† that expansions in Jacobi series be used for the representation of a certain class of frequency functions in statistics. The question then is not primarily that of approximating to the given function point by point, but of approximating to the value of its definite integral over an arbitrary subinterval of the range under consideration. Convergence in this sense is readily proved under fairly general hypotheses.

Let the range once more be the interval $(-1,1)$, and let $\rho(x)$ be a function summable over this interval—not necessarily of the particular form leading to the Jacobi polynomials. Let $F(x)$ be another function defined over the interval. It will ordinarily be enough in the way of generality if both $\rho(x)$ and $F(x)$ are thought of as continuous, except that they may become infinite at isolated points, subject to conditions of integrability. The function $F(x)$ is to be developed in a series of the form

$$c_0 p_0(x)\rho(x) + c_1 p_1(x)\rho(x) + c_2 p_2(x)\rho(x) + \cdots,$$

or, if $F(x)/\rho(x)$ is denoted by $f(x)$, the latter function is to be expanded in a series of the form (2). *Let $f(x)$ be supposed continuous for* $-1 \leqq x \leqq 1$, after elimination of removable singularities. By Weierstrass's theorem it is possible to define a sequence of polynomials $q_n(x)$, which may be thought of as having the degree indicated by the subscript in each case, so that $f(x) - q_n(x)$ approaches zero uniformly throughout the interval. If $|f(x) - q_n(x)| \leqq \epsilon_n$,

$$\int_{-1}^{1} \rho(x)[f(x) - q_n(x)]^2 dx \leqq \epsilon_n^2 \int_{-1}^{1} \rho(x)\, dx,$$

* See Colloquium, Chapter I, Theorem VIII.

† See e. g. V. Romanovsky, *Generalisation of some types of the frequency curves of Professor Pearson*, Biometrika, vol. 16 (1924), pp. 106–116; E. H. Hildebrandt, *Systems of polynomials connected with the Charlier expansions and the Pearson differential and difference equations*, Annals of Mathematical Statistics, vol. 2 (1931), pp. 379–439.

and this approaches zero as n becomes infinite when ε_n is made to approach zero. By the least-square property, the integral on the left is greater than or equal to that obtained if $q_n(x)$ is replaced by $s_n(x)$, the partial sum of the series, and consequently, if γ_n denotes the integral

$$\int_{-1}^{1}\rho(x)[f(x)-s_n(x)]^2dx,$$

$\lim_{n\to\infty}\gamma_n=0$. This conclusion holds under much more general conditions, but the greater generality is of secondary interest for the interpretation in terms of frequency functions. For an application of Schwarz's inequality, let

$$u=[\rho(x)]^{-1/2}|F(x)-\rho(x)s_n(x)|=[\rho(x)]^{1/2}|f(x)-s_n(x)|,$$

$$v=[\rho(x)]^{1/2},$$

$$\delta_n=\int_{-1}^{1}|F(x)-\rho(x)s_n(x)|\,dx=\int_{-1}^{1}uv\,dx.$$

Then

$$\delta_n^2\leqq\int_{-1}^{1}u^2dx\int_{-1}^{1}v^2dx=\gamma_n\int_{-1}^{1}\rho(x)\,dx,$$

and as γ_n approaches zero, δ_n approaches zero likewise. If x_1 and $x_2(>x_1)$ are any two points in the interval,

$$\left|\int_{x_1}^{x_2}[F(x)-\rho(x)s_n(x)]\,dx\right|\leqq\int_{x_1}^{x_2}|F(x)-\rho(x)s_n(x)|\,dx\leqq\delta_n,$$

and from the approach of δ_n to zero it follows that

$$\lim_{n\to\infty}\int_{x_1}^{x_2}\rho(x)s_n(x)\,dx=\int_{x_1}^{x_2}F(x)\,dx,$$

which expresses the desired property of convergence.

ORTHOGONAL TRIGONOMETRIC SUMS*

DUNHAM JACKSON, University of Minnesota

1. Introduction. The purpose of this paper is to develop the beginnings of a theory analogous to that of systems of polynomials orthogonal with respect to a given weight function, for the corresponding case of trigonometric sums.**

It is well known that if a set of functions $\psi_1(x)$, $\psi_2(x),\ldots,$ is given in an interval (a,b), each of the ψ's being summable over the interval together with its square, and any finite number of them being properly independent, in the sense that every linear combination with coefficients that do not all vanish is different from zero over a set of positive measure, then it is possible to construct a normalized orthogonal system $\varphi_1(x),\varphi_2(x),\ldots,$ in which the general function $\varphi_n(x)$ is a linear combination of $\psi_1(x),\ldots,\psi_n(x)$, so that

$$\int_a^b \varphi_m(x)\varphi_n(x)\,dx=0 \qquad (m\neq n),$$

$$\int_a^b [\varphi_n(x)]^2\,dx=1.$$

Let $\rho(x)$ be a non-negative summable function of period 2π, different from zero over a set of positive measure in any interval of length 2π, and let $\psi_1(x),\psi_2(x),\ldots,$ be the set of functions

$$[\rho(x)]^{1/2},\ [\rho(x)]^{1/2}\cos x,\ [\rho(x)]^{1/2}\sin x,\ [\rho(x)]^{1/2}\cos 2x,\ [\rho(x)]^{1/2}\sin 2x,\ldots.$$

The corresponding φ's for a period interval will be a sequence†

$$[\rho(x)]^{1/2}u_0(x),\ [\rho(x)]^{1/2}u_1(x),\ [\rho(x)]^{1/2}v_1(x),\ [\rho(x)]^{1/2}u_2(x),\ [\rho(x)]^{1/2}v_2(x),\ldots,$$

in which $u_n(x)$ and $v_n(x)$ are trigonometric sums of the nth order, the former containing no term in $\sin nx$, and the characteristic properties of the sequence are expressed by the relations

$$\int_{-\pi}^{\pi}\rho(x)u_m(x)u_n(x)\,dx=0,\quad \int_{-\pi}^{\pi}\rho(x)v_m(x)v_n(x)\,dx=0, \qquad (m\neq n),$$

$$\int_{-\pi}^{\pi}\rho(x)u_m(x)v_n(x)\,dx=0 \text{ for all } m \text{ and } n,$$

$$\int_{-\pi}^{\pi}\rho(x)[u_n(x)]^2\,dx=1,\quad \int_{-\pi}^{\pi}\rho(x)[v_n(x)]^2\,dx=1. \tag{1}$$

The content of the equations with right-hand member zero can be expressed by

* Received February 21, 1933.—Presented to the American Mathematical Society, August 30, 1932, and April 15, 1933.

**The treatment is to a considerable extent parallel to that in the writer's expository paper entitled *Series of orthogonal polynomials*, Annals of Mathematics, vol. 34 (1933), pp.527–545.

† See e.g. the writer's *Theory of Approximation*, American Mathematical Society Colloquium Publications, vol. XI, New York,1930, pp. 89–91.

saying that any two functions of the sequence of u's and v's are orthogonal to each other with respect to $\rho(x)$ as weight function. The following discussion is concerned with further properties of the u's and v's belonging to a given weight function $\rho(x)$, and in particular with the convergence of series expansions in terms of them.

It is apparent from the manner of construction of the orthogonal system that, in the general formulation, $\varphi_n(x)$ as written in terms of the ψ's actually contains a term in $\psi_n(x)$, with non-vanishing coefficient, for each value of n, and hence that each $\psi_n(x)$ is linearly expressible in terms of $\varphi_1(x),\ldots,\varphi_n(x)$. Consequently any linear combination of $\psi_1,\ldots,\psi_{n-1}$ is a linear combination of $\varphi_1,\ldots,\varphi_{n-1}$, and $\varphi_n(x)$ is orthogonal to any such combination. In the special case under consideration, $u_n(x)$ actually has a term in $\cos nx$, and $v_n(x)$ actually has a term in $\sin nx$. Any trigonometric sum of the nth order containing no term in $\sin nx$ is linearly expressible in terms of $u_0(x), u_1(x), v_1(x),\ldots,u_{n-1}(x), v_{n-1}(x), u_n(x)$, and any trigonometric sum of the nth order is expressible in terms of $u_0(x),\ldots,u_n(x), v_n(x)$. With respect to the weight function $\rho(x)$, $u_n(x)$ and $v_n(x)$ are orthogonal to any trigonometric sum of order lower than the nth, and $v_n(x)$ is orthogonal also to any sum of the nth order lacking the term in $\sin nx$.

2. Recursion formulas. The u's and v's are connected by recursion formulas analogous to that of Darboux* for orthogonal polynomials, though somewhat less simple in form. Let μ be any constant. (The essence of the method would be sufficiently illustrated by taking $\mu=0$, but it is desirable for the applications to have the degree of generality conferred by the presence of the arbitrary constant.) The product $\cos(x-\mu)u_n(x)$ is a trigonometric sum of order $n+1$, and as such can be expressed in terms of $u_0(x),\ldots,u_{n+1}(x), v_{n+1}(x)$:

$$\begin{aligned}\cos(x-\mu)u_n(x) = {} & \alpha_n u_{n+1}(x) + A_n u_n(x) + A_n' u_{n-1}(x) + A_n'' u_{n-2}(x) + \cdots \\ & + \beta_n v_{n+1}(x) + B_n v_n(x) + B_n' v_{n-1}(x) + B_n'' v_{n-2}(x) + \cdots. \end{aligned} \quad (2)$$

The coefficients of course depend on μ, but it is not necessary to indicate that in the notation. For any value of k from 0 to n, multiplication by $\rho(x)u_{n-k}(x)$ and integration from $-\pi$ to π, with attention to the relations (1), gives

$$\int_{-\pi}^{\pi} \rho(x)\cos(x-\mu)u_n(x)u_{n-k}(x)\,dx = A_n^{(k)}. \quad (3)$$

If the trigonometric factors under the integral sign are regarded as forming the product of $u_n(x)$ and $\cos(x-\mu)u_{n-k}(x)$, the latter is a trigonometric sum of order lower than n if k has any value greater than 1, and the integral consequently vanishes: $A_n^{(k)}=0$ for $k \geqq 2$. Similarly

$$\int_{-\pi}^{\pi} \rho(x)\cos(x-\mu)u_n(x)v_{n-k}(x)\,dx = B_n^{(k)} \quad (4)$$

---

* G. Darboux, *Sur l'approximation des fonctions de très-grands nombres*... Journal de mathématiques pures et appliquées (3), vol. 4 (1878), pp. 377–416, see pp. 411–416.

for $k=0,1,\ldots,n-1$ (there is no $v_0(x)$), and the integral vanishes for $k \geqq 2$. Replacement of n by $n-1$ in (2) gives an identity of the form

$$\cos(x-\mu)u_{n-1}(x)=\alpha_{n-1}u_n(x)+\beta_{n-1}v_n(x)+\cdots. \tag{5}$$

If this expression for $\cos(x-\mu)u_{n-1}(x)$ is substituted in (3), for $k=1$, the value of the integral defining A_n' is seen to be α_{n-1}. So (2) may be rewritten in the form

$$\begin{aligned}\cos(x-\mu)u_n(x)=&\ \alpha_n u_{n+1}(x)+A_n u_n(x)+\alpha_{n-1}u_{n-1}(x)\\&+\beta_n v_{n+1}(x)+B_n v_n(x)+B_n' v_{n-1}(x).\end{aligned} \tag{6}$$

Further specifications with regard to the coefficients will be brought out later.

After the manner of (2) there may be written down a corresponding identity

$$\begin{aligned}\cos(x-\mu)v_n(x)=&\ \gamma_n u_{n+1}(x)+C_n u_n(x)+C_n' u_{n-1}(x)+C_n'' u_{n-2}(x)+\cdots\\&+\delta_n v_{n+1}(x)+D_n v_n(x)+D_n' v_{n-1}(x)+D_n'' v_{n-2}(x)+\cdots.\end{aligned} \tag{7}$$

The coefficients are given by the integrals

$$\begin{aligned}&\int_{-\pi}^{\pi}\rho(x)\cos(x-\mu)v_n(x)u_{n-k}(x)\,dx=C_n^{(k)},\\&\int_{-\pi}^{\pi}\rho(x)\cos(x-\mu)v_n(x)v_{n-k}(x)\,dx=D_n^{(k)},\end{aligned} \tag{8}$$

and it is seen that $C_n^{(k)}=D_n^{(k)}=0$ for $k \geqq 2$. Substitution from the identity

$$\cos(x-\mu)v_{n-1}(x)=\gamma_{n-1}u_n(x)+\delta_{n-1}v_n(x)+\cdots \tag{9}$$

gives $D_n'=\delta_{n-1}$. Also, (9) may be used in (4) and (5) in (8) to show that $B_n'=\gamma_{n-1}$ and $C_n'=\beta_{n-1}$. And it appears from comparison of (4) and (8) for $k=0$ that $B_n=C_n$. So the identities (6) and (7) take the form

$$\begin{aligned}\cos(x-\mu)u_n(x)=&\ \alpha_n u_{n+1}(x)+A_n u_n(x)+\alpha_{n-1}u_{n-1}(x)\\&+\beta_n v_{n+1}(x)+B_n v_n(x)+\gamma_{n-1}v_{n-1}(x),\end{aligned} \tag{10}$$

$$\begin{aligned}\cos(x-\mu)v_n(x)=&\ \gamma_n u_{n+1}(x)+B_n u_n(x)+\beta_{n-1}u_{n-1}(x)\\&+\delta_n v_{n+1}(x)+D_n v_n(x)+\delta_{n-1}v_{n-1}(x).\end{aligned} \tag{11}$$

These hold for $n \geqq 1$, provided that when $n=1$ it is understood that $v_0(x)$ is to be replaced by 0. For $n=0$ no calculation is required to justify the formula

$$\cos(x-\mu)u_0(x)=\alpha_0 u_1(x)+\beta_0 v_1(x)+A_0 u_0(x).$$

3. Christoffel-Darboux formula. From the recursion formulas it is possible to obtain an analogue of the Christoffel-Darboux formula* in the theory of orthogonal polynomials. It is a question of an identity for the sum

$$\begin{aligned}K_n(x,t)\equiv&\ u_0(x)u_0(t)+u_1(x)u_1(t)+\cdots+u_n(x)u_n(t)\\&+v_1(x)v_1(t)+\cdots+v_n(x)v_n(t).\end{aligned} \tag{12}$$

* Darboux, loc. cit.

Let (10) and (11) be multiplied by $u_n(t)$ and $v_n(t)$ respectively, and let the corresponding formulas for $\cos(t-\mu)u_n(x)u_n(t)$ and $\cos(t-\mu)v_n(x)v_n(t)$ be written down by interchange of t and x. Combination of the resulting identities by the indicated additions and subtractions yields

$$\begin{aligned}&\big[\cos(t-\mu)-\cos(x-\mu)\big]\big[u_n(x)u_n(t)+v_n(x)v_n(t)\big]\\ &\quad=\alpha_n\big[u_n(x)u_{n+1}(t)-u_{n+1}(x)u_n(t)\big]-\alpha_{n-1}\big[u_{n-1}(x)u_n(t)-u_n(x)u_{n-1}(t)\big]\\ &\quad+\beta_n\big[u_n(x)v_{n+1}(t)-v_{n+1}(x)u_n(t)\big]-\beta_{n-1}\big[u_{n-1}(x)v_n(t)-v_n(x)u_{n-1}(t)\big]\\ &\quad+\gamma_n\big[v_n(x)u_{n+1}(t)-u_{n+1}(x)v_n(t)\big]-\gamma_{n-1}\big[v_{n-1}(x)u_n(t)-u_n(x)v_{n-1}(t)\big]\\ &\quad+\delta_n\big[v_n(x)v_{n+1}(t)-v_{n+1}(x)v_n(t)\big]-\delta_{n-1}\big[v_{n-1}(x)v_n(t)-v_n(x)v_{n-1}(t)\big].\quad(13)\end{aligned}$$

These relations hold for $n \geqq 1$, with the previous understanding as to v_0. For $n=0$,

$$\begin{aligned}&\big[\cos(t-\mu)-\cos(x-\mu)\big]u_0(x)u_0(t)\\ &\qquad=\alpha_0\big[u_0(x)u_1(t)-u_1(x)u_0(t)\big]+\beta_0\big[u_0(x)v_1(t)-v_1(x)u_0(t)\big].\quad(14)\end{aligned}$$

The desired identity for $K_n(x,t)$ is derived by writing (13) with n replaced by k for the values $k=1,2,\dots,n$ successively, summation over these values of the index, and addition of the terms obtained from (14):

$$\begin{aligned}&\big[\cos(t-\mu)-\cos(x-\mu)\big]K_n(x,t)\\ &\qquad=\alpha_n\big[u_n(x)u_{n+1}(t)-u_{n+1}(x)u_n(t)\big]+\beta_n\big[u_n(x)v_{n+1}(t)-v_{n+1}(x)u_n(t)\big]\\ &\qquad+\gamma_n\big[v_n(x)u_{n+1}(t)-u_{n+1}(x)v_n(t)\big]+\delta_n\big[v_n(x)v_{n+1}(t)-v_{n+1}(x)v_n(t)\big],\end{aligned}$$

or, if the long right-hand member is denoted by $N_n(x,t)$,

$$K_n(x,t)=\frac{N_n(x,t)}{\big[\cos(t-\mu)-\cos(x-\mu)\big]}. \tag{15}$$

The constants $\alpha_n,\beta_n,\gamma_n,\delta_n$ can be evaluated in terms of the leading coefficients in the expressions of $u_n(x)$ and $v_n(x)$ as combinations of sines and cosines. More important for the subsequent applications, however, is the fact that *their absolute values are less than unity*. Multiplication of (10) by $\rho(x)u_{n+1}(x)$ and integration from $-\pi$ to π gives

$$\alpha_n=\int_{-\pi}^{\pi}\rho(x)\cos(x-\mu)u_n(x)u_{n+1}(x)\,dx.$$

Hence

$$|\alpha_n \leqq \int_{-\pi}^{\pi}\rho(x)|\cos(x-\mu)u_n(x)u_{n+1}(x)|\,dx.$$

Let Schwarz's inequality be applied to this integral, the integrand being regarded as the product of the factors $[\rho(x)]^{1/2}|\cos(x-\mu)u_n(x)|$ and $[\rho(x)]^{1/2}|u_{n+1}(x)|$:

$$\begin{aligned}\alpha_n^2 &\leqq \int_{-\pi}^{\pi}\rho(x)\cos^2(x-\mu)\big[u_n(x)\big]^2dx\int_{-\pi}^{\pi}\rho(x)\big[u_{n+1}(x)\big]^2dx\\ &<\int_{-\pi}^{\pi}\rho(x)\big[u_n(x)\big]^2dx\int_{-\pi}^{\pi}\rho(x)\big[u_{n+1}(x)\big]^2dx=1;\end{aligned}$$

the replacement of $\cos^2(x-\mu)$ by unity actually increases the value of the integrand over a set of positive measure and increases the value of the integral, the alternative of equality being ruled out. A corresponding calculation gives a similar result for β_n, γ_n, and δ_n (and incidentally for each of the coefficients denoted by A_n, B_n, D_n in the preceding section; in the case of A_n and D_n the fact can also be recognized directly from the form of the integrals defining them, without the use of Schwarz's inequality).

If the terms of the nth order in $u_n(x)$ and $v_n(x)$ are denoted by $a_n \cos nx$, $c_n \cos nx$, and $d_n \sin nx$ respectively, the explicit formulas for $\alpha_n, \beta_n, \gamma_n, \delta_n$ are found to be as follows:

$$\alpha_n = \frac{1}{2}\frac{a_n}{a_{n+1}}\left(\cos\mu - \frac{c_{n+1}}{d_{n+1}}\sin\mu\right),$$

$$\beta_n = \frac{1}{2}\frac{a_n}{d_{n+1}}\sin\mu,$$

$$\gamma_n = \frac{1}{2a_{n+1}}\left[(c_n\cos\mu - d_n\sin\mu) - \frac{c_{n+1}}{d_{n+1}}(c_n\sin\mu + d_n\cos\mu)\right],$$

$$\delta_n = \frac{1}{2d_{n+1}}(c_n\sin\mu + d_n\cos\mu),$$

for $n \geqq 1$, while

$$\alpha_0 = \frac{a_0}{a_1}\left(\cos\mu - \frac{c_1}{d_1}\sin\mu\right), \quad \beta_0 = \frac{a_0}{d_1}\sin\mu.$$

These are set down merely as a matter of record, and will not be used further.

4. Formal expansion of an arbitrary function. An arbitrary function $f(x)$ of period 2π, subject to the condition that $\rho(x)f(x)$ be summable over a period, can be formally expanded in a series

$$a_0u_0(x) + a_1u_1(x) + a_2u_2(x) + \cdots + b_1v_1(x) + b_2v_2(x) + \cdots \tag{16}$$

by the procedure that is usual in connection with series of orthogonal functions; the meaning of a_n here of course has no connection with the use of the same symbol in the preceding paragraph. The general coefficients are

$$a_k = \int_{-\pi}^{\pi}\rho(x)f(x)u_k(x)\,dx, \quad b_k = \int_{-\pi}^{\pi}\rho(x)f(x)v_k(x)\,dx.$$

The partial sum of the series through terms of the nth order is

$$s_n(x) = \int_{-\pi}^{\pi}\rho(t)f(t)K_n(x,t)\,dt, \tag{17}$$

where $K_n(x,t)$ is defined as above. In particular,

$$1 \equiv \int_{-\pi}^{\pi}\rho(t)K_n(x,t)\,dt;$$

multiplication of this identity by $f(x)$, which is independent of the variable of integration, and subtraction from (17) gives

$$s_n(x)-f(x)=\int_{-\pi}^{\pi}\rho(t)[f(t)-f(x)]K_n(x,t)\,dt. \tag{18}$$

5. Analogue of the Riemann-Lebesgue theorem. If $\rho(x)[f(x)]^2$ as well as $\rho(x)f(x)$ is summable, further calculation which is well known as applied to orthogonal functions in general yields the relation

$$\int_{-\pi}^{\pi}\rho(x)[f(x)-s_n(x)]^2dx=\int_{-\pi}^{\pi}\rho(x)[f(x)]^2dx-a_0^2-\sum_{k=1}^{n}(a_k^2+b_k^2).$$

As the magnitude of the sum on the right can never exceed that of the integral on the right as n increases, by reason of the non-negative character of the left-hand member, the corresponding infinite series is convergent, and the general term approaches zero:

$$\lim_{k\to\infty}a_k=0,\quad \lim_{k\to\infty}b_k=0.$$

If the sums $u_k(x), v_k(x)$ *are uniformly bounded*, as will be true at least in particular cases, this conclusion can be made more general by omission of the requirement that ρf^2 be summable, the hypothesis of summability being imposed merely on ρ and ρf. Suppose $|u_k(x)|\leqq H$, $|v_k(x)|\leqq H$, for all k and all x. For arbitrary positive N, let $f_N(x)=f(x)$ when $|f(x)|\leqq N$, $f_N(x)=0$ when $|f(x)|>N$. If ε is any positive number,

$$\int_{-\pi}^{\pi}\rho(x)|f(x)-f_N(x)|\,dx<\frac{\varepsilon}{2H}$$

when N is sufficiently large, and hence

$$\left|\int_{-\pi}^{\pi}\rho(x)f(x)u_k(x)\,dx-\int_{-\pi}^{\pi}\rho(x)f_N(x)u_k(x)\,dx\right|<\frac{1}{2}\varepsilon,$$

and similarly with $u_k(x)$ replaced by $v_k(x)$, for all values of k. Let N be given a fixed value so large that these inequalities hold. Since ρf_N^2 is summable,

$$\left|\int_{-\pi}^{\pi}\rho(x)f_N(x)u_k(x)\,dx\right|<\frac{1}{2}\varepsilon,\quad \left|\int_{-\pi}^{\pi}\rho(x)f_N(x)v_k(x)\,dx\right|<\frac{1}{2}\varepsilon,$$

for all values of k from a certain point on, and when this is true

$$|a_k|<\varepsilon,\quad |b_k|<\varepsilon,$$

if a_k and b_k are the coefficients defined for the original function $f(x)$.

It will be desirable subsequently to have the conclusion formulated for a function other than the one whose series expansion is under investigation, in the form of a

LEMMA. *If $\rho(t)\varphi(t)$ and $\rho(t)[\varphi(t)]^2$ are summable, or if $\rho(t)\varphi(t)$ is summable and the functions $u_k(t), v_k(t)$ are uniformly bounded, the integrals*

$$\int_{-\pi}^{\pi} \rho(t)\varphi(t)u_k(t)\,dt, \quad \int_{-\pi}^{\pi} \rho(t)\varphi(t)v_k(t)\,dt$$

approach zero as k becomes infinite.

(It is understood throughout the paper that the weight function itself is summable.)

6. Special uniformly bounded systems. As there has been occasion above to suppose the functions $u_k(x), v_k(x)$ uniformly bounded, and as this hypothesis will be fundamental in a large part of what follows, it is worth while, pending investigation of general conditions under which the property in question is realized, to point out that in a particular class of cases its validity can be recognized immediately. In other cases it can be seen that the u's and v's are bounded for specified values of x, or uniformly bounded over a part of a period, and this also is significant for the theory of convergence.

Let $\rho(x)$ be any weight function for which the u's and v's are bounded for a particular value of x as k becomes infinite, or uniformly bounded over a designated range of values of x:

$$|u_k(x)| \leqq H, \quad |v_k(x)| \leqq H, \tag{19}$$

for values of x belonging to a point set E. Certainly this condition is fulfilled everywhere in the special case $\rho(x) \equiv 1$, the functions of the normalized system for $k \geqq 1$ being $(1/\pi^{1/2})\cos kx$ and $(1/\pi^{1/2})\sin kx$. Let $U_k(x), V_k(x)$ be the sums of the kth order in the normalized orthogonal system corresponding to the weight function $\rho(x)\tau(x)$, where $\tau(x)$ is any non-negative trigonometric sum. It will appear that $U_k(x), V_k(x)$ are bounded at any point of E where $\tau(x) \neq 0$, and uniformly bounded over any part of E where $\tau(x)$ has a positive lower bound.

Let p be the order of the sum $\tau(x)$. Since $\tau(x)U_n(x)$ is a trigonometric sum of order $n+p$ it can be expressed in the form

$$\begin{aligned}\tau(x)U_n(x) \equiv A_0u_0(x) + A_1u_1(x) + \cdots + A_{n+p}u_{n+p}(x)\\ + B_1v_1(x) + \cdots + B_{n+p}v_{n+p}(x),\end{aligned}$$

the coefficients depending on n, to be sure, as well as on the index which appears explicitly. (The present A's and B's have no relation to those used previously.) By the property of orthogonality of the sums $u_k(x), v_k(x)$,

$$A_k = \int_{-\pi}^{\pi} \rho(x)\tau(x)U_n(x)u_k(x)\,dx.$$

But $U_n(x)$ is orthogonal to any trigonometric sum of lower order with respect to the weight function $\rho(x)\tau(x)$ for which the U's and V's are constructed, and hence

$A_k=0$ for $k<n$. Similarly, $B_k=0$ for $k<n$. So

$$U_n(x)\equiv\frac{1}{\tau(x)}\left[A_n u_n(x)+\cdots+A_{n+p}u_{n+p}(x)+B_n v_n(x)+\cdots+B_{n+p}v_{n+p}(x)\right],$$

the number of terms in the bracket being $2p+2$, and remaining finite as n increases. In the same way $V_n(x)$ has the form

$$V_n(x)\equiv\frac{1}{\tau(x)}\left[C_n u_n(x)+\cdots+C_{n+p}u_{n+p}(x)+D_n v_n(x)+\cdots+D_{n+p}v_{n+p}(x)\right].$$

For x in E, if $\tau(x)\neq 0$,

$$|U_n(x)|\leqq\frac{H}{\tau(x)}(|A_n|+\cdots+|A_{n+p}|+|B_n|+\cdots+|B_{n+p}|),$$

whence, as any one of the absolute values in the parenthesis is certainly not greater than the square root of the sum of the squares of all of them,

$$|U_n(x)|\leqq\frac{H}{\tau(x)}(2p+2)\left(A_n^2+\cdots+A_{n+p}^2+B_n^2+\cdots+B_{n+p}^2\right)^{1/2}. \tag{20}$$

(By Schwarz's inequality the last relation holds in fact with $(2p+2)$ replaced by $(2p+2)^{1/2}$.) If M_0 is the maximum of $\tau(x)$,

$$A_n^2+\cdots+A_{n+p}^2+B_n^2+\cdots+B_{n+p}^2=\int_{-\pi}^{\pi}\rho(x)\left[\tau(x)U_n(x)\right]^2dx$$

$$\leqq M_0\int_{-\pi}^{\pi}\rho(x)\tau(x)\left[U_n(x)\right]^2dx=M_0,$$

the last equality resulting from the normalization of the U's with respect to $\rho(x)\tau(x)$ as weight function. So the whole right-hand member of (20) has an upper bound independent of n. Similar reasoning applies to $V_n(x)$. So the assertion with regard to the boundedness of the U's and V's is justified. In particular, if (19) holds for all values of x and if $\tau(x)$ has a positive minimum *the U's and V's are uniformly bounded everywhere.*

7. First convergence theorem. Let $\rho(x)$ be a non-negative summable weight function of period 2π for which the sums $u_k(x), v_k(x)$ are everywhere uniformly bounded. Let $f(x)$ be a function of period 2π such that $\rho(x)f(x)$ is summable. Let $s_n(x)$, as above, be the partial sum of the series (16) through terms of the nth order. This section is concerned with the setting up of conditions for the convergence of $s_n(x)$ toward $f(x)$ for an arbitrarily specified value of x, to be held fast throughout the discussion. The question of uniformity of convergence will be reserved for treatment by a different method in the next section.

The value of x being given, let $\psi_1(t)$ and $\psi_2(t)$ be functions of period 2π, defined in the interval $x-(\pi/2)\leqq x<x+(3\pi/2)$ as follows:

$$\psi_1(t)=f(t)-f(x)\quad\text{for } |t-x|\leqq\frac{\pi}{2},\quad \psi_1(t)=0\qquad\text{for } |t-x|>\frac{\pi}{2},$$

$$\psi_2(t)=0\qquad\text{for } |t-x|\leqq\frac{\pi}{2},\quad \psi_2(t)=f(t)-f(x)\quad\text{for } |t-x|>\frac{\pi}{2}.$$

Then $\psi_1(t)+\psi_2(t)=f(t)-f(x)$ everywhere, and by (18)

$$s_n(x)-f(x)=\int_{-\pi}^{\pi}\rho(t)\psi_1(t)K_n(x,t)\,dt+\int_{-\pi}^{\pi}\rho(t)\psi_2(t)K_n(x,t)\,dt$$
$$=I_1+I_2.$$

It will be recalled that in the expression (15) for $K_n(x,t)$ the constant μ is arbitrary, the coefficients $\alpha_n,\beta_n,\gamma_n,\delta_n$ which enter into the numerator depending on μ, but being always less than 1 in absolute value. For substitution of (15) in the integral I_1, let $\mu=x+(\pi/2)$. Then

$$\cos(x-\mu)=\cos\left(-\frac{\pi}{2}\right)=0,\quad \cos(t-\mu)=\cos\left(t-x-\frac{\pi}{2}\right)=\sin(t-x),$$
$$\cos(t-\mu)-\cos(x-\mu)=\sin(t-x).$$

So I_1 is the sum of

$$\alpha_n\int_{-\pi}^{\pi}\rho(t)\frac{\psi_1(t)}{\sin(t-x)}\left[u_n(x)u_{n+1}(t)-u_{n+1}(x)u_n(t)\right]dt \tag{21}$$

and three other integrals of similar form. Let

$$\Phi(t)\equiv\frac{f(t)-f(x)}{t-x},$$

and let it be supposed that the function $f(t)$ is such that there is an interval of values of t about the point x over which $\rho(t)\Phi(t)$ is summable. Then $\rho(t)\phi(t)$ is summable over any finite interval, being the product of the summable function $\rho(t)[f(t)-f(x)]$ by the continuous function $1/(t-x)$ over any interval which does not approach the point $t=x$. If

$$\varphi_1(t)\equiv\frac{\psi_1(t)}{\sin(t-x)},$$

$\rho(t)\varphi_1(t)$ is summable over the interval $x-(\pi/2)\leqq t\leqq x+(\pi/2)$, being the product of $\rho(t)\Phi(t)$ by the continuous function $(t-x)/\sin(t-x)$; it is summable from $x+(\pi/2)$ to $x+(3\pi/2)$, being identically zero except at the point $t=x+\pi$ where it is not defined; and therefore it is summable over the entire period from $x-(\pi/2)$ to $x+(3\pi/2)$, or equally well over any other period interval, in particular that from $-\pi$ to π. The expression (21) is equal to

$$\alpha_n\left[u_n(x)\int_{-\pi}^{\pi}\rho(t)\varphi_1(t)u_{n+1}(t)\,dt-u_{n+1}(x)\int_{-\pi}^{\pi}\rho(t)\varphi_1(t)u_n(t)\,dt\right].$$

Here each integral approaches zero as n becomes infinite, by the Lemma of Section 5; $u_n(x)$ and $u_{n+1}(x)$ are bounded by hypothesis; and $|\alpha_n|<1$. So the whole expression approaches 0. Similar reasoning, with a similar conclusion, is applicable to each of the other three parts of I_1. Hence $\lim_{n\to\infty}I_1=0$.

For the application of (15) in I_2, let $\mu=x$,

$$\cos(t-\mu)-\cos(x-\mu)=\cos(t-x)-1,$$

and let $\varphi_2(t) \equiv \psi_2(t)/[\cos(t-x)-1]$. With new values of $\alpha_n, \ldots, \delta_n$, the integral I_2 is the sum of

$$\alpha_n\left[u_n(x)\int_{-\pi}^{\pi}\rho(t)\varphi_2(t)u_{n+1}(t)\,dt - u_{n+1}(x)\int_{-\pi}^{\pi}\rho(t)\varphi_2(t)u_n(t)\,dt\right]$$

and three other expressions of like character. The product $\rho(t)\varphi_2(t)$ is summable from $x-(\pi/2)$ to $x+(\pi/2)$, where it is identically zero except for $t=x$, and from $x+(\pi/2)$ to $x+(3\pi/2)$, where $|\cos(t-x)-1| \geqq 1$, and so is summable over any period interval. By use of the Lemma again it is seen that $\lim_{n\to\infty} I_2 = 0$. The fact that I_1 and I_2 both approach zero means that

$$\lim_{n\to\infty} s_n(x) = f(x).$$

The proof of convergence goes equally well without the requirement that the u's and v's be uniformly bounded, if it is assumed that they are bounded at the point x and that $\rho(t)[f(t)]^2$ and $\rho(t)[\Phi(t)]^2$ are summable. These results are expressed in

THEOREM I. *The series* (16) *will converge to the value $f(x)$ for a specified value of x if the functions $u_k(t)$ and $v_k(t)$ are uniformly bounded, if $\rho(t)f(t)$ is summable, and if there is an interval about the point x over which $\rho(t)\Phi(t)$ is summable, where $\Phi(t)$ denotes the difference quotient $[f(t)-f(x)]/(t-x)$; or if $u_k(x)$ and $v_k(x)$ are bounded for the particular value of x in question, if $\rho(t)[f(t)]^2$ is summable, and if $\rho(t)[\Phi(t)]^2$ is summable over an interval about the point x.*

The conditions on $\Phi(t)$ can of course be interpreted in more specific terms* with reference to properties of $f(t)$. They will be more than satisfied, for example, if $f(t)$ satisfies a Lipschitz condition in the neighborhood of $t=x$.

More particularly still, the condition on $\Phi(t)$ will be satisfied if $f(t)$ vanishes identically throughout an interval about the point x. If $f_1(t)$ and $f_2(t)$ are two functions which are equal throughout such an interval, the series for their difference, which is the difference of their respective series, will converge toward zero at the point x; if the series for f_1 converges, the series for f_2 will converge likewise, regardless of any difference between the functions outside the interval designated, provided ρf_1 and ρf_2 are summable, and, in case of need, ρf_1^2 and ρf_2^2 also. *The convergence of the series for a function $f(x)$ at a specified point depends only on the behavior of the function in the neighborhood of the point*, if the u's and v's are uniformly bounded and ρf is summable, or if the u's and v's are bounded at the point in question and ρf^2 is summable.

8. Second convergence theorem. Let $T_n(x)$ be any trigonometric sum of the nth order. It can be expressed as a linear combination of $u_0(x), u_1(x), v_1(x), \ldots, u_n(x), v_n(x)$, and this linear expression is at the same time the series development of $T_n(x)$ in terms of the u's and v's. The partial sum of the series through terms of

* Cf. e.g. H. Lebesgue, *Leçons sur les séries trigonométriques*, Paris, 1906, Chapter III.

order n is identical with $T_n(x)$ itself. This means that

$$T_n(x) \equiv \int_{-\pi}^{\pi} \rho(t) T_n(t) K_n(x,t)\, dt.$$

Let $f(x) - T_n(x) \equiv r_n(x)$. Then $f(x) = T_n(x) + r_n(x)$, and if $s_n(x)$ is the partial sum of the series for $f(x)$,

$$\begin{aligned} s_n(x) &= \int_{-\pi}^{\pi} \rho(t) f(t) K_n(x,t)\, dt \\ &= \int_{-\pi}^{\pi} \rho(t) T_n(t) K_n(x,t)\, dt + \int_{-\pi}^{\pi} \rho(t) r_n(t) K_n(x,t)\, dt \\ &= T_n(x) + \int_{-\pi}^{\pi} \rho(t) r_n(t) K_n(x,t)\, dt, \end{aligned}$$

$$f(x) - s_n(x) = r_n(x) - \int_{-\pi}^{\pi} \rho(t) r_n(t) K_n(x,t)\, dt.$$

If $|r_n(x)| \leqq \varepsilon_n$,

$$\left| \int_{-\pi}^{\pi} \rho(t) r_n(t) K_n(x,t)\, dt \right| \leqq \varepsilon_n \int_{-\pi}^{\pi} \rho(t) |K_n(x,t)|\, dt,$$

and consequently

$$|f(x) - s_n(x)| \leqq \varepsilon_n [1 + \lambda_n(x)],$$

if

$$\lambda_n(x) = \int_{-\pi}^{\pi} \rho(t) |K_n(x,t)|\, dt.$$

So the convergence of $s_n(x)$ toward $f(x)$ can be discussed in terms of the order of magnitude of ε_n and the order of magnitude of $\lambda_n(x)$. Information with regard to the former, when the sums $T_n(x)$ are appropriately chosen, is given by known theorems on degree of approximation. The latter will be examined here.*

Let it be supposed that $\rho(x)$ *is such that* $u_k(x)$, $v_k(x)$ *are uniformly bounded, and furthermore*, for simplicity at least, *that* $\rho(x)$ *itself is bounded.* Let $\rho(x) \leqq G$, $|u_k(x)| \leqq H$, $|v_k(x)| \leqq H$. In view of the periodicity of the functions involved, the integral defining $\lambda_n(x)$, instead of being taken from $-\pi$ to π, may be extended over any other interval of length 2π, say that from $x - (\pi/2)$ to $x + (3\pi/2)$. Let the integrals from $x - (\pi/2)$ to $x + (\pi/2)$ and from $x + (\pi/2)$ to $x + (3\pi/2)$ be represented by J_1 and J_2 respectively, and let J_1 be further subdivided into the integrals from $x - (\pi/2)$ to $x - (1/n)$, from $x - (1/n)$ to $x + (1/n)$, and from $x + (1/n)$ to $x + (\pi/2)$, denoted respectively by J_1', J_1'', and J_1''', so that

$$\lambda_n(x) = J_1 + J_2 = J_1' + J_1'' + J_1''' + J_2,$$

the integrand each time being $\rho(t)|K_n(x,t)|$.

* For a corresponding treatment of the problem of polynomial approximation see J. Shohat, *On the development of continuous functions in series of Tchebycheff polynomials*, Transactions of the American Mathematical Society, vol. 27 (1925), pp. 537–550.

In J_1' and J_1''' let $K_n(x,t)$ be represented by (15) with $\mu = x + (\pi/2)$, whereby the denominator takes the form

$$\cos(t-\mu) - \cos(x-\mu) = \sin(t-x).$$

Since $|t-x| \leqq \pi/2$ over the range in question, $|\sin(t-x)| \geqq (2/\pi)|t-x|$. In the first of the four parts of the numerator, the relations

$$|\alpha_n| < 1, \quad |u_n(x)| \leqq H, \quad |u_{n+1}(t)| \leqq H, \quad |u_{n+1}(x)| \leqq H, \quad |u_n(t)| \leqq H$$

give $|\alpha_n[u_n(x)u_{n+1}(t) - u_{n+1}(x)u_n(t)]| \leqq 2H^2$. As corresponding inequalities hold for the other three parts, and as $\rho(t) \leqq G$,

$$\rho(t)|K_n(x,t)| \leqq \frac{8\pi GH^2}{2|t-x|}.$$

By explicit integration,

$$\int_{x-(\pi/2)}^{x-(1/n)} \frac{dt}{|t-x|} = \int_{x+(1/n)}^{x+(\pi/2)} \frac{dt}{|t-x|} = \log\left(\frac{\pi}{2}\right) + \log n,$$

and J_1' and J_1''' have each an upper bound of the order of $\log n$. In J_1'' it suffices to read off from the definition of $K_n(x,t)$ in (12), together with the hypothesis on the u's and v's, that $|K_n(x,t)| \leqq (2n+1)H^2$; the length of the interval being $2/n$, it is seen that $J_1'' \leqq 2[2+(1/n)]GH^2 \leqq 6GH^2$. Consequently the whole of J_1 does not exceed a constant multiple of $\log n$.

In J_2 the formula (15) is to be used with $\mu = x$. As $t-x$ ranges from $\pi/2$ to $3\pi/2$,

$$|\cos(t-\mu) - \cos(x-\mu)| = |\cos(t-x) - 1| \geqq 1,$$

and $\rho(t)|K_n(x,t)| \leqq 8GH^2, J_2 \leqq 8\pi GH^2$.

By combination of these results it is recognized that

$$\lambda_n(x) \leqq C \log n$$

for $n > 1$, where C is independent of n and x.

The conclusions with regard to convergence and degree of convergence are expressed in

THEOREM II. *If $\rho(x)$ is bounded, if $u_k(x)$ and $v_k(x)$ are uniformly bounded, and if trigonometric sums $T_n(x)$ of the nth order exist* so that*

$$|f(x) - T_n(x)| \leqq \varepsilon_n,$$

then

$$|f(x) - s_n(x)| \leqq C_1 \varepsilon_n \log n$$

* For the requisite theorems on degree of approximation see e.g. the writer's Colloquium, cited above, Chapter I.

*for $n>1$, where C_1 is independent of n and x; in particular,*** *the series converges uniformly to the value $f(x)$ for all values of x if $f(x)$ has a modulus of continuity $\omega(\delta)$ such that $\lim_{\delta\to 0}\omega(\delta)\log\delta=0$.*

9. General weight functions. Some information as to the order of magnitude of the u's and v's, when they are not known to remain bounded, is readily accessible in the case of weight functions of considerable generality.

Suppose that $\rho(x)$ is summable and has a positive lower bound for all values of x:

$$\rho(x) \geqq v > 0. \tag{22}$$

Let $T_n(x)$ be any trigonometric sum of the nth order (not necessarily one of those in the orthogonal sequence) satisfying the condition that

$$\int_{-\pi}^{\pi} \rho(x)[T_n(x)]^2 dx = 1. \tag{23}$$

In consequence of (22),

$$\int_{-\pi}^{\pi} [T_n(x)]^2 dx \leqq \frac{1}{v}. \tag{24}$$

Let c_0 be the constant term in the trigonometric sum $[T_n(x)]^2$, and let $S_n(x) = [T_n(x)]^2 - c_0$, so that $S_n(x)$ is a trigonometric sum of order $2n$ without constant term. The integral (24) has the value $2\pi c_0$, and hence $|c_0| \leqq 1/(2\pi v)$. For any value of x between $-\pi$ and π,

$$\int_{-\pi}^{x} [T_n(x)]^2 dx \leqq \int_{-\pi}^{\pi} [T_n(x)]^2 dx \leqq \frac{1}{v},$$

and

$$\left|\int_{-\pi}^{x} S_n(x)\,dx\right| \leqq \int_{-\pi}^{x} [T_n(x)]^2 dx + \int_{-\pi}^{x} |c_0|\,dx \leqq \frac{2}{v}.$$

The integral between bars in the left-hand member is itself a trigonometric sum of order $2n$, having $S_n(x)$ for its derivative. Hence, by Bernstein's theorem,

$$|S_n(x)| \leqq \frac{4n}{v}.$$

It follows that $[T_n(x)]^2 \equiv c_0 + S_n(x)$ has an upper bound of the order of n, and

**For the case of Fourier series, and for the special theorem on degree of approximation needed here, see also Lebesgue, *Sur la représentation trigonométrique approchée des fonctions satisfaisant à une condition de Lipschitz*, Bulletin de la Société Mathématique de France, vol. 38 (1910), pp. 184–210; pp. 201–202, 209.

$|T_n(x)|$ has an upper bound of the order of $n^{1/2}$. In particular, $|u_n(x)|$ and $|v_n(x)|$ can not exceed a constant multiple of $n^{1/2}$.

Now let the requirement that $\rho(x)$ have a positive lower bound be replaced by a less restrictive one: let it be assumed merely that $[\rho(x)]^{-1}$ as well as $\rho(x)$ is summable over a period. Let $T_n(x)$ again be a trigonometric sum of the nth order satisfying (23). Let Schwarz's inequality be applied to the integral of the product of the factors $[\rho(x)]^{-1/2}$ and $[\rho(x)]^{1/2}|T_n(x)|$:

$$\left[\int_{-\pi}^{\pi}|T_n(x)|\,dx\right]^2 \leqq \int_{-\pi}^{\pi}[\rho(x)]^{-1}dx\int_{-\pi}^{\pi}\rho(x)[T_n(x)]^2dx=\int_{-\pi}^{\pi}[\rho(x)]^{-1}dx.$$

The square root of the value of the last integral being denoted by I,

$$\int_{-\pi}^{\pi}|T_n(x)|\,dx \leqq I.$$

Let c_0 this time be the constant term in $T_n(x)$, and $S_n(x)$ the sum of the remaining terms. It is seen that

$$2\pi|c_0|=\left|\int_{-\pi}^{\pi}T_n(x)\,dx\right|\leqq I,\quad |c_0|\leqq\frac{I}{2\pi}.$$

Furthermore,

$$\int_{-\pi}^{\pi}|S_n(x)|\,dx\leqq\int_{-\pi}^{\pi}|T_n(x)|\,dx+\int_{-\pi}^{\pi}|c_0|\,dx\leqq 2I,$$

$$\left|\int_{-\pi}^{x}S_n(x)\,dx\right|\leqq\int_{-\pi}^{x}|S_n(x)|\,dx\leqq\int_{-\pi}^{\pi}|S_n(x)|\,dx\leqq 2I.$$

Since $\int S_n(x)\,dx$ is a trigonometric sum of the nth order, Bernstein's theorem gives for $|S_n(x)|$, and consequently for $|T_n(x)|$, an upper bound of the order of n. Specifically, $|u_n(x)|$ and $|v_n(x)|$ can not exceed a constant multiple of n.

These bounds for the u's and v's, however, do not seem to lead immediately to any better theorems on convergence than can be obtained by direct use of the least-square property of the series.* Consider for example the case of the last paragraph, under which $\rho(x)$ and $[\rho(x)]^{-1}$ are summable. It follows from well-known properties of orthogonal functions in general that the integral

$$\gamma_n=\int_{-\pi}^{\pi}\rho(x)[f(x)-s_n(x)]^2dx$$

has a smaller value than the corresponding integral with $s_n(x)$ replaced by any other trigonometric sum $t_n(x)$ of the nth order. Suppose sums $t_n(x)$ exist so that

$$|f(x)-t_n(x)|\leqq\varepsilon_n$$

* See e.g. the writer's Colloquium, already cited, Chapter III; D. Jackson, *A generalized problem in weighted approximation*, Transactions of the American Mathematical Society, vol. 26 (1924), pp. 133–154; J. Shohat, *On the polynomial and trigonometric approximation of measurable bounded functions on a finite interval*, Mathematische Annalen, vol. 102 (1930), pp. 157–175.

for all values of x. Then it is certain that γ_n does not exceed a constant multiple of ε_n^2, say $\gamma_n \leqq h^2\varepsilon_n^2$. Let Schwarz's inequality be applied with $[\rho(x)]^{-1/2}$ and $[\rho(x)]^{1/2}|f(x)-s_n(x)|$ as factors under the integral sign:

$$\left[\int_{-\pi}^{\pi} |f(x)-s_n(x)|\,dx\right]^2 \leqq I^2\gamma_n,$$

I having the same meaning as before. It follows that

$$\int_{-\pi}^{\pi} |f(x)-s_n(x)|\,dx \leqq Ih\varepsilon_n. \tag{25}$$

If this integral is denoted by g_n it is easily shown* that

$$|f(x)-s_n(x)| \leqq 4ng_n + 5\varepsilon_n,$$

which means, in consequence of (25), that $s_n(x)$ converges uniformly toward $f(x)$ if $\lim_{n\to\infty} n\varepsilon_n = 0$. And it will be possible to construct sums $t_n(x)$ so that this condition is satisfied, if $f(x)$ has a continuous derivative.** The proof of convergence thus obtained applies even at points where $\rho(x)$ vanishes.

THE CONVERGENCE OF FOURIER SERIES[1]

DUNHAM JACKSON, University of Minnesota

1. *Introduction.* Considerable parts of the theory of Fourier series have an interest, both for their mathematical content and by reason of the importance of their applications, for students whose experience of mathematics in general is only moderately advanced. The writer has had occasion more than once to give an introductory course on Fourier series and related topics for classes whose mathematical preparation was not assumed to extend beyond a first course in the calculus. The question has arisen each time how far it is possible to go beyond merely formal relationships, and to give such a class a genuine appreciation of some of the properties of convergence, even the most elementary of which are so characteristic of the type of series in question and have had so profound an influence on the course of modern mathematical development. This paper is an outline of the writer's most recent attempt in that direction. No part of the treatment is new, and most parts have been used long since for purposes of elementary exposition. The object of this account is merely to suggest one way of cutting the whole picture into pieces of convenient size, and arranging them in

* See Colloquium, pp. 87–88. It is to be observed that the function denoted by $\rho(x)$ in the passage cited is not the $\rho(x)$ of the present problem, but is to be taken as 1, while the sum $T_n(x)$ there is to be taken as the present $s_n(x)$, and $m=1$.

**See Colloquium, p. 12, Theorem IV, Corollary.

[1] Presented to the Mathematical Association of America at New Orleans, December 31, 1931.

order so that at any stage the next piece is not far to seek.[2]

Specifically, the essential technicalities are brought within easy reach by the following devices:

(a) The main convergence proof is made to depend on nothing more abstruse than the fact that the general term of a convergent series approaches zero.

(b) Certain theorems relating to an arbitrary continuous function are made to depend on acceptance of the proposition that the graph of such a function can be approximated with any desired accuracy by a broken line. Elsewhere a reader unfamiliar with the precise definition of the word "continuous" may take it as self-explanatory, as far as an understanding of the main features of the argument is concerned.

(c) The phrase "uniformly convergent" is introduced as naturally descriptive of a type of convergence already characterized in quantitative terms.

Such facts as the integrability of a continuous function, or of a function which is continuous except for a finite number of finite jumps, will be considered "obvious" on the basis of the interpretations that are customary in a first study of the calculus. An attempt is made throughout to give an exposition to which more critical study may have something to add, but from which it will have nothing to retract. In actual presentation to a class the less familiar ideas are naturally explained and illustrated at greater length than in the text.

2. *Formulas for the coefficients.* The Fourier series for a given function has the form

$$\begin{aligned} &a_0/2 + a_1 \cos x + a_2 \cos 2x + \cdots \\ &\quad + b_1 \sin x + b_2 \sin 2x + \cdots, \end{aligned} \tag{1}$$

in which the coefficients are given by the formulas

$$a_k = \frac{1}{\pi}\int_{-\pi}^{\pi} f(t)\cos kt\,dt, \qquad b_k = \frac{1}{\pi}\int_{-\pi}^{\pi} f(t)\sin kt\,dt. \tag{2}$$

These are found by setting $f(t)$ equal to the series (1), written in terms of the variable t, multiplying by $\cos kt$ or $\sin kt$, and integrating term by term, with the use of the relations

$$\begin{aligned} &\int_{-\pi}^{\pi} \cos jt \cos kt\,dt = 0, \qquad \int_{-\pi}^{\pi} \sin jt \sin kt\,dt = 0 \qquad (j \neq k), \\ &\int_{-\pi}^{\pi} \sin jt \cos kt\,dt = 0 \qquad (\text{for all } j \text{ and } k), \\ &\int_{-\pi}^{\pi} \cos^2 kt\,dt = \int_{-\pi}^{\pi} \sin^2 kt\,dt = \pi \qquad (k \geqq 1). \end{aligned} \tag{3}$$

[2] Among the numerous more or less extensive presentations of the theory of Fourier series the following may be specially mentioned in connection with the present paper: M. Bôcher, *Introduction to the Theory of Fourier's Series*, Annals of Mathematics, (2), vol. 7 (1905–06), pp. 81–152; H. Lebesgue, *Leçons sur les séries trigonométriques*, Paris, 1906.

The representation of the constant term by $a_0/2$ rather than a_0 is an artifice to make the general formula for a_k applicable without change when $k=0$.

If the calculation leading to the equations (2) were to be regarded as a "proof," some inquiry would be necessary as to the validity of the processes involved. In the discussion to be presented here the formulas (2), which are suggested by the calculation as at least presumptively important, and which have a meaning whenever $f(t)$ is an integrable function, will be taken outright as starting point. A series (1) will be written down with these coefficients, and it will be inquired whether the series does in fact converge and represent $f(x)$. It will be assumed throughout that $f(x)$ is periodic, with period 2π.

3. *Order of magnitude of the coefficients in the case of certain continuous functions.* If $f(t)$ has a continuous derivative, integration by parts gives

$$\pi a_k = \left[\frac{1}{k} f(t) \sin kt\right]_{-\pi}^{\pi} - \frac{1}{k}\int_{-\pi}^{\pi} f'(t) \sin kt\, dt = -\frac{1}{k}\int_{-\pi}^{\pi} f'(t) \sin kt\, dt.$$

If there is a continuous second derivative,

$$\int_{-\pi}^{\pi} f'(t) \sin kt\, dt = \left[-\frac{1}{k} f'(t) \cos kt\right]_{-\pi}^{\pi} + \frac{1}{k}\int_{-\pi}^{\pi} f''(t) \cos kt\, dt$$
$$= \frac{1}{k}\int_{-\pi}^{\pi} f''(t) \cos kt\, dt.$$

Let M be the maximum of $|f''(t)|$. Then

$$|f''(t) \cos kt| \leqq M, \qquad \left|\int_{-\pi}^{\pi} f''(t) \cos kt\, dt\right| \leqq 2M\pi,$$

and $|a_k| \leqq 2M/k^2$. Similarly $|b_k| \leqq 2M/k^2$. It follows that

$$|a_k \cos kx + b_k \sin kx| \leqq 4M/k^2,$$

and as $1/k^2$ is the general term of a well known convergent series, the series (1) is also certainly convergent.

The same conclusion can be reached with a somewhat less restrictive hypothesis on $f(t)$, and this will be important for an application later. Let $f(t)$ still be continuous everywhere, and let it be supposed that any period interval can be divided into a finite number of subintervals throughout each of which $f(t)$ has continuous first and second derivatives, but that the derivatives may not be continuous in passing from one subinterval to the next. The graph of $f(t)$ over a period is then made up of a finite number of pieces, each having continuous curvature, but there may be corners (or, as an admissible alternative, abrupt changes of curvature without change of direction) at the points where two pieces come together. Let the successive points of division marking the subintervals of the period from $-\pi$ to π be $x_1, x_2, \cdots, x_{p-1}$, and for uniformity of notation let $x_0 = -\pi$, $x_p = \pi$. The derivatives may have different values from the right and from the left at these points, but the function $f(t)$ itself has a determinate value at each of them. For each value of i from 0 to $p-1$,

$$\int_{x_i}^{x_{i+1}} f(t) \cos kt\, dt = \left[\frac{1}{k} f(t) \sin kt\right]_{x_i}^{x_{i+1}} - \frac{1}{k}\int_{x_i}^{x_{i+1}} f'(t) \sin kt\, dt.$$

When equations of this form are written for all p subintervals and added, the terms $(1/k)f(x_i) \sin kx_i$ cancel, each occurring once with a plus and once with a minus sign, and

$$\int_{-\pi}^{\pi} f(t) \cos kt\, dt = -\frac{1}{k}\sum_{i=0}^{p-1} \int_{x_i}^{x_{i+1}} f'(t) \sin kt\, dt.$$

Another integration by parts gives

$$\int_{x_i}^{x_{i+1}} f'(t) \sin kt\, dt = \left[-\frac{1}{k} f'(t) \cos kt\right]_{x_i}^{x_{i+1}} + \frac{1}{k}\int_{x_i}^{x_{i+1}} f''(t) \cos kt\, dt.$$

When these expressions are added for the various intervals the terms outside the signs of integration do not cancel, since $f'(x_i)$ at the left-hand end of one interval does not in general mean the same thing as $f'(x_i)$ at the right-hand end of the preceding interval. But under the hypotheses $f'(t)$ and $f''(t)$ remain finite everywhere, in spite of their discontinuities; if M and M_1 are numbers such that $|f''(t)| \leqq M$, $|f'(t)| \leqq M_1$, for all values of t, then

$$|f'(x_{i+1}) \cos kx_{i+1} - f'(x_i) \cos kx_i| \leqq 2M_1$$

in each case, and

$$\left|\int_{x_i}^{x_{i+1}} f''(t) \cos kt\, dt\right| \leqq M(x_{i+1} - x_i).$$

Hence

$$|\pi a_k| = \left|\int_{-\pi}^{\pi} f(t) \cos kt\, dt\right| \leqq \frac{2pM_1}{k^2} + \frac{2\pi M}{k^2}.$$

A similar calculation applies to b_k. With a readily intelligible abbreviation of the hypothesis, the result may be stated as follows:

THEOREM I. *If $f(x)$ is a function which has a continuous second derivative except for a finite number of corners in a period, and if a_k, b_k are the coefficients in its Fourier series, there is a number C, independent of k, such that*

$$|a_k| \leqq \frac{C}{k^2}, \qquad |b_k| \leqq \frac{C}{k^2}.$$

The convergence of the series is an immediate corollary. The special importance of the generalized hypothesis is that it applies in particular to a function whose graph over any period is made up of a finite number of straight line segments of finite slope joined end to end, or, as it may be described for brevity, *a function whose graph is a broken line.*

It must be recognized however that the series has not yet been proved to converge *to the value* $f(x)$. The preceding convergence proof would apply equally well, as far as it goes, to a series of cosines alone, without sine terms; but a cosine series is not adequate for the representation of an arbitrary periodic function. From the point of view of completeness of demonstration the question still remains whether the cosines and sines together are sufficient in all cases, or whether still other terms may sometimes be needed. An answer to this question will be found later, after some further preliminaries.

4. *Approach of the coefficients to zero in general (Riemann's Theorem).* Let $f(x)$ now be any function of period 2π which (with sufficient generality for the purposes of this paper) is continuous except for a finite number of finite jumps in a period. Let a_k, b_k be its Fourier coefficients (2), and let $S_n(x)$ be the partial sum of the series (1) through terms of the nth order:

$$\begin{aligned} S_n(x) = a_0/2 + a_1 \cos x + \cdots + a_n \cos nx \\ + b_1 \sin x + \cdots + b_n \sin nx. \end{aligned} \tag{4}$$

By the use of the relations (3) it is seen that

$$\int_{-\pi}^{\pi} S_n(t) \cos kt \, dt = \pi a_k = \int_{-\pi}^{\pi} f(t) \cos kt \, dt,$$

$$\int_{-\pi}^{\pi} S_n(t) \sin kt \, dt = \pi b_k = \int_{-\pi}^{\pi} f(t) \sin kt \, dt$$

for values of $k \leqq n$. If these equations, read from right to left, are multiplied by $a_0/2, a_1, \cdots, a_n, b_1, \cdots, b_n$ for the successive values of k respectively and added, it is found that

$$\int_{-\pi}^{\pi} f(t)S_n(t)dt = \pi\left[\frac{a_0^2}{2} + \sum_{k=1}^{n} (a_k^2 + b_k^2)\right] = \int_{-\pi}^{\pi} [S_n(t)]^2 dt.$$

Hence

$$\begin{aligned} \int_{-\pi}^{\pi} [f(t) - S_n(t)]^2 dt &= \int_{-\pi}^{\pi} [f(t)]^2 dt - 2\int_{-\pi}^{\pi} f(t)S_n(t)dt + \int_{-\pi}^{\pi} [S_n(t)]^2 dt \\ &= \int_{-\pi}^{\pi} [f(t)]^2 dt - \int_{-\pi}^{\pi} [S_n(t)]^2 dt \\ &= \int_{-\pi}^{\pi} [f(t)]^2 dt - \pi\left[\frac{a_0^2}{2} + \sum_{k=1}^{n} (a_k^2 + b_k^2)\right]. \end{aligned}$$

As the first member can not be negative, it must be that

$$\frac{a_0^2}{2} + \sum_{k=1}^{n} (a_k^2 + b_k^2) \leqq \frac{1}{\pi} \int_{-\pi}^{\pi} [f(t)]^2 dt.$$

The fact that this is true for all values of n, while the last integral does not depend on n, means that the infinite series

$$\frac{a_0^2}{2} + \sum_{k=1}^{\infty} (a_k^2 + b_k^2)$$

is convergent. And since the general term of a convergent series approaches zero it must be that

$$\lim_{k\to\infty} a_k = 0, \quad \lim_{k\to\infty} b_k = 0.$$

It will be convenient to have this result stated for reference with another notation for the arbitrary function, and with the index k replaced by n:

THEOREM II. *If $\phi(t)$ is any function of period 2π which is continuous except for a finite number of finite jumps in a period,*

$$\lim_{n\to\infty} \int_{-\pi}^{\pi} \phi(t) \cos nt\, dt = \lim_{n\to\infty} \int_{-\pi}^{\pi} \phi(t) \sin nt\, dt = 0.$$

A simple consequence of this theorem, of no conspicuous interest in itself, will be required later as a lemma. It is clear that the hypothesis of periodicity is not essential, since the integrals to which the theorem relates do not involve values of the function outside the interval $(-\pi, \pi)$. If any function whatever is given over this interval, a periodic function can be constructed from it by suitable repetition of its values in successive intervals of length 2π. If the original function as defined from $-\pi$ to π approaches different limits at the two ends of this interval the corresponding periodic function will have a finite jump, to be sure, in passing from one interval to the next, but such a discontinuity is admissible under the hypothesis. If $\phi(x)$ is any function continuous from $-\pi$ to π except for a finite number of finite jumps the same will be true of the functions $\phi(x) \cos (x/2)$ and $\phi(x) \sin (x/2)$, and the theorem can be applied to these functions, regardless of the fact that $\cos (x/2)$ and $\sin (x/2)$ do not of themselves have the period 2π when considered for unrestricted values of x. With a change of notation for the independent variable, application of the theorem to the combination

$$\phi(u) \sin (n + \tfrac{1}{2})u \equiv [\phi(u) \sin \tfrac{1}{2}u] \cos nu + [\phi(u) \cos \tfrac{1}{2}u] \sin nu$$

gives the

COROLLARY. *If $\phi(u)$ is any function which is continuous from $-\pi$ to π except for a finite number of finite jumps,*

$$\lim_{n\to\infty}\int_{-\pi}^{\pi}\phi(u)\sin(n+\tfrac{1}{2})u\,du = 0.$$

5. *Integral formula for the partial sum of the series.* Further study of convergence depends on a trigonometric identity. Since

$$\begin{aligned}\sin\tfrac{1}{2}v[\tfrac{1}{2}+\cos v+\cos 2v+\cdots+\cos nv] &= \tfrac{1}{2}\sin\tfrac{1}{2}v+\sum_{k=1}^{n}\sin\tfrac{1}{2}v\cos kv\\ &= \tfrac{1}{2}\sin\tfrac{1}{2}v+\tfrac{1}{2}\sum_{k=1}^{n}[\sin(k+\tfrac{1}{2})v-\sin(k-\tfrac{1}{2})v]\\ &= \tfrac{1}{2}\sin(n+\tfrac{1}{2})v\end{aligned}$$

it appears that

$$\tfrac{1}{2}+\cos v+\cdots+\cos nv = \frac{\sin(n+\tfrac{1}{2})v}{2\sin\tfrac{1}{2}v}. \tag{5}$$

If the values of a_k, b_k given by (2) are substituted explicitly in (4) and the resulting expression written out at length it is found that

$$\begin{aligned}S_n(x) &= \frac{1}{\pi}\int_{-\pi}^{\pi} f(t)\left[\tfrac{1}{2}+\sum_{k=1}^{n}(\cos kt\cos kx+\sin kt\sin kx)\right]dt\\ &= \frac{1}{\pi}\int_{-\pi}^{\pi} f(t)\left[\tfrac{1}{2}+\sum_{k=1}^{n}\cos k(t-x)\right]dt,\end{aligned}$$

which by the identity just obtained reduces to

$$S_n(x) = \frac{1}{\pi}\int_{-\pi}^{\pi} f(t)\,\frac{\sin(n+\tfrac{1}{2})(t-x)}{2\sin\tfrac{1}{2}(t-x)}\,dt. \tag{6}$$

With the substitution $t-x=u$ this becomes

$$S_n(x) = \frac{1}{\pi}\int_{-\pi-x}^{\pi-x} f(x+u)\,\frac{\sin(n+\tfrac{1}{2})u}{2\sin\tfrac{1}{2}u}\,du.$$

The integrand has the period 2π when considered as a function of u. (Addition of 2π to u reverses the signs of numerator and denominator in the fraction, but leaves the fraction as a whole unchanged.) It is a general fact that the integral of a periodic function over any interval whose length is a period is the same as the integral over any other interval of equal length. This is evident from the interpretation of the integral as the area under a curve, and is readily proved analytically with the aid of a suitable change of variable. In the present instance the integral from $-\pi-x$ to $\pi-x$ is the same as that from $-\pi$ to π, so that

$$S_n(x) = \frac{1}{\pi}\int_{-\pi}^{\pi} f(x+u)\frac{\sin(n+\frac{1}{2})u}{2\sin\frac{1}{2}u}\,du. \tag{7}$$

By integration of the identity (5) it is seen that

$$1 = \frac{1}{\pi}\int_{-\pi}^{\pi} \frac{\sin(n+\frac{1}{2})u}{2\sin\frac{1}{2}u}\,du. \tag{8}$$

(This is in fact merely the form taken by the general expression (7) for the special case $f(x)\equiv 1$, since the Fourier series for any constant reduces to the constant itself.) If (8) is multiplied by $f(x)$, for any particular value of x, the factor $f(x)$, being independent of the variable of integration, may be placed under the integral sign:

$$f(x) = \frac{1}{\pi}\int_{-\pi}^{\pi} f(x)\frac{\sin(n+\frac{1}{2})u}{2\sin\frac{1}{2}u}\,du.$$

Hence

$$S_n(x) - f(x) = \frac{1}{\pi}\int_{-\pi}^{\pi} [f(x+u)-f(x)]\frac{\sin(n+\frac{1}{2})u}{2\sin\frac{1}{2}u}\,du. \tag{9}$$

The problem of convergence of the series is thus reduced to the problem of showing that the last integral approaches zero, under suitable hypotheses with regard to the function $f(x)$ under consideration.

6. *Convergence at a point of continuity.* Attention will be restricted for the present to the question of convergence at a point where $f(x)$ is continuous. *Let it be supposed that $f(x)$ is continuous everywhere, or continuous except for a finite number of finite jumps in a period, and that at the particular point where convergence is to be proved it is continuous and has a finite right-hand derivative and a finite left-hand derivative,* which may or may not be equal. This means that its graph is continuous at the point in question, and may be smooth or may have a corner there, with finite slopes from both sides.

Analytically the hypothesis implies that the difference quotient

$$\frac{f(x+u)-f(x)}{u}$$

approaches a definite limit as u approaches zero through positive values, and approaches the same or a different limit as u approaches zero through negative values. *Considered as a function of u, the quotient has at most a finite jump for* $u=0$. The same is true of the function $\phi(u)$ defined by the formula

$$\phi(u) = \frac{f(x+u)-f(x)}{2\sin\frac{1}{2}u} = \frac{f(x+u)-f(x)}{u}\cdot\frac{\frac{1}{2}u}{\sin\frac{1}{2}u},$$

since $(\frac{1}{2}u)/(\sin \frac{1}{2}u)$ has the limit 1. For any other value of u between $-\pi$ and π this $\phi(u)$, considered as a function of u for a fixed value of x, is continuous if $f(x+u)$ is continuous, and has a finite jump if $f(x+u)$ has a finite jump. *The function $\phi(u)$ is continuous from $-\pi$ to π except for a finite number of finite jumps.* For the value of x in question, by (9)

$$S_n(x) - f(x) = \frac{1}{\pi}\int_{-\pi}^{\pi} \phi(u) \sin (n + \tfrac{1}{2})u \, du.$$

Hence, by the Corollary of Theorem II, $S_n(x)-f(x)$ approaches zero as n becomes infinite.[1] This proves

THEOREM III. *If $f(x)$ is continuous except for a finite number of finite jumps in a period, its Fourier series converges to the value $f(x)$ at every point where $f(x)$ is continuous and has a finite right-hand derivative and a finite left-hand derivative.*

7. *Uniform convergence.* Theorem III applies in particular to any function satisfying the hypotheses of Theorem I, and it is now established without question that under the conditions of Theorem I the series converges *to the value $f(x)$* for all values of x. To that extent Theorem I is superseded by Theorem III. The earlier theorem, however, gives important additional information with regard to the manner of convergence, under its own more restrictive hypotheses.

Let $f(x)$ be a function for which the conditions of Theorem I are satisfied By the conclusion of that theorem, together with the fact that the sum of the series is $f(x)$,

$$|f(x) - S_n(x)| = \left| \sum_{k=n+1}^{\infty} (a_k \cos kx + b_k \sin kx) \right| \leqq \sum_{k=n+1}^{\infty} \frac{2C}{k^2}.$$

It is clear that

$$\frac{1}{k^2} = \int_{k-1}^{k} \frac{du}{k^2} < \int_{k-1}^{k} \frac{du}{u^2},$$

since $u<k$ throughout the interior of the interval of integration, and hence

$$\sum_{k=n+1}^{\infty} \frac{1}{k^2} < \sum_{k=n+1}^{\infty} \int_{k-1}^{k} \frac{du}{u^2} = \int_{n}^{\infty} \frac{du}{u^2} = \frac{1}{n}.$$

So

$$|f(x) - S_n(x)| \leqq \frac{2C}{n},$$

for all values of x. The fact that the remainder does not exceed a quantity which is *independent of* x and which approaches zero as n becomes infinite is expressed

[1] For graphs illustrating the convergence in special cases see, e.g., Byerly, *Fourier's Series and Spherical Harmonics*, Boston, 1895, pp. 62–64.

by saying that the series is *uniformly convergent.* The conclusion thus noted may be recorded as

THEOREM IV. *If $f(x)$ is a function which has a continuous second derivative except for a finite number of corners in a period, its Fourier series converges uniformly to the value $f(x)$ for all values of x.*

8. *Weierstrass's Theorem for trigonometric approximation.* The preceding, like Theorem I, holds in particular for a broken-line function. Its application to such a function is not merely of interest in itself, but can be used to prove an important general theorem, known as Weierstrass's theorem, which says that any continuous function of period 2π can be approximately represented with any assigned degree of accuracy by a suitably constructed trigonometric sum. By a trigonometric sum is meant an expression of the form

$$\alpha_0/2 + \alpha_1 \cos x + \alpha_2 \cos 2x + \cdots + \alpha_n \cos nx$$
$$+ \beta_1 \sin x + \beta_2 \sin 2x + \cdots + \beta_n \sin nx,$$

with any constant coefficients α_k, β_k.

Let $f(x)$ be any function which has the period 2π and is continuous for all values of x. It is clear that it can be approximated by a broken-line function as closely as may be desired. In terms of geometric representation this can be accomplished simply by marking points close enough together on the graph of $f(x)$ and joining them in succession by line segments. Let ϵ be any positive number, arbitrarily small, and let $g(x)$ be a broken-line function constructed so that the difference $|f(x)-g(x)|$ is not merely less than ϵ, but less than $\epsilon/2$, for all values of x. Let $T_n(x)$ be the partial sum of the Fourier series *for the function* $g(x)$, through the terms involving $\cos nx$ and $\sin nx$. Since the series converges uniformly to the value $g(x)$, it will be possible to take n so large that $|g(x)-T_n(x)|<\epsilon/2$ for all values of x; if C_0 is the constant given by Theorem I, and entering into the proof of Theorem IV, as applied to $g(x)$, it is sufficient to take $n>4C_0/\epsilon$, so that $2C_0/n<\epsilon/2$. Then

$$|f(x) - T_n(x)| < \epsilon$$

for all values of x, and this is the essence of the conclusion to be proved. It may be stated as

THEOREM V (*Weierstrass's Theorem*). *If $f(x)$ is any continuous function of period 2π, and if ϵ is any positive number, arbitrarily small, it is possible to construct a trigonometric sum $T_n(x)$ so that*

$$|f(x) - T_n(x)| < \epsilon$$

for all values of x.

If the Fourier series for $f(x)$ itself were known to converge uniformly to the right value there would of course be no need of bringing in the auxiliary function $g(x)$; but the Fourier series for a given function is *not* necessarily convergent if the function is merely assumed to be continuous. While continuous functions having divergent Fourier series are of complicated structure, and not likely to be encountered except when cited expressly for purposes of illustration, the fact that such functions exist gives significance to the general theorem of Weierstrass, which holds for all continuous functions without exception.

9. *Completeness of the series.* An important consequence for the theory of Fourier series, which will be stated first and then proved, is

THEOREM VI. *If $f(x)$ is a continuous function of period 2π whose Fourier coefficients are all zero, then $f(x)=0$ identically.*

The hypothesis means that

$$\int_{-\pi}^{\pi} f(x) \cos kx\,dx = \int_{-\pi}^{\pi} f(x) \sin kx\,dx = 0$$

for all integral values of k. It follows that

$$\int_{-\pi}^{\pi} f(x)T_n(x)dx = 0$$

if $T_n(x)$ is any trigonometric sum whatever. Let M be the maximum of $|f(x)|$; the conclusion to be proved means of course that $M=0$, but that is not assumed for the time being. Let ϵ be any positive quantity. Corresponding to the positive quantity $\epsilon/[2\pi(M+1)]$ let a trigonometric sum $T_n(x)$ be constructed according to Weierstrass's theorem so that

$$|f(x) - T_n(x)| \leqq \frac{\epsilon}{2\pi(M+1)}$$

for all values of x. Then

$$\int_{-\pi}^{\pi} [f(x)]^2dx = \int_{-\pi}^{\pi} f(x)[f(x) - T_n(x)]dx$$

$$\leqq \int_{-\pi}^{\pi} M\cdot\frac{\epsilon}{2\pi(M+1)}\,dx = \frac{M\epsilon}{M+1} < \epsilon.$$

Since this is true no matter how small ϵ is taken, it must be that

$$\int_{-\pi}^{\pi} [f(x)]^2dx = 0,$$

from which it follows that $f(x)\equiv 0$, as the theorem asserts.

When the Fourier coefficients for a function are all zero, the corresponding series is of course convergent, and has zero for its sum. The theorem may be regarded as a statement that under these conditions, if the function is continuous, it is identical with the sum of the series. It is a special case of the general proposition, not yet proved in this paper, that any continuous function is equal to the sum of its Fourier series if the series is uniformly convergent. From the point of view of demonstration the special case is not trivial; on the contrary, it contains the essence of the general theorem, which can be deduced from it almost immediately with the aid of certain standard theorems on uniformly convergent series. These theorems, which will not be proved here, are to the effect that a uniformly convergent series of continuous functions represents a continuous function and can be integrated term by term. If they are assumed as known the reasoning proceeds as follows. Let $f(x)$ be the given continuous function, let a_k and b_k be its Fourier coefficients, and let $h(x)$ be the sum of the series, supposed uniformly convergent:

$$h(x) \equiv \frac{a_0}{2} + \sum_{k=1}^{\infty} (a_k \cos kx + b_k \sin kx).$$

By one of the theorems cited $h(x)$ also is continuous. The series will still be uniformly convergent, and so integrable term by term, if multiplied through by $\cos kx$ or $\sin kx$. When the integration is performed it is seen by means of (3) that

$$\int_{-\pi}^{\pi} h(x) \cos kx \, dx = \pi a_k, \qquad \int_{-\pi}^{\pi} h(x) \sin kx \, dx = \pi b_k.$$

Comparison of these formulas with (2), by means of which a_k and b_k are defined, shows that

$$\int_{-\pi}^{\pi} [f(x) - h(x)] \cos kx \, dx = \int_{-\pi}^{\pi} [f(x) - h(x)] \sin kx \, dx = 0$$

for all values of k, and application of Theorem VI to the difference $f(x)-h(x)$ gives $f(x)-h(x)\equiv 0$.

10. *Convergence at a point of discontinuity.* Throughout the discussion of convergence so far it has been assumed that the function is continuous at least at the point where convergence is to be proved, though in Theorem III it may have discontinuities elsewhere. By way of introduction to a treatment of convergence at a point of discontinuity let $F_0(x)$ be the particular function obtained by the following construction: it is defined by the formula $(\pi-x)/2$ for $0<x<2\pi$, is made periodic by repetition of these values in successive intervals of length 2π, and at the points of discontinuity $x=2k\pi$, $k=0, \pm 1, \pm 2, \cdots$, it is expressly given the value zero. Its graph, except for isolated points, is thus made up of an infinite succession of straight line segments, each with slope $-\frac{1}{2}$, the right-hand end of each segment being π units below the left-hand end of the next, while for the values of x corresponding to breaks in the graph the value of the function is not the end value belonging to either of the segments concerned, but is half-way between them. Let a_k, b_k be the Fourier coefficients of this function $F_0(x)$. Since the function satisfies the identity $F_0(-x)\equiv -F_0(x)$, the integral of $F_0(t) \cos kt$ from $-\pi$ to 0 cancels the integral from 0 to π, and $a_k=0$. It

is readily found by explicit calculation that $b_k = 1/k$. So $F_0(x)$ has the Fourier series

$$(10) \qquad \sin x + \tfrac{1}{2}\sin 2x + \tfrac{1}{3}\sin 3x + \cdots .$$

It is known from Theorem III, without further inquiry, that the series must converge to $F_0(x)$ at all points where $F_0(x)$ is continuous.[1] And when x is zero or any integral multiple of 2π the convergence of the series to the prescribed value is immediately apparent, since each term then reduces to zero separately. So $F_0(x)$ is represented by the series for *all* values of x. The value assigned to the function at the points of discontinuity of course has no influence on the determination of the coefficients; the essential observation is that the series (10) does converge at these points, and the value to which it converges is the one designated.

Now let $f(x)$ be any function of period 2π which is continuous except for a finite number of finite jumps in a period, which actually does have a finite jump for $x=0$, and which has a derivative from the right and a derivative from the left there; the last requirement is naturally interpreted to mean that the function has a derivative from the right if defined for $x=0$ by the limit approached from the right, and a left-hand derivative if defined by the limit from the left, or in other words that each segment of the graph has a determinate finite slope where the break occurs. Let the limits approached by $f(x)$ from the right and from the left be denoted by $f(0+)$ and $f(0-)$, and let $D=f(0+)-f(0-)$. Let $f_0(x)=f(x)-(D/\pi)F_0(x)$. This function approaches the limit $[f(0+)+f(0-)]/2$ as x approaches zero from the right, and has the same limit for approach from the left; moreover it has a right-hand and a left-hand derivative for $x=0$, by the corresponding hypothesis on $f(x)$. So its Fourier series converges for $x=0$ to the value $[f(0+)+f(0-)]/2$, by direct application of Theorem III. As the Fourier series for $f(x)$ can be obtained by adding the series for $f_0(x)$ and the series for $(D/\pi)F_0(x)$, and as the latter converges to the value zero, it is seen that the series for $f(x)$ converges to the mean value $[f(0+)+f(0-)]/2$.

Similar reasoning is applicable in the case of a finite jump for any other value of x. The question is merely that of convergence at the point where the discontinuity occurs; convergence at points of continuity is already taken care of by Theorem III. If carried through in detail, the proof would involve something amounting to explicit verification of the fact that the Fourier series for

[1] In particular, setting $x=\pi/2$ gives

$$\frac{\pi}{4} = 1 - \frac{1}{3} + \frac{1}{5} - \frac{1}{7} + \cdots,$$

in agreement with the result obtained by setting $x=1$ in the power series for arc tan x. So the present reasoning incidentally gives a proof of the validity of the power series for $x=1$.

$F_0(x-c)$ as a function of x is

$$\sum_{k=1}^{\infty}\left(\frac{\cos kc}{k}\sin kx - \frac{\sin kc}{k}\cos kx\right) = \sum_{k=1}^{\infty}\frac{\sin k(x-c)}{k},$$

if c is any constant. One way of accomplishing this is as follows: As the Fourier coefficients for $F_0(x)$ are known to be 0 and $1/k$,

$$\frac{\sin kx}{k} = \frac{1}{\pi}\int_{-\pi}^{\pi} F_0(t)\cos k(t-x)dt.$$

Let $F_0(x-c)$ be denoted by $F(x)$, and let A_k, B_k be its Fourier coefficients. Then

$$\begin{aligned} A_k \cos kx + B_k \sin kx &= \frac{1}{\pi}\int_{-\pi}^{\pi} F(u)\cos k(u-x)du \\ &= \frac{1}{\pi}\int_{-\pi}^{\pi} F_0(u-c)\cos k(u-x)du. \end{aligned}$$

By the substitution $u-c=t$ the last expression becomes

$$\frac{1}{\pi}\int_{-\pi-c}^{\pi-c} F_0(t)\cos k[t-(x-c)]dt,$$

and as integration over any period interval gives the same result this reduces to

$$\frac{1}{\pi}\int_{-\pi}^{\pi} F_0(t)\cos k[t-(x-c)]dt = \frac{\sin k(x-c)}{k}.$$

The conclusion with regard to convergence at a point of discontinuity may be summarized in

THEOREM VII. *If $f(x)$ is continuous except for a finite number of finite jumps in a period, if it has a finite jump for $x=x_0$, the limits approached from the right and from the left being $f(x_0+)$ and $f(x_0-)$, and if it has a derivative from the right and a derivative from the left at this point, its Fourier series converges for $x=x_0$ to the value $[f(x_0+)+f(x_0-)]/2$.*

11. *Least-square property.* An important characteristic of the Fourier series is the least-square property, according to which the integral

$$\int_{-\pi}^{\pi} [f(x) - S_n(x)]^2 dx,$$

where $S_n(x)$ is the partial sum of the series, has a smaller value than that which is obtained if $S_n(x)$ is replaced by any other trigonometric sum of the nth order. This property however is not needed for the present discussion of convergence, and it is not necessary to repeat a proof of it here.

12. *Summation by the first arithmetic mean.* A matter which does have an intimate connection with the preceding work is the *summation* of Fourier series by the method of the arithmetic mean. The mean in question is the quantity

$$\sigma_n(x) \equiv \frac{S_0(x) + S_1(x) + \cdots + S_{n-1}(x)}{n}.$$

Although, as has been stated, there exist continuous functions for which the sums $S_n(x)$ do not give a convergent approximation, Fejér[1] proved the striking theorem that $\sigma_n(x)$ *always converges uniformly to the value* $f(x)$ *if* $f(x)$ *is continuous.* A proof[2] will be given here which is in part different in arrangement from that of Fejér.

A preliminary observation is that the arithmetic mean converges uniformly in the case of any broken-line function of the sort previously considered. Let $g(x)$ be any such function, let $T_n(x)$ be the partial sum of its Fourier series, and let

$$\tau_n(x) \equiv (1/n)[T_0(x) + T_1(x) + \cdots + T_{n-1}(x)].$$

Let ϵ be any positive quantity, and in accordance with the uniform convergence of the series let p be a number so large that $|g(x) - T_k(x)| < \epsilon/2$ everywhere for $k = p$ and for all larger values of k. For $n > p$,

$$\begin{aligned} g(x) - \tau_n(x) &= \frac{1}{n}\sum_{k=0}^{n-1}[g(x) - T_k(x)] \\ &= \frac{1}{n}\sum_{k=0}^{p-1}[g(x) - T_k(x)] + \frac{1}{n}\sum_{k=p}^{n-1}[g(x) - T_k(x)]. \end{aligned}$$

Let the sums from 0 to $p-1$ and from p to $n-1$ be denoted respectively by Σ_1 and Σ_2, so that $g(x) - \tau_n(x) = (1/n)(\Sigma_1 + \Sigma_2)$. Then

$$|g(x) - \tau_n(x)| \leqq \frac{1}{n}|\Sigma_1| + \frac{1}{n}|\Sigma_2|.$$

In Σ_2 each difference $g(x) - T_k(x)$ is less than $\epsilon/2$ in absolute value, and as the number of terms in the summation is not greater than n it is certain that $|\Sigma_2| < n\epsilon/2$ and $(1/n)|\Sigma_2| < \epsilon/2$. The sum Σ_1 does not depend on n; if G is the maximum of its absolute value, $(1/n)|\Sigma_1| < \epsilon/2$ as soon as $n > 2G/\epsilon$. When the last condition is satisfied,

$$|g(x) - \tau_n(x)| < \epsilon$$

[1] L. Fejér, *Untersuchungen über Fouriersche Reihen,* Mathematische Annalen, vol. 58 (1904), pp. 51–69.

[2] For the method cf. A. Haar, *Zur Theorie der orthogonalen Funktionensysteme,* Dissertation, Göttingen, 1909, reprinted in Mathematische Annalen, vol. 69 (1910), pp. 331–371.

for all values of x. When any positive ϵ is chosen, no matter how small, the inequality is satisfied for all values of n from a certain point on, and $\tau_n(x)$ thus converges uniformly toward $g(x)$.

It is readily seen (though this is not necessary for present purposes) that the method of proof is of more general applicability, and that if any series whatever is convergent the corresponding means will converge to the same value. The significance of the process of "summation" lies in the fact that the means will sometimes converge when the original series does not.

From the identity (6), written with k in place of n, it is seen that

$$\sigma_n(x) = \frac{1}{n} \sum_{k=0}^{n-1} S_k(x) = \frac{1}{n\pi} \int_{-\pi}^{\pi} f(t) \left[\sum_{k=0}^{n-1} \frac{\sin (k + \frac{1}{2})(t - x)}{2 \sin \frac{1}{2}(t - x)} \right] dt.$$

The relations

$$\begin{aligned} \sin \tfrac{1}{2}v \sum_{k=0}^{n-1} \sin (k + \tfrac{1}{2})v &= \tfrac{1}{2} \sum_{k=0}^{n-1} [\cos kv - \cos (k + 1)v] \\ &= \tfrac{1}{2}(1 - \cos nv) = \sin^2 \tfrac{1}{2}nv \end{aligned}$$

give

$$\frac{1}{\sin \frac{1}{2}v} \sum_{k=0}^{n-1} \sin (k + \tfrac{1}{2})v = \frac{\sin^2 \frac{1}{2}nv}{\sin^2 \frac{1}{2}v}$$

and hence

$$\sigma_n(x) = \frac{1}{n\pi} \int_{-\pi}^{\pi} f(t) \frac{\sin^2 \frac{1}{2}n(t - x)}{2 \sin^2 \frac{1}{2}(t - x)} dt.$$

For the study of convergence an important difference between $\sigma_n(x)$ and $S_n(x)$ is that the trigonometric factor in the last integral is always positive or zero, while the corresponding factor in (6) is of variable sign. This is the underlying reason for the greater simplicity of some of the properties of $\sigma_n(x)$. (It is not implied that the present formulas are in any sense to be regarded as superseding the earlier ones; the original sums $S_n(x)$ continue to be of more fundamental significance and have important advantages of their own, and in particular are likely to converge more rapidly when they do converge.)

In the case of a function $f(x)$ which is identically 1 each $S_k(x)$ reduce to 1, and $\sigma_n(x)$ consequently is identically 1 also, for any value of n, so that

$$1 \equiv \frac{1}{n\pi} \int_{-\pi}^{\pi} \frac{\sin^2 \frac{1}{2}n(t - x)}{2 \sin^2 \frac{1}{2}(t - x)} dt. \tag{11}$$

An inference from (11) is that for any $f(x)$, if M is a number such that $|f(x)| \leqq M$ for all values of x,

$$|\sigma_n(x)| \leqq \frac{1}{n\pi} \int_{-\pi}^{\pi} M \frac{\sin^2 \frac{1}{2}n(t - x)}{2 \sin^2 \frac{1}{2}(t - x)} dt = M. \tag{12}$$

It is possible now to proceed to the proof of the main convergence theorem for $\sigma_n(x)$. Let $f(x)$ be an arbitrary continuous function of period 2π, and $\sigma_n(x)$ the arithmetic mean of the first n partial sums of its Fourier series. Let ϵ be any positive quantity. Let $g(x)$ be a broken-line function constructed so that

$$|f(x) - g(x)| < \epsilon/3$$

for all values of x, and let $r(x) = f(x) - g(x)$. Let $\tau_n(x)$ and $\rho_n(x)$ be the arithmetic means pertaining to $g(x)$ and $r(x)$ respectively. Then

$$\sigma_n(x) = \tau_n(x) + \rho_n(x), \qquad \sigma_n(x) - f(x) = [\tau_n(x) - g(x)] + [\rho_n(x) - r(x)],$$

and

$$|\sigma_n(x) - f(x)| \leqq |\tau_n(x) - g(x)| + |r(x)| + |\rho_n(x)|.$$

The definition of $r(x)$ makes

$$|r(x)| < \epsilon/3.$$

By application of (12) to the function $r(x)$, whose maximum absolute value is less than $\epsilon/3$, it is seen that

$$|\rho_n(x)| < \epsilon/3$$

for all x and all n. As it has been shown that $\tau_n(x)$ converges uniformly toward $g(x)$,

$$|\tau_n(x) - g(x)| < \epsilon/3$$

for all values of n from a certain point on. For such values of n it follows that

$$|\sigma_n(x) - f(x)| < \epsilon,$$

and this relation, satisfied for all values of x, expresses the property of uniform convergence. This completes the proof of

THEOREM VIII. *If $f(x)$ is any continuous function of period 2π, and $\sigma_n(x)$ the arithmetic mean of the partial sums $S_0(x), S_1(x), \cdots, S_{n-1}(x)$ of its Fourier series, $\sigma_n(x)$ converges uniformly toward $f(x)$ for all values of x.*

13. *Weierstrass's Theorem for polynomial approximation.* Inasmuch as $\sigma_n(x)$ is a trigonometric sum, Theorem V is incidentally obtained again as a corollary of Theorem VIII, though with the order of presentation followed here the first proof is simpler.

In conclusion it may be noted that the better known theorem of Weierstrass which relates to polynomial approximation is also immediately obtainable from the work that has been done. The earlier form of proof will be preferred again.

Let $f(x)$ be a function of x which is continuous for $-1 \leqq x \leqq 1$. Let $x = \cos\theta$, and let $f(x) = f(\cos\theta) = \phi(\theta)$. This is a function of period 2π which is defined and continuous for all values of θ, and which furthermore satisfies the relation

$\phi(-\theta) \equiv \phi(\theta)$. Let ϵ be any positive quantity, and let $g(\theta)$ be defined for $0 \leqq \theta \leqq \pi$ as a broken-line function such that $|\phi(\theta) - g(\theta)| < \epsilon/2$ throughout the interval. Then if $g(\theta)$ is defined for $-\pi \leqq \theta \leqq 0$ by the relation $g(-\theta) = g(\theta)$, and for values of θ outside the interval $(-\pi, \pi)$ by the requirement that it shall have the period 2π, it is a function to which Theorem IV is applicable, and $|\phi(\theta) - g(\theta)| < \epsilon/2$ for all values of θ. Let $T_n(\theta)$ be the partial sum of the Fourier series for $g(\theta)$, and let n be taken so large that by virtue of the uniform convergence $|g(\theta) - T_n(\theta)| < \epsilon/2$ everywhere, and hence $|\phi(\theta) - T_n(\theta)| < \epsilon$. Since $g(-\theta) \equiv g(\theta)$ each sine coefficient b_k in $T_n(\theta)$ is zero; the cosine of $k\theta$ is expressible as a polynomial of the kth degree in $\cos \theta$ for each value of k; and hence $T_n(\theta)$ is a polynomial in $\cos \theta$, which may be denoted by $P_n(\cos \theta)$ or $P_n(x)$. As $\phi(\theta) \equiv f(x)$, an approximating polynomial has been constructed for $f(x)$ so that

$$|f(x) - P_n(x)| < \epsilon$$

for $-1 \leqq x \leqq 1$.

This result can be extended to any interval (a, b) by the change of variable $y = (2x - a - b)/(b - a)$. Any continuous function of x for $a \leqq x \leqq b$ is a continuous function of y for $-1 \leqq y \leqq 1$; an approximating polynomial in terms of y can be found for this function; and any polynomial in y is a polynomial of the same degree in x. The general conclusion can be stated as

THEOREM IX (*Weierstrass's Theorem for polynomial approximation*). *If $f(x)$ is any continuous function for $a \leqq x \leqq b$, and if ϵ is any positive number, arbitrarily small, it is possible to construct a polynomial $P_n(x)$ so that*

$$|f(x) - P_n(x)| < \epsilon$$

for $a \leqq x \leqq b$.

The discussion of convergence of Fourier series given in the earlier sections can be carried over in part to the case of Legendre series and to more general developments in series of polynomials. This is done, together with an extension to less elementary parts of the theory, in a paper in the Annals of Mathematics.[1]

[1] D. Jackson, *Series of orthogonal polynomials*, Annals of Mathematics, vol. 34 (1933), pp. 527–545.

APPENDIX TO "SERIES OF ORTHOGONAL POLYNOMIALS," "ORTHOGONAL TRIGONOMETRIC SUMS," AND "THE CONVERGENCE OF FOURIER SERIES"

GUIDO WEISS, Washington University, St. Louis

Jackson's papers deal with the convergence of Fourier series and with some basic properties of series of orthogonal polynomials. The literature involving these subjects is so vast that it would take a serious scholarly effort to include an appendix that would bring these papers "up to date" (and not offend various researchers in this field).

Concerning convergence, one should mention Lusin's conjecture that Fourier series of a continuous function (more generally, an L^2 function) converge almost everywhere. This conjecture, made in 1913 in his paper, "Sur la convergence des séries trigonométriques de Fourier," C. R. Acad. Sci. Paris, 156 (1913) 1655–1658, was a long outstanding problem and was settled by L. Carleson in his paper, "On convergence and growth of partial sums of Fourier series," Acta Math., 116 (1966) 135–157. R. A. Hunt has extended Carleson's result and showed that the Fourier series of any L^p function, $p>1$, converges almost everywhere. Hunt's article appeared in the volume, *Orthogonal Expansions and Their Continuous Analogous*, D. T. Haimo, Editor, Southern Illinois University Press, Carbondale, 1968, 235–255. A recent work that brings some of the subject of orthogonal expansions up to date is a book by A. M. Oleveskii, *Fourier Series with Respect to General Orthogonal Systems*, (translated by B. P. Marshall and H. J. Christoffers), Ergebnisse der Mathematik und ihrer Grenzgebiete, Band 86, Springer-Verlag, Berlin, Heidelberg, New York, 1975. Finally, we think it is appropriate to mention that a standard reference for the development of Fourier series up to the early '60's is Zygmund's work *Trigonometric Series*, Cambridge University Press (the second edition is dated 1959 and the last printing was in 1968).

5

GORDON THOMAS WHYBURN

Gordon Thomas Whyburn was born on January 7, 1904, in Lewisville, Texas. He received a B.A. majoring in Chemistry in 1925 from the University of Texas followed by an M.A. in 1926 and Ph.D. in 1927. He was a Guggenheim Fellow at the University of Vienna, Austria in 1929–30. He received a star in American Men of Science, 1933, an honorary doctorate from Washington and Lee University in 1949, and The Thomas Jefferson award in 1968. He held a Ford Foundation Fellowship at Stanford University, 1952–53. In 1926–27 he was an instructor of mathematics at Texas and an adjunct professor, 1927–29. At Johns Hopkins University he became an associate professor in mathematics for the period 1929–1934 and was professor at Virginia, 1934–57. He was named Alumni professor at Virginia in 1957, where he remained until an untimely death in September 1969.

He was department chairman from 1935–66. He was a member of the Center for Advanced Studies at Virginia, 1966–69, a visiting professor at Stanford University in the summers 1929, 32, 38 and 41, at the University of California at Los Angeles, summer 1940, University of California, summer 1947, University of Colorado, summer 1956. He was also a member of the War Policy Commission for Mathematics, the National Research Council Fellowship Board, the National Academy of Sciences, the American Mathematical Society (President, 1952–54, trustee 1948–66, Colloquium Speaker, 1940), the Mathematical Association of America, managing editor of the *Transactions*, 1942–52, and a Vice-President of the American Association for the Advancement of Science. He was a member of the Conference for the formation of the International Mathematical Union, 1950.

His principal mathematical interest was in topology in which field he wrote the books *Analytic Topology* (1942) and *Topological Analysis* (1958). In addition he published over one hundred and forty research papers. He was a product of R. L. Moore and many of his methods were passed on to his students so that many of today's point set topologists are known as of the "Whyburn-Moore School." His original doctoral thesis on continua in the plane set the tone for later studies of locally connected continua, cut points, and end points of curves. His work was influenced by that of H. Hahn, S. Mazurkiewicz, and W. Sierpinski on local connectedness. His work in dynamic topology was useful in characterizing two-dimensional manifolds. In addition he applied purely topological methods to the study of analyticity of functions of a complex variable, leading to a topological solution of the Cauchy-Goursat problem.

Professor Whyburn was known not only as a leading research mathematician, but also for his teaching skills. He used the method of involvement of his students in open discussion of problems rather than simply lectures. These methods have since been used by students such as J. L. Kelley as standard methods for teaching topology. His influence on his students was strong, producing a great sense of loyalty to the Whyburn methods.

ON THE STRUCTURE OF CONTINUA*

G. T. WHYBURN, University of Virginia

1. Introduction. The notion of a continuum, according to present usage of the term, is both an abstraction and a generalization of the more classical concept of the continuum of real numbers. The essential properties taken from the line or interval are its closedness and its connectedness. A set of points is said to be closed if it contains all of its limiting points; and, following the Lennes-Hausdorff definition, a set is connected provided that however it be divided into two disjoint subsets, one of these must contain a limiting point of the other. Loosely speaking, a set is connected provided it all hangs together in one piece. Thus an interval, a line, or a circle is connected, whereas a set consisting of two distinct intervals, or of a circle and an interval not intersecting it, or of an interval and a point not on it would not be connected. Combining these two properties, then, we understand by a *continuum* any set which is both closed and connected. Thus all properties of the linear continuum which have to do with its linearity or with the order and arrangement of its points have been rejected or, rather, reserved for more specialized topological concepts. Hence not only are sets such as an interval, circle, lemniscate, and sphere included among the continua but also many less orthodox sets such as the curve $y = \sin(1/x)$ together with the interval $(-1, 1)$ of the y axis, or the set consisting of a circle and a spiral approaching it asymptotically, and a host of even more complicated figures.

I shall not attempt to bring in review all of the many valuable results proved in recent years which may be relevant to the title I have selected for my address. Instead I shall limit my remarks to certain lines of investigation which have been most closely related to my own work and which lead to some of the most interesting and important unsolved problems in the field of topology.

Inasmuch as the principal interest of the results to be discussed lies in their concrete visual character rather than in the great generality or abstractness of the spaces in which they hold, for simplicity I shall suppose once for all that the sets I consider all lie in a compact metric space, or even in a bounded portion of a euclidean space, although in nearly all cases this is an unnecessary restriction. A continuum, then, will be a closed, connected, and compact set of points.

2. Cut points. My paper is devoted largely to theorems grouped around the notion of a cut point of a continuum and the key which this notion gives to the structure of a continuum through its decomposition into cyclic elements. The point p of a continuum M is called a *cut point* of M provided the set $M - p$ is not connected. Thus $M - p$ falls into at least two disjoint sets neither of which contains a limiting point of the other. For example, any interior point of an arc is a cut point, the double point of a lemniscate is a cut point but is its only cut point, all points on the continuum constructed above using the curve $y = \sin(1/x)$ except the

* An address delivered at the meeting of the AMS on September 13, 1935, in Ann Arbor, by invitation of the Committee on Program.

end point and points on the limiting set are cut points, and no point of a simple closed curve is a cut point.

This type of point was first studied by R. L. Moore [1]* and S. Mazurkiewicz [1],* each of whom proved theorems from which it follows that no simple closed curve can contain more than a countable number of cut points of any locally connected continuum† containing it. Moore's result is considerably more general and states that if K is any subcontinuum of an arbitrary continuum M, then all save at most a countable number of the cut points of M which belong to K are cut points also of K. This theorem in turn has been shown to hold for arbitrary connected sets M and K by Zarankiewicz [1]. Moore also obtained a characterization in terms of cut points of the class of continua called *dendrites*, that is, the locally connected continua containing no simple closed curves. As a result of a theorem proved later by R. L. Wilder [1], this may be stated as follows: *In order for a continuum D to be a dendrite it is necessary and sufficient that D consist wholly of cut points and end points.*‡

Now an excellent account of the results concerning cut points which were known up to that time was given in 1927 by J. R. Kline in an address before this Society which has subsequently been published in the Bulletin (see Kline [1]). Hence I shall content myself with mentioning only two more recent results which seem to be of fundamental and central significance. The first of these might well be called the *Cut Point-Order Theorem.* It states that *all save a countable number of the cut points of any continuum whatever must be points of Menger-Urysohn order*§ 2 *of that continuum* (see G. T. Whyburn [1]). Thus in particular it follows that the branch points of a dendrite must be countable; and, indeed, the points of any continuum which cut it into more than two pieces must be countable. Also, since any continuum is locally connected at every point of finite order, it follows that any continuum is locally connected at all except possibly a countable number of its cut points.

The second result has to do with the Borel classification of the set G of all cut points of a continuum M. Zarankiewicz [1] has shown that if M is locally connected, the set G of all of its cut points is an F_σ-set, that is, a set which is the sum of a countable number of closed sets. Later it was shown in one of my own papers (see Whyburn [2]) that the set of all cut points of an arbitrary continuum is a Borel set of the class $G_{\delta\sigma}$, that is, a set which is the sum of a countable number of G_δ-sets, where a G_δ-set in turn is one which is the product of a countable number of

* Numbers in brackets refer to the bibliography at the end of the paper.

† A continuum M is locally connected provided each point of M is contained in arbitrarily small neighborhoods whose common part with M is connected.

‡ By an end point of a continuum M we will understand a point $p \in M$ which is contained in arbitrarily small neighborhoods whose boundaries intersect M in just one point, that is, a point of M of Menger-Urysohn order 1. That this definition is equivalent to earlier definitions used by various authors was proved by H. M. Gehman [1].

§ The point p of a continuum M is a point of Menger-Urysohn order n of M provided n is the least integer such that p is contained in arbitrarily small neighborhoods whose boundaries intersect M in at most n points. See Menger [1] and Urysohn [1].

open sets. From these results it follows that if the set G of all cut points of a continuum M is uncountable, it necessarily contains a perfect set; and thus in any case the set G is either vacuous, finite, countable, or of the power of the continuum.

3. Cyclic elements. The principal feature of cut points with which I shall deal is the insight they give into the structure of a locally connected continuum through the decomposition of such a continuum into cyclic elements.

Let S designate any definite locally connected continuum. Then the *cyclic elements* of S are (a) the cut points of S, (b) the end points of S (that is, the points of order 1) and (c) the subsets C of S which are connected and have no cut point and are maximal in S relative to this property (see Whyburn [3] and [4]). The cyclic elements of type (c) are called *true cyclic elements* or *non-degenerate cyclic elements*, since the elements of the other two types reduce to single points. There are other equivalent ways of defining the cyclic elements of S (see Moore [2], and Kuratowski [1] and Whyburn), but I have adopted this definition here since in this form it admits of extension to what we may call n-dimensional or nth-order cyclic elements of an arbitrary continuum.

For example, in the continuum consisting of two disjoint circles joined by an arc, the two circles are true cyclic elements and every other remaining point is itself a degenerate cyclic element. A sphere has only one cyclic element, namely, the sphere itself. A lemniscate has two true cyclic elements and one cut point. In the continuum consisting of an infinite sequence of tangent spheres converging to a single point, the spheres are true cyclic elements, the points of tangency are cut points (and thus are cyclic elements) and the point to which the spheres converge is an end point (also a cyclic element). The cyclic elements of a dendrite are all individual points, so that it has no true cyclic elements.

Now let us consider briefly some of the general properties of the cyclic elements of a locally connected continuum S. In the first place, S is the sum of its cyclic elements, that is, every point of S is either a cut point, an end point, or a point of at least one true cyclic element. Furthermore, the true cyclic elements of S are countable, in fact there are at most a finite number of them of diameter greater than any preassigned positive quantity, and no two of them intersect in more than one point. If two of them do have a common point, this point must be a cut point of S, so that the intersection of two cyclic elements is either vacuous or is itself a cyclic element. Also each true cyclic element C is itself a locally connected continuum and moreover C is cyclicly connected (see Whyburn [4] and Ayres [1]) in the sense that any two points of C lie together on a simple closed curve in C. If Z is an arbitrary connected subset of S, the intersection $Z \cdot C$ of Z with an arbitrary cyclic element is either vacuous or connected. Thus it follows in particular that any arc in S whose end points belong to C must itself lie wholly in C. From this property it results that there can be no closed ring of cyclic elements.

Hence, with respect to its cyclic elements, the continuum S has a structure analogous to that of a dendrite. There are many other properties of cyclic elements and of S relative to its cyclic elements which strengthen this analogy. For example, a dendrite is characterized among the locally connected continua by the property of containing one and only one simple arc joining any two of its points. Correspondingly, for any locally connected continuum S, there is one and only one simple chain (see Whyburn [3]) of cyclic elements joining any two given elements. Also, any connected subset of a dendrite is arcwise connected; and correspondingly, any connected set of cyclic elements of S is arcwise connected and also cyclic chainwise connected.

It is notable that sets of this type, that is, connected sets of cyclic elements, as well as closed and connected sets of cyclic elements, present many interesting phenomena. For example, Ayres has shown (see Ayres [2]) that they may be identified with the sets which he has called arc-curves. If K is any subset of S, then by the *arc-curve* $M(K)$ is meant the set of all points x of S which lie on arcs axb in S joining some two points $a, b \in K$. Ayres shows that any arc-curve is a connected set of cyclic elements and conversely any connected set H of cyclic elements is an arc-curve, in fact we have $M(H) \equiv H$.

4. Cyclicly extensible and reducible properties. It may well be said that it is to this same "dendritic" structure of an arbitrary locally connected continuum relative to its cyclic elements that we owe the principal applications of the cyclic element notion. The fruitfulness of this notion in the study of the structure of locally connected continua is due in large measure to the fact that so many questions concerning locally connected continua of various types can be reduced to the same questions concerning the individual cyclic elements of those continua. In other words, there are a large number of properties P which are *cyclicly extensible* in the sense that if each cyclic element has property P, then the whole continuum S has property P. Similarly a property P is *cyclicly reducible* provided that if S has property P, so also does each cyclic element of S. Thus P is cyclicly extensible if

$$(P \text{ in each } C) \rightarrow (P \text{ in } S);$$

P is cyclicly reducible if

$$(P \text{ in } S) \rightarrow (P \text{ in each } C).$$

Let us consider some properties of this sort. First, for a continuum M, let P be the property of having all of its connected subsets arcwise connected. Clearly this is a property of simple sets such as arcs, simple closed curves, θ-curves, that is, curves like the letter θ, and in fact of linear graphs in general. Furthermore, this property is cyclicly extensible and reducible. Consequently any locally connected continuum S every true cyclic element of which is, say, a simple closed curve, has this property. Now suppose we had this question: Given a locally connected continuum M in the plane, let S be the boundary of any complementary domain of M; then is it true that every connected subset of S is arcwise connected? This question was answered in the affirmative in 1923 by R. L. Wilder (see Wilder [1]) by other methods. However, we can answer it readily in the following way. First it is known that S is locally connected and that every true cyclic element of S is a simple closed curve. Then since each cyclic element of S has the property in question and this property is cyclicly extensible, it results that S itself has this property. The same reasoning would apply, of course, to the case where every true cyclic element is a θ-curve or in fact is a linear graph of any sort. This illustrates the usefulness of the cyclic element decomposition. For we are here able to reduce the problem to the same problem for the cyclic elements by virtue of the cyclic extensibility of the property involved, and then to solve it for the cyclic elements since their structure is markedly simpler than is that of the whole continuum.

Cyclicly extensible properties, then, seem to be of fundamental importance. In a report on cyclic elements published in 1930 by C. Kuratowski and myself there were listed some fourteen non-trivial properties of this sort, and since that time at least as many more significant ones have been discovered. I shall discuss briefly here only three or four which seem particularly interesting.

(a) *Separation of the plane and of n-space.* In the first place let us consider the property P of failing to separate the plane, and let us suppose our continuum S lies in a plane. This property is cyclicly extensible and reducible as was shown in a paper of mine published in 1928 (see Whyburn [3]). Thus if no cyclic element of S separates the plane, neither does S; and conversely if S does not separate the plane, neither does any cyclic element of S. Now any locally connected plane continuum C which has no cut point and fails to separate its plane must be a closed 2-cell, that is, it is homeomorphic with a circle plus its interior. Thus we have the result, that *in order that a locally connected plane continuum S should fail to separate its plane it is necessary and sufficient that every true cyclic element of S be a closed* 2-*cell* (see Whyburn [3]). This type of continuum S has been called a *base set* by C. B. Morrey [1] who has used it extensively in his recent work on the theory of surfaces.

Now it is well known that the property of not separating the plane may be stated as an intrinsic property of S by saying that the one-dimensional connectivity or Betti number of S is 0. This raises the question as to whether the property of having a vanishing first Betti number is cyclicly extensible in an arbitrary locally connected continuum S, whether S lies in a plane or not. This is indeed true, as was shown by Borsuk (see Borsuk [1]). I shall consider this result in more detail later since it also appears as a special case of a much more general formula for Betti numbers of all dimensions. Borsuk also showed (Borsuk [2]) that the property of failing to separate n-space is cyclicly extensible when S is imbedded in the euclidean n-dimensional space, for any n.

(b) *Fixed point property.* Consider next the property of a continuum A to have a fixed point under every continuous transformation of A into a subset of itself. A point x is said to be a fixed point of a transformation $T(A)=B$ provided $T(x)=x$; and if for every single-valued [though not necessarily (1-1)] and continuous transformation $T(A)=B$, where $B \subset A$, there is at least one fixed point, then A has our property. Now a set such as an interval, a 2-cell, or a T-shaped curve will have this property; and it was shown by Brouwer that an n-cell has it. On the other hand a circle would not have it, since a simple rotation of a circle through an angle of π leaves no point fixed.

Now it was shown by Sherrer [1] that every dendrite has the fixed point property. This result in a way suggests that the fixed point property might be cyclicly extensible; and it was shown by W. L. Ayres [3] that if we restrict ourselves to topological transformations in defining the fixed point property (that is, to transformations which are both (1-1) and continuous), then this property is extensible. In fact Ayres proves in general that if $T(S)=R$ is any topological transformation of a locally connected continuum S into a subset R of itself, then there is at least one cyclic element C of S which maps into a subset of itself under S, that is, we have $T(C)\subset C$. From this it results of course that if each such element C has a fixed point under every homeomorphism of C into a subset of itself, then so also does the whole continuum S. It was shown independently by Borsuk [2] that even without restricting it to topological transformations, the fixed point property is cyclicly extensible. Thus not only is it a property of dendrites as shown by Sherrer but also of a continuum such as two tangent circles with their interiors or a set consisting of a circle together with its interior plus a perpendicular through its center. In fact, since by the Brouwer fixed point theorem it follows that every n-cell has this property, it results that any locally connected continuum every true cyclic element of which is a k-cell, where k may vary from element to element, has the fixed point property. Thus, in particular, any base-set has this property.

(c) *Curve types.* As a final example, let us consider some curve types occurring in the curve theory and dimension theory as developed extensively by Urysohn [1] and Menger [1, 2]. A continuum M is said to be a *regular curve* provided each point of M is contained in arbitrarily small neighborhoods whose boundaries intersect M in only a finite set of points, and M is a *rational curve* if these neighborhoods can be chosen so that they intersect M in only a countable set of points. For example, a circle plus two perpendicular diameters is regular and hence also rational; a curve consisting of an interval AB together with a sequence of intervals $AB_1, AB_2, \ldots$, converging to AB is rational but not regular. Similarly, M is said to be of dimension $\leqq n$ provided each point of M has arbitrarily small neighborhoods whose boundaries intersect M in a set of dimension $\leqq n-1$, where the null set has dimension -1. Now the property of being a regular curve (Whyburn [5]), of being rational (Whyburn [6]), or of being of dimension $\leqq n$ for $n>1$ (Kuratowski [2]) are all cyclicly extensible and reducible. Thus if each true cyclic element of S is a regular curve, so also is S; hence, in particular, if the boundary of a plane domain is locally connected, it is a regular curve, since each of its true cyclic elements is a simple closed curve. Likewise the Menger-Urysohn dimensionality of S may be found by taking the maximum of the dimensionalities of the true cyclic elements.

Another important curve type is the so called *hereditarily locally connected continuum*, that is, a continuum M having the property that every subcontinuum of M is locally connected. The property of being hereditarily locally connected also is cyclicly extensible and reducible (Whyburn [3]). In fact, as Zippin [1] has noted, we may say even more, namely, that if some subcontinuum N of S fails to be locally connected, then there exists a true cyclic element C of S such that the continuum $C \cdot N$ fails to be locally connected.

5. Cyclic elements of higher orders. So much for the cyclic element decomposition of a locally connected continuum. The usefulness of this decomposition leads us naturally to ask (see Wilder [2]) whether it is not possible in the first place to extend it to arbitrary continua, locally connected or not, and in the second place to carry it further and obtain a finer decomposition of a continuum in some analogous fashion into elements with respect to which the structure of the continuum is simple, though not so simple as the dendrite, and which may yield an even deeper light into the structure of the continuum. This is indeed the case; for as we shall see, it is possible to obtain, for each integer $r \geqq 0$, a decomposition of this sort of any continuum into elements which we will call E_r which enjoy properties analogous to those of the cyclic elements and reduce, in case $r=0$ and in case the continuum is locally connected, to the cyclic elements themselves.

The key to these finer decompositions is furnished by the notion of a *complete cycle* as introduced originally by Vietoris [1] and subsequently extensively developed by Vietoris, Alexandroff, Čech, Borsuk, Lefschetz, R. L. Wilder, Pontrajagn, Kuratowski, and others. For convenience and to distinguish them from the ordinary geometric combinatorial cycles Γ^r, we shall follow the usual practice of calling these cycles "Vietoris cycles" and we shall designate them by γ^r, where r denotes the dimensionality of the cycle. For our purposes we will understand by a cycle of dimension 0, either geometric Γ^0 or Vietoris γ^0, any *even number* of points (0-cells). A Vietoris cycle is said to be *essential* if it has at least one carrier C in which it is not ~ 0. A closed set C "carries" a γ^r provided all vertices of all cycles in γ^r belong to C.

We shall call a closed set of points which carries no essential r-dimensional Vietoris cycle a *T_r-set* or simply a T_r. Thus a single point is a T_0, an arc or dendrite is a T_1, a 2-cell is a T_2, and so on; whereas a point pair is not a T_0, a circle is not a T_1, but is a T_r for $r > 1$, a sphere or torus is not a T_2, and so on.

Let M denote any compact continuum. A non-degenerate subset X of M will be called an *E_r-set* in M or merely an E_r, provided X is not separated by any T_r in X and X is maximal in M relative to this property. In other words, if no T_r in X cuts X and if, when Y is any larger set containing X, some T_r cuts Y, then X is an E_r. The sets E_r may be called the rth order true cyclic elements of M for each $r \geqq 0$. It is to be noted first, that in case $r=0$, the sets E_r are the maximal sets in M which are not separated by any T_0, that is, by any single point. Thus the sets E_0 are the maximal sets in M having no cut point, so that if M is locally connected they are identically the ordinary true cyclic elements of M. For $r=1$ consider the following example. Let W be the set consisting of a torus together with a disk just fitting into it. Let C be a cube or sphere attached to W along a simple arc, let Q be a 2-cell attached to C along an arc and intersecting W along another arc having nothing in common with C, and let $M = W + C + Q +$any finite collection of arcs joined on in an arbitrary manner. Then the sets E_1 in M are W and C, whereas the sets W and C lie in the same E_0 (cyclic element), of M. For any set cutting C or W must contain a ring-shaped figure and thus carries a 1-cycle, whereas any larger continuum in M containing C or W is cut by a T_1.

Now the existence of the sets E_r for any continuum M is an easy consequence of a general theorem of mine [7] on the existence of maximal sets which I shall not take time to discuss here. Suffice it to say that any irreducible carrier K of an essential γ^r in M is contained in some E_{r-1} and the decomposition into sets E_r is always possible. Furthermore, the sets E_r are continua and the intersection of any two of them is always a T_r-set. For example, two E_0's have at most one point in common, two E_1's in the above example have only an arc (which is a T_1) in common. It is no longer true that the E_r's are countable, even for $r=0$. For we are not supposing M locally connected, so that M might consist of a non-enumerable family of concentric circles connected up by a radius; and in this case each of the circles is an E_0 of M. However, corresponding to the property of cyclic elements that the product of each cyclic element C of S by any continuum N in S is itself a continuum, we have an analogous property of the E_r's for any r. In the language of homologies this property states that any γ^0 in $C \cdot N$ which is ~ 0 in N is ~ 0 in $C \cdot N$. Correspondingly (see Whyburn [8]), if E_r is any E_r-set in any continuum M and N is any closed subset of M, then any γ^r in $E_r \cdot N$ which is ~ 0 in N is ~ 0 in $E_r \cdot N$. Thus in the above example it will be noted that a 1-cycle in C or W which is ~ 0 in M is ~ 0 in C or W, respectively. Also it is interesting to note that for increasing values of r the decompositions into sets E_r are monotone decreasing, that is, each E_r is $\subset$ some $E_{r-1} \subset$ some $E_{r-2} \cdots \subset$ some E_0, so that we really do have finer decompositions as r increases. We have seen already that any irreducible carrier of a γ^{r+1} is $\subset$ some E_r. Thus there exists no closed $(r+1)$-dimensional ring of elements E_r just as there existed no closed 1-ring of cyclic elements. Hence, relative to the elements E_r, the structure of M is "like a T_{r+1}-set" just as, relative to its cyclic elements, S was "dendritic" or "like a T_1."

6. Applications. We have seen that for the cyclic elements, that is, the E_0's, we have cyclicly extensible and reducible properties. Similarly for the E_r's we have a group of properties which are *E_r-extensible* and *E_r-reducible*, that is, properties P such that

$$(P \text{ in each } E_r) \rightarrow (P \text{ in } M),$$

and conversely.

For example, we saw that if $S \subset R^2$ the property of not separating R^2 is cyclicly extensible and reducible, that is, E_0-extensible and reducible. Similarly, for any $r \geqq 0$, if $M \subset R^{r+2}$, then the property of not separating R^{r+2} is E_r-extensible and reducible. Also the property of being locally γ^s-connected for any $s > r$ is E_r-extensible and reducible. A compact set N is locally γ^s-connected provided that if $\varepsilon > 0$, a $\delta_\varepsilon > 0$ exists such that any Vietoris cycle γ^s in N of diameter $< \delta_\varepsilon$ in ~ 0 is a subset of N of diameter $< \varepsilon$. To say that this property is E_r-extensible and reducible means of course that if each E_r is locally γ^s-connected, so also is M, and conversely.

Finally, the property of having a vanishing s-dimensional Betti number, for any $s>r$, is E_r-extensible and reducible. This result generalizes the theorem of Borsuk mentioned earlier to the effect that the property of having a zero first Betti number is cyclicly extensible (E_0-extensible) in locally connected continua. However, this result in turn is a consequence of a much more inclusive formula by means of which it is possible to express the s-dimensional Betti number of any continuum M in terms of the corresponding numbers of the sets E_r, where $r<s$. In fact, if we denote the s-dimensional Betti number of a set X by $p^s(X)$, we have simply

$$p^s(M)=\sum p^s(E_r), \quad \text{for any } r<s,$$

the summation being extended over all sets E_r in M (see Whyburn [8]). Thus, in particular, we have $p^s(M)=\sum p^s(E_{s-1})$. To see how this formula works, let us take the simple case where $s=1$, and where M is a lemniscate plus a cross bar on one loop. The sets $E_0=E_{s-1}$ in M are the loop L and the θ-curve. The formula gives us

$$p^1(M)=p^1(L)+p^1(\theta)=1+2=3.$$

Thus we can obtain the kth Betti numbers of any continuum M simply by adding together the kth Betti numbers of the sets E_{k-1}. Now clearly the E_r-extensibility and reducibility of the property of having a zero Betti number of dimension s is obtained merely by setting all the numbers on one side or the other of this equation equal to 0.

7. A problem. In connection with the higher order cyclic elements there remains a very fundamental problem which is as yet unsolved, namely, whether or not in a locally γ^r-connected continuum M every element E_r carries an essential Vietoris cycle γ^{r+1} of dimension $r+1$. In case $r=0$ and in case M is locally γ^0-connected (that is, locally connected in the ordinary sense), the elements E_r reduce to the ordinary cyclic elements of M and we have seen that each cyclic element is cyclicly connected. Hence each E_0 certainly contains a simple closed curve and hence carries an essential γ^1. In the light of known results, this problem may be stated as follows: If a continuum M is locally γ^r-connected and is not separated by any T_r, does M necessarily carry an essential γ^{r+1}? This is no longer necessarily true even in case $r=0$ if we leave off the condition that M shall be locally connected. For it is well known (see Knaster [1]) that there exists an indecomposable continuum C in the plane having a vanishing first Betti number; and since no indecomposable continuum can have a cut point, C would have no cut point and hence would have only one set E_0, namely, C itself; and clearly C carries no essential γ^1.

8. Separating points and local separating points. Returning to the notion of a cut point, let us consider briefly what happens when we localize this concept. A natural way to do this is simply to say that a point p of a continuum M is a local cut point or a local separating point of M, provided p is a cut point of any sufficiently small neighborhood of p in M. However, this definition leads to difficulties for it would require that we be able to find arbitrarily small neighborhoods of p such that the part of M in these neighborhoods is connected. This, of course, requires that M be locally connected, and for such sets this definition is perfectly good. For continua in general, however, we need first to extend the notion of a cut point or a separating point of a continuum to an arbitrary set K, which we do as follows.

If K is any set, connected or not, a point p of K is called a *separating point* of K, provided that p separates K between some two points in the component* C of K containing p, that is, we have a separation $K-p=K_1+K_2$, where $K_1 \cdot C \neq 0 \neq K_2 \cdot C$ (compare with Menger [4]). In case K is connected, of course, $C=K$ and the separating points of K are merely the cut points of K. It is always true that a separating point p of K is a cut point of the component C of K containing p. However, it is not generally true that every cut point of a component C of K is a separating point of K. For in the example K consisting of a sequence of intervals converging to a limiting interval, every inner point of the limiting interval is a cut point of that interval but no such point is a separating point of K.

Now the concept of a separating point localizes directly and without difficulty. A point p of a continuum M (or of any set M) is a *local separating point* of M provided that p is a separating point of some open subset of M. From this definition it results easily that any local separating point p of M is a separating point of the part of M in any sufficiently small neighborhood of p. For example, any point of a circle or a lemniscate is a local separating point. Any linear graph consists entirely of local separating points plus a finite number of end points.

Since obviously any cut point of a continuum is also a local separating point but not conversely, it follows that the chance of existence of local separating points is much greater than for cut points. It can be proved easily, for example, that in any regular curve, rational curve, or hereditarily locally connected continuum, the local separating points must be everywhere dense (Whyburn [9]), whereas, obviously, there exist curves of these types which have no cut point. There exist very simple curves, for example, a circle, having no cut point, whereas, any locally connected continuum having no local separating point must contain, for any two of its points a and b, a set of arcs $[axb]$ of the power of the continuum each pair of which intersect in just $a+b$ (see Whyburn [11] and Zippin [2]).

The more fundamental theorems on cut points extend with little or no modification to local separating points. Thus the cut point-order theorem extends to give the following theorem.

LOCAL SEPARATING POINT-ORDER THEOREM. *All save possibly a countable number of the local separating points of any continuum M are points of order 2 of M* (Whyburn [11]).

---

* By the component of K containing p is meant the maximal connected subset of K containing p.

Also, the local separating points of any locally connected set form a Borel set of the class F_σ and the set of all such points of any continuum is a set of class $G_{\delta\sigma}$ just as was the case with cut points (Whyburn [2]).

So far as has yet been discovered, the local separating points of a continuum M do not yield a useful decomposition of M which is strictly analogous to the cyclic element decomposition of a locally connected continuum. In case they exist in sufficient numbers, however (that is, if they are uncountable), the local separating points of M do yield quite useful decompositions of M of a slightly different sort. To obtain such a decomposition, let G denote the set of all local separating points of M and, for each $p \in M$, let $C(p)$ denote the maximal subcontinuum of M containing p and such that $G \cdot C(p)$ is at most countable. Then the sets $C(p)$ always exist and no two of them which are different can intersect at all. Thus we obtain a decomposition of M into disjoint continua $[C(p)]$. This decomposition is upper semi-continuous in the sense of R. L. Moore [3, 4] and its hyperspace H, that is, the space whose points are the sets $[C(p)]$, is a regular curve of simple structure. In fact, every subcontinuum of H must contain uncountably many local separating points of H and H is a continuum of finite degree in the sense of Kamiya [1] (see Whyburn [12], [13]).

Still other decompositions by means of local separating points are possible under suitable restrictions on M. For example, we may decompose M into maximal subcontinua $D(p)$ of M which have only a countable number of local separating points. However, the possibilities in this direction have not been extensively investigated and herein lies a most interesting and what promises to be a very fruitful unexplored realm of topology, namely, to study intensely the possible decompositions of continua by means of local separating points. It would be most desirable to develop a group of properties which would be extensible from the sets $C(p)$ to the whole continuum M in the same sense that the cyclicly extensible properties extend from the cyclic elements to the whole continuum.

9. Applications. Among the applications of the local separating point notion to problems in continuum structure, I shall mention three which seem particularly far-reaching.

(i) First we mention an application to the problem of arcwise connectivity of all connected subsets of a continuum. It was shown independently by Moore [4, 5] and by Menger [3] that any connected and locally connected G_δ-set is arcwise connected. This result lends particular significance to the problem of finding necessary and sufficient conditions in order that every connected subset of a continuum M should be a G_δ-set. The solution to this problem is easily given in terms of local separating points. It is embodied simply in the condition that all except a countable number of the points of M shall be local separating points of M (see Whyburn [14]). In fact we may state a more general theorem as follows.

THEOREM. *If every connected subset of a continuum M is a Borel set (of any class whatever), then the non-local separating points of M are countable. Conversely, if the non-local separating points of M are countable, then every connected subset of M is locally connected and is both a* G_δ *and an* F_σ.

In conjunction with the Moore-Menger result and the cyclic extensibility of the property of having all connected subsets arcwise connected, this gives the theorem that *if the non-local separating points of each true cyclic element of a locally connected continuum S are countable, then every connected subset of S is arcwise connected.* Another interesting result of the above theorem is that *if each connected subset of M is a Borel set of some class, then every such connected set must be both an* F_σ *and a* G_δ.

(ii) Secondly, the local separating points of a continuum M have a deciding position relative to the existence in M of totally imperfect connected subsets, that is, connected subsets in M which contain no perfect subsets. It has been shown by F. Bernstein [1] and Sierpinski [1] that such sets exist in any euclidean space of dimension greater than 1 and by Knaster [2] and Kuratowski that they even exist in the Sierpinski regular curve. This naturally raises the question: Under what conditions will such sets exist in a continuum M? Again the answer is readily provided in terms of local separating points. *For, in order that M contain a totally imperfect connected subset it is necessary and sufficient that the local separating points of some subcontinuum be countable* (Whyburn [15]). In particular it follows from this that the hyperspace H of the decomposition of M above into sets $C(p)$ can contain no totally imperfect connected subset.

(iii) As a final application I will call attention to the natural order basis which the local separating points provide in any regular curve. If K is a regular curve in the Menger-Urysohn sense, a subset B of K is called an *order basis* for K provided each point p of K is contained in arbitrarily small neighborhoods whose boundaries intersect K in only a finite number of points all of which lie in B and the number of which does not exceed the order of p in K. Now we may assert that *if Q is any set of local separating points of K which includes all of the at most countable number of local separating points of K of order* >2 *and which is dense in the set of all local separating points of K, then Q is an order basis for K.* Clearly, by virtue of the local separating point-order theorem, such a set Q can always be chosen so that it is countable; and hence *every regular curve has a countable order basis consisting entirely of local separating points.*

10. A general problem on continuous transformations. I shall devote the time that remains to a consideration of the following general problem concerning the preservation of the structure of a continuum when the continuum undergoes a continuous transformation. Let A and B be continua, and let $T(A)=B$ be a single-valued continuous transformation of A into B. The problem is to find conditions on the transformation T and its inverse which will insure that B will be topologically equivalent to A, that is, homeomorphic with A. In other words, we are asking what sort of continuous transformation will preserve or leave invariant all topological properties of a given continuum A. I shall consider only those conditions on T which concern the inverse sets $T^{-1}(b)$ of points on B. Thus we are seeking conditions on the sets $T^{-1}(b)$ which will insure that B be topologically equivalent to A.

Now the problem has the obvious and trivial solution embodied in the condition that the inverse T^{-1} should also be single-valued, that is, that T be (1-1). For, in the case of compact sets A and B, this merely makes T itself a topological transformation. However, in a number of important cases this condition is known to be stronger than necessary. For example, in case A is a simple arc, it is enough, to insure that B also will be an arc, to assume merely that the inverse set $T^{-1}(b)$ for each point $b \in B$ should be connected.

A satisfactory general solution to this problem as I have stated it seems as yet to be considerably beyond our reach. As we shall see later, no such solution may be expected so long as we limit ourselves to conditions on the sets $T^{-1}(b)$ alone; but when we allow conditions both on these sets and on their complements in A, the outlook is considerably more hopeful. The principal result up to date in this direction has been obtained by James F. Wardwell. In his dissertation he obtained a condition which is sufficient to make B topologically equivalent to A provided the number of and the condensation of those sets $[T^{-1}(b)]$ which are non-degenerate also are suitably restricted.

Aside from Wardwell's work, practically all other progress which has been made on this problem is confined to particular types of sets A; and concerning these types a rich collection of theorems has been proved. I have already mentioned the case where A is a simple arc, and the condition in this case is simply that the inverse sets $\left[T^{-1}(b)\right]$ should all be continua. Now a continuous transformation satisfying this condition, that is, such that $T^{-1}(b)$ is connected for every $b \in B$, has been called (see Morrey [1]) a *monotone transformation*. The term seems appropriate because of the analogy with the case of a real function $y=f(x)$; for if $f(x)$ is monotone in the usual sense, then for each value y_1 of y the set of values x_1 of x such that $f(x_1)=y_1$ is always connected. It is interesting to note that a continuous transformation $T(A)=B$, where A and B are compact, will be monotone if and only if the property of connectedness is invariant under T^{-1}.

In case A is an arc, then the solution to our problem is embodied in the condition that the transformation T be monotone. The same is true in case A is a circle or any simple closed curve. Thus we have the following result.

THEOREM. *If A is a simple arc {simple closed curve} and $T(A)=B$ is monotone, then B is a simple arc {simple closed curve} or a single point.*

11. Monotone transformations on the sphere. Cactoids. Now when we take A to be a sphere (that is, the surface of a sphere), it appears at once that the condition that T be a monotone is no longer sufficient to make B homeomorphic with A. For we can transform a sphere A into a set B consisting of two tangent spheres by sending the equator on A into the point of tangency and mapping each of the two hemispheres onto the two spheres minus one point. Similarly, by sending two circles on A into points, we can map it onto three tangent spheres, or we can map it into a diameter by a simple projection which clearly is a monotone transformation. Thus, if we are to have B topologically equivalent to A, extra conditions must be added. Now it will be noted that in each of the cases here illustrated where B is not a topological sphere, some of the sets $T^{-1}(b)$ separate A. In the first case, the inverse of the point of tangency of the two spheres is the equator of the given sphere and hence cuts it into two parts. In the last case, all sets $T^{-1}(b)$ except two cut A. Thus we are led to the following theorem due to R. L. Moore [3]; it solves our problem in the case of the sphere.

THEOREM. *If A is a sphere, if T is monotone and no set $T^{-1}(b)$ separates A, and if B contains more than one point, then B is homeomorphic with A.*

Moore states this theorem in terms of upper semi-continuous decompositions of a sphere (or plane) into continua rather than in terms of a continuous transformation as I have done. It was shown later by Alexandroff [1] and by Kuratowski [3] that any upper semi-continuous decomposition of a compact space is equivalent to a continuous transformation defined on that space; and hence we can use the language of continuous transformations to describe this and other results some of which were originally stated in terms of other notions.

The case of the sphere (or plane) is of such interest and usefulness in connection with the study of surfaces both from a topological standpoint and an analytical standpoint that it may be worthwhile to consider briefly some results which have been found in connection with this theorem. In the first place we have already seen that assuming T monotone was not sufficient, in the case of the sphere, to make B topologically equivalent to A. However, suppose we investigate the possible images of A when we do just assume T monotone. We have seen that we may get images such as a string of tangent spheres, an arc, two spheres joined by an arc. Now from the point of view of cyclic elements, these various possibilities are very similar. In fact in each case every true cyclic element of B is a topological sphere. This is not accidental; indeed it is a characteristic property for images of a sphere under monotone transformations. Thus we have the following theorem which is due to R. L. Moore [2].

THEOREM. *If A is a sphere and T monotone, then every true cyclic element of B is a topological sphere.*

The very picturesque name *cactoid* has been used to describe such a set B, that is, a cactoid is a locally connected continuum every true cyclic element of which is a topological sphere. Now the cactoids form a well defined mathematical class which is equivalent to the class of all monotone transformations definable on the sphere. That is, not only is the image of any sphere under any monotone transformation always a cactoid, but also any cactoid is always the image under some monotone transformation of the sphere (see Moore [2]). Also this class is closed under the operation of taking images under monotone transformations, by the following theorem. (Whyburn [16].)

THEOREM. *The image of any cactoid under any monotone transformation is itself a cactoid.*

12. Non-alternating transformations on the circle. Boundary curves. The relation of the cactoids to the sphere suggests that there should be a class of one-dimensional curves which is analogously related to the circle. This is indeed true as we shall soon see. Since a cactoid is a locally connected continuum every true cyclic element of which is a topological sphere, its one-dimensional analog, therefore, is a locally connected continuum every true cyclic element of which is a topological circle. We have called such a curve a *boundary curve* due to the fact (see Ayres [4]) that it is also characterized by the property that it is always homeomorphic with the boundary of a plane domain. We now ask: *What kind of a transformation will produce a boundary curve from a circle*? Of course, a monotone transformation will do so, since it always produces a topological circle from a circle; but given a boundary curve, in general it cannot of course be obtained from a circle by a monotone transformation.

The answer is given by the condition on the sets $T^{-1}(b)$ that they not separate each other on the circle A. This means that for no two points b_1 and b_2 of B will $T^{-1}(b_1)$ separate any two points of $T^{-1}(b_2)$ on A. A transformation having this property is called a *non-alternating* transformation (Whyburn [16]), since in the case of the circle A it simply means that as we move around the circle A we never can meet alternately points of two distinct sets $T^{-1}(b)$. Thus we have the following theorem.

THEOREM. *The image of every circle under any non-alternating transformation is a boundary curve.*

The converse of this is also true, namely, *any boundary curve B is the image under some non-alternating transformation of a circle*. Thus, given in particular any locally connected continuum B bounding a plane domain, we can map the circle A onto B by a non-alternating transformation. It is interesting to note that this can always at the same time be done in a certain minimal way from the standpoint of multiplicity. That is, we can map the circle A onto any boundary curve B by a non-alternating transformation T in such a way that for each $b \in B$, the number of points in $T^{-1}(b)$ is exactly the same as the number of components into which p cuts B, provided either of these numbers is finite. Finally, just as the class of cactoids is closed under monotone transformations, so also is the class of boundary curves closed under non-alternating transformations, which means simply that the image of any boundary curve under any non-alternating transformation is itself a boundary curve.

13. The 3-space. Returning to our problem of finding conditions on the sets $T^{-1}(b)$ which will make B homeomorphic with A, we have seen that the solution has been found for the case of the circle and of the sphere. It is clear that the results above stated could easily be modified so as to yield the solution when A is the ordinary line or plane. Let us next consider the case of the 3-dimensional space. We may expect of course that extra conditions on the sets $T^{-1}(b)$ may be necessary. Just how much more may be necessary, however, no one is able at present to say, since the problem for this case is still unsolved. The difficulties met here may be illustrated by some simple considerations. Suppose we let A be the 3-space and let us take the simple case where one single arc xy in A goes into a point b of B but where every other point of B comes from a single point of A, that is, $T(xy)=b$ but T is (1-1) on $A-xy$. Even in this case we cannot say that B is homeomorphic with A. If xy is a linear interval this will indeed be true. But Antoine [1] has shown that there are arcs xy in 3-space A which are knotted in the sense that $A-xy$ is not homeomorphic with A minus a linear interval. Thus in the example just given, if xy is such an arc, then since $B-b$ is homeomorphic with $A-xy$, we see that B cannot be a topological 3-space; because if it were, $B-b$ would be homeomorphic with B minus a linear interval.

14. A trial condition. The example just given shows clearly that if we limit ourselves to topological conditions on the sets $T^{-1}(b)$ alone, no satisfactory solution to our problem is possible for the case where A is a euclidean space of dimension three or greater. The condition imposed above in the case of the sphere, that the sets not separate A, may of course be stated as an intrinsic property of the sets by using the notion of higher connectivity.

However, when the conditions on both the sets $T^{-1}(b)$ and their complements are allowed, the outlook is more hopeful. Suppose for instance we impose the condition that for each $b \in B$ the complement of every set $T^{-1}(b)$ in A be homeomorphic with the complement of every single point in A. This condition will rule out at once the possibility encountered in the example and described in §13, that is, if any set $T^{-1}(b)$ is an arc in A, where A is the 3-space, then this arc cannot be knotted in the sense described above. However, in the cases of the circle and the sphere it is to be noted that it reduces to exactly the conditions appropriate to these respective cases. For, if A is a circle, then to say that the complement of a closed subset K of A is homeomorphic with the complement of a point is exactly the same as saying that K is connected. Also, if A is a sphere, then the condition that a closed set on A be connected and not separate A is equivalent to the condition that its complement be homeomorphic with the complement of a point. Whether this condition actually will be sufficient to make B homeomorphic with A in general or in the case of the 3-space, is not yet known. It is being investigated at present and some results have been obtained using it; so that one can at least hope that it or some of its many possible modifications may lead to a solution to the problem, if not in general, at least in a number of the more interesting particular cases such as the ones I have indicated.

Bibliography

ALEXANDROFF, P.

1. Über stetige Abbildungen kompakter Räume, Mathematische Annalen, vol. 96 (1926), pp. 555–571.

ANTOINE, L.

1. Sur les voisinages de deux figures homéomorphes, Fundamenta Mathematicae, vol. 5 (1924), pp. 265–287.

AYRES, W. L.

1. Concerning continuous curves in metric space, American Journal of Mathematics, vol. 51 (1929), pp. 577–594.

2. Concerning the arc-curves and basic sets of a continuous curve, Trans. AMS, vol. 30 (1928), pp. 567–578, and vol. 31 (1929), pp. 595–612.

3. Some generalizations of the Sherrer fixed-point theorem, Fundamenta Mathematicae, vol. 16 (1930), pp. 332–336.

4. Continuous curves homeomorphic with the boundary of a plane domain, Fundamenta Mathematicae, vol. 14 (1929), pp. 92–95.

BERNSTEIN, F.

1. Leipziger Berichte, vol. 60 (1908), p. 329.

BORSUK, K.

1. Über eine Klasse von lokal zusammenhängenden Räumen, Fundamenta Mathematicae, vol. 19 (1932), p. 230.

2. Einige Sätze über stetige Streckenbilder, Fundamenta Mathematicae, vol. 18 (1932), pp. 198–213.

GEHMAN, H. M.

1. Concerning end points of continuous curves and other continua, Trans. AMS, vol. 30 (1928), pp. 63–84.

KAMIYA, H.

1. Tôhoku Mathematical Journal, vol. 36 (1933), pp. 58–72.

KLINE, J. R.

1. Separation theorems and their relation to recent developments in analysis situs, Bull. AMS, vol. 34 (1928), pp. 155–192.

KNASTER, B.

1. Un continu dont tout sous-continu est indécomposable, Fundamenta Mathematicae, vol. 3 (1922), pp. 247–286.

2. (With C. Kuratowski) A connected and connected im kleinen point set which contains no perfect subset, Bull. AMS, vol. 33 (1927), pp. 106–109.

KURATOWSKI, C.

1. (With G. T. Whyburn) Sur les éléments cycliques et leurs applications, Fundamenta Mathematicae, vol. 16 (1930), pp. 305–331.

2. Quelques applications d'éléments cycliques de M. Whyburn, Fundamenta Mathematicae, vol. 14 (1929), pp. 138–144.

3. Sur les décompositions semi-continues d'espaces métriques compacts, Fundamenta Mathematicae, vol. 11 (1928), pp. 169–185.

MAZURKIEWICZ, S.

1. Un théorème sur les lignes de Jordan, Fundamenta Mathematicae, vol. 2 (1921), pp. 119–130.

MENGER, K.

1. Grundzüge einer Theorie der Kurven, Mathematische Annalen, vol. 95 (1925), pp. 272–306.

2. Kurventheorie, B. G. Teubner, 1932.

3. Zur Begründung einer axiomatischen Theorie der Dimension, Monatshefte für Mathematik und Physik, vol. 36 (1926), pp. 193–218.

4. Remarks concerning the paper of W. L. Ayres on the regular points of a continuum, Trans. AMS, vol. 33 (1931), pp. 663–667.

MOORE, R. L.

1. Concerning cut points of continuous curves and of other closed and connected point sets, Proceedings of the National Academy of Sciences, vol. 9 (1923), pp. 101–106.

2. Concerning upper semi-continuous collections, Monatshefte für Mathematik und Physik, vol. 36 (1929), pp. 81–88.

3. Concerning upper semi-continuous collections of continua, Trans. AMS, vol. 27 (1925), pp. 416–428.

4. Foundations of Point Set Theory, American Mathematical Society Colloquium Publications, 1932.

5. Abstract sets and the foundations of analysis situs (abstract), Bull. AMS, vol. 33 (1927), p. 141.

MORREY, C. B.

1. The topology of (path) surfaces, American Journal of Mathematics, vol. 57 (1935), pp. 17–50.

SHERRER, W.

1. Über ungeschlossene stetige Kurven, Mathematische Zeitschrift, vol. 24 (1925), pp. 125–130.

SIERPINSKI, W.

1. Sur un ensemble punctiforme connexe, Fundamenta Mathematicae, vol. 1 (1920), pp. 7–10.

URYSOHN, P.

1. Mémoire sur les multiplicités cantoriennes. Deuxième partie Verhandelingen der koninklijke Akademie van Wetenshappen te Amsterdam, vol. 13 (1928), No. 4.

VIETORIS, L.

1. Über den höheren Zusammenhang kompakter Räume und eine Klasse von zusammenhangstreuen Abbildungen, Mathematische Annalen, vol. 97 (1927), pp. 454–572.

WHYBURN, G. T.

1. Concerning cut points of continua, Trans. AMS, vol. 30 (1928), pp. 597–609.

2. Cut points of connected sets and of continua, ibid., vol. 32 (1930), pp. 147–154.

3. Concerning the structure of a continuous curve, American Journal of Mathematics, vol. 50 (1928), pp. 167–194.

4. Cyclicly connected continuous curves, Proceedings of the National Academy of Sciences, vol. 13 (1927), pp. 31–38.

5. Concerning Menger regular curves, Fundamenta Mathematicae, vol. 12 (1928), pp. 264–294.

6. The rationality of certain continuous curves, Bull. AMS, vol. 36 (1930), pp. 522–524.

7. Concerning maximal sets, Bull. AMS, vol. 40 (1934), pp. 159–164.

8. Cyclic elements of higher orders, American Journal of Mathematics, vol. 56 (1934), pp. 133–146.

9. On regular points of continua and regular curves of at most order n, Bull. AMS, vol. 35 (1929), pp. 218–224.

10. Continuous curves without local separating points, American Journal of Mathematics, vol. 53 (1931), pp. 163–166.

11. Local separating points of continua, Monatshefte für Mathematik und Physik, vol. 36 (1929), pp. 305–314.

12. Decompositions of continua by means of local separating points, American Journal of Mathematics, vol. 55 (1933), pp. 437–457.

13. Concerning continua of finite degree and local separating points, ibid, vol. 57 (1935), pp. 11–16.

14. Sets of local separating points of a continuum, Bull. AMS, vol. 39 (1933), pp. 97–100.

15. On the existence of totally imperfect and punctiform connected subsets in a given continuum, American Journal of Mathematics, vol. 55 (1933), pp. 146–152.

16. Non-alternating transformations, ibid, vol. 56 (1934), pp. 294–302.

WILDER, R. L.

1. Concerning continuous curves, Fundamenta Mathematicae, vol. 7 (1925), pp. 340–377.

2. Point sets in three and higher dimensions and their investigation by means of unified analysis situs, Bull. AMS, vol. 38 (1932), pp. 649–692.

ZARANKIEWICZ, C.

1. Sur les points de division dans les ensembles connexes, Fundamenta Mathematicae, vol. 9 (1927), pp. 124–171.

ZIPPIN, L.

1. On continuous curves and the Jordan curve theorem, American Journal of Mathematics, vol. 52 (1930), pp. 331–350.

2. Note on locally cyclicly connected continua (abstract), Bull. AMS, vol. 36 (1930), p. 805.

APPENDIX TO "ON THE STRUCTURE OF CONTINUA"

F. B. JONES, University of California, Riverside

A *continuum* shall be defined as above, i.e., a connected, compact subset of a metric space. No attempt will be made to document the generalizations to Hausdorff spaces. A continuum M is *semi-locally connected* (s-l-c) if for each point p of M there exists an arbitrarily small open neighborhood of p whose complement (in M) has a finite number of components **[22]**. A complementary (or dual) notion states that a continuum M is *aposyndetic* at the point p of M if for each point q of $M-\{p\}$ some closed connected neighborhood of p misses q **[8]**.

2. Cut points. If a continuum M is aposyndetic at no point of M, then M must contain at least one point c such that $M-\{c\}$ is not continuum-wise connected (and c is called a (*weak*) *cut point*) **[8]**. If p and q are distinct points of a non-separating subcontinuum of the plane and no point (weakly) cuts p from q in M, then M contains a simple closed curve from p to q **[10]**. Consequently every non-separating subcontinuum of the plane which contains no (weak) cut point is cyclicly connected.

3. Cyclic elements. Whyburn generalized his cyclic element theory to s-l-c continua **[22]**. For a history and extensive bibliography see **[11]**. By introducing the notion of contiguous points, Moore gets the ultimate decomposition theorem: the space whose points are the cyclic elements of a locally connected continuum *is* a dendrite, not just "analogous to a dendrite" **[15]**. If one decomposes a lemniscate into its cyclic elements, one obtains an arc consisting of three (or two, depending upon the definition of "contiguous") points. Also of interest is another approach to "cyclic elements" by McAllister and McAuley **[12]**.

4, 5, 6. For many new results concerning these topics the reader is referred to Simon **[17]**, Gary **[7]** and Wilder **[23]**.

7. This question appears not to have been answered.

8. Whyburn has generalized the cut point order theorems by using the notion: potential order (a generalization of Menger order) **[22]**.

10, 11. Monotone transformations. By generalizing the sets $C(p)$ [or Moore's $M(P)$] McAuley has shown how to construct a (usually non-trivial) monotone transformation T of a continuum M such that $T(M)$ is aposyndetic ($\equiv$s-l-c) **[13]**. In fact, this can be done so that the decomposition of M is atomic **[14]**. Swingle and FitzGerald have gotten a similar (possibly the same) decomposition by an alternate procedure **[20]**. Thomas has exhaustively studied analogous decompositions of irreducible continua **[19]**. Vought **[21]**, Stratton **[18]**, and many others have made contributions to this form of decomposition theory.

12, 13. 3-space. Moore's theorem that the monotone image of S^2 is topologically S^2 (if for each p in $T(S^2)$, $T^{-1}(p)$ is point like) does not generalize to S^3. Bing's "dog bone" space demonstrates this fact [**4**], i.e., for each point p of $T(S^3)$, $T^{-1}(p)$ is an arc (or a point) such that $S^3 - T^{-1}(p)$ is homeomorphic with the complement of a single point, yet $T(S^3)$ is *not* homeomorphic with S^3. This problem of Whyburn's as much as anything has inspired the tremendous output of the Bing School of 3-space (and geometric) topology [**1**, **2**, **3**, **5**, **6**].

References

1. Steve Armentrout, Monotone decompositions of E^3, Annals of Mathematical Studies, Princeton University Press, 60 (1966) 1–26.

2. ———, A decomposition of E^3 into straight lines and singletons, Dissertationes Math., 73 (1970) 1–49.

3. ———, Cellular decompositions of 3-manifolds that yield 3-manifolds, Memoirs of Amer. Math. Soc., 107 (1971).

4. R. H. Bing, A decomposition of E^3 into points and tame arcs such that the decomposition space is topologically different from E^3, Ann. of Math., 65 (1957) 484–500.

5. C. E. Burgess and J. W. Cannon, Embedding of Surfaces in E^3, Rocky Mountain J. Math., 1 (1971) 259–344.

6. C. E. Burgess, Embedding of surfaces in Euclidean three-space, Bull. Amer. Math. Soc., 81 (1975) 795–818.

7. John Gary, Higher dimensional cyclic elements, Pacific J. Math., 9 (1959) 1061–1070.

8. F. B. Jones, Concerning aposyndetic and non-aposyndetic continua, Bull. Amer. Math. Soc., 58 (1952) 137–151.

9. ———, Aposyndetic continua, Coll. Math. Soc. Janos Bolyai, Topics in topology, Keszthely (Hungary), 8 (1972) 437–447.

10. ———, The cyclic connectivity of plane continua, Pacific J. Math., 11 (1961) 1013–1016.

11. B. L. McAllister, Cyclic elements in topology—a history, Amer. Math. Monthly, 73 (1966) 337–350.

12. ———, and L. F. McAuley, A new "cyclic" element theory, Math. Zeitschr., 101 (1967) 152–164.

13. L. F. McAuley, On decomposition of continua into aposyndetic continua, Trans. Amer. Math. Soc., 81 (1956) 74–91.

14. ———, On atomic decomposition of continua into aposyndetic continua, *ibid.*, 88 (1958) 1–11.

15. R. L. Moore, Foundations of a point set theory of spaces in which some points are contiguous to others, The Rice Institute Pamphlet, 23 (1936) 1–41.

16. ———, Foundations of point set theory, Amer. Math. Soc. Coll. Pub., 13 (revised edition), Providence, R.I., 1962.

17. A. B. Simon, n-cyclic elements I, Duke Math. J., 24 (1957) 1–7.

18. H. H. Stratton, On continua which resemble simple closed curves, Fund. Math., 68 (1970) 121–128.

19. E. S. Thomas, Jr., Monotone decomposition of irreducible continua, Rasprawy Math., 50 (1966) 1–74.

20. P. M. Swingle and R. W. FitzGerald, Core decompositions of continua, Fund. Math., 61 (1967) 33–50.

21. E. J. Vought, Monotone decompositions of continua into arcs and simple closed curves, Fund. Math.

22. G. T. Whyburn, Analytic Topology, Amer. Math. Soc. Coll. Publ., 28, Providence, R.I., 1942.

23. R. L. Wilder, Topology of Manifolds, Amer. Math. Soc. Coll. Publ., 32, Providence, R.I., 1949.

6

SAUNDERS MAC LANE

Saunders Mac Lane was born in Norwich, Connecticut, August 4, 1909. He received a Ph.B. degree from Yale in 1930, an M.A. from Chicago in 1931, and a Ph.D. from Göttingen in 1934. He was a Sterling Fellow at Yale in 1933–34, a Peirce Instructor at Harvard in 1934–36, an instructor at Cornell, 1936–37, and at Chicago 1937–38. He was an assistant professor at Harvard, 1938–41, associate professor, 1941–46, and professor, 1946–47 after which he went to Chicago as professor where he has remained, having become Max Mason Distinguished Professor in 1965. He received an honorary D.Sc., Purdue, 1965, Yale, 1969 and LL.D., Glasgow, 1971, and Honorary D.Sc., Coe College, 1974. He was a Guggenheim Fellow at the Swiss Federal Institute of Technology and at Columbia 1947–48, a member of the executive committee of the International Mathematical Union, 1954–58, a Visiting Professor, Heidelberg, 1958 and again, 1976. He was a member of the council of the National Academy of Sciences, 1959–62 and 1967–72, Chairman of the Editorial Board of the *Proceedings*, 1966, Visiting Professor, Frankfurt, 1966, Tulane, 1969, Fulbright Fellow, Canberra, Australia, 1969. He was a civilian with the Office of Research and Development, 1943–45. He is a member of the National Academy of Sciences, the American Mathematical Society (Vice-President, 1946–48, President, 1973), American Philosophical Society (Vice-President, 1948–49, President, 1951–52), the American Academy of Arts and Sciences, Vice-President, National Academy of Sciences, 1973–1977, and a member of the National Science Board, 1974–1980.

His early mathematical interests were in abstract algebra, especially groups, rings, and field theory. Later at Chicago he turned to topology and with S. Eilenberg was responsible for the development of the new field of category theory which grew out of algebraic topology. His recent interests have been in the theory of topos. In addition he has shown strong interest in logic and foundations. His wide range of mathematical interests is illustrated by his paper "Hamiltonian Mechanics and Geometry" at the Chauvenet Symposium.

Professor Mac Lane, besides being a top flight research mathematician, has always been an enthusiastic teacher, devoted to the cause of better education at all levels. His early collaboration with G. Birkhoff in 1939–41 in writing the *Survey of Modern Algebra* touched off the expansion of abstract algebra into colleges during the last thirty-five years. R. P. Boas, Jr., writes that Mac Lane "has not only been influential in the cause of better education in mathematics, he has set a good personal example. I know of no more enthusiastic teacher or skillful expositor," [1]. His entire career is the embodiment of the principles on which Coolidge established the Chauvenet Prize.

Reference

1. R. P. Boas, Jr., Professor Saunders Mac Lane, The Monthly, 82 (1975) 107–8.

SOME RECENT ADVANCES IN ALGEBRA

SAUNDERS MAC LANE, University of Chicago

The rapid exploitation of the new techniques of abstract algebra and the introduction of many new types of algebraic systems, though perhaps superficially confusing, has actually centered about several well defined lines of investigation: function-fields, linear algebras, p-adic fields, Lie algebras, matrices. The present direction and the close interrelation of these fields of investigation was clearly indicated in a conference on algebra recently held at the University of Chicago. We here attempt to summarize some of the ideas and inter-connections brought out at this conference, not in the fashion of a handbook or monograph, but rather as a survey for the generally interested mathematical public. After the necessary background has been filled in, we state the sorts of problems considered and the types of answers obtained. For detailed statements of results we refer to the skeleton bibliography at the end. In the last paragraph (§12) we attempt to state a general definition of algebra and a summary of its fundamental problems. To the authors whose ideas are here assembled we apologize for the omission, inevitably attendant upon such a summary, of the many difficulties and subtleties of their work.

1. Algebraic geometry, power series, and valuations. The relation of algebra to algebraic geometry was a lively topic of debate, stimulated by a paper of Lefschetz on the use of formal power series in algebraic geometry **[21]**.* The origin of this algebraic-geometric connection might be described thus: the geometry of an algebraic curve can be reduced to the algebra of a certain corresponding field;† specifically, if k denotes the field of all complex numbers, and if a curve is defined in the plane by an irreducible polynomial equation $f(x, y)=0$, then the corresponding field K is the totality $k(x, y)$ of all rational functions $z=g(x, y)/h(x, y)$ with complex coefficients of the two quantities x and y. It should be emphasized that x and y do not figure in this field as variables taking on values in the sense of analysis, but merely as quantities on which rational operations of addition, multiplication, *etc.*, can be performed, subject always to the proviso that the rational combination $f(x, y)$ is to be 0. The use of this "function field" $k(x, y)$ corresponding to the curve has one considerable advantage: a birational transformation of the curve will leave the corresponding field invariant.

Near a point $x=a$, $y=b$ of the curve $f(x, y)=0$, the variable y can be expanded in one or more series of the form

$$(1) \qquad y = b + b_1 t + b_2 t^2 + \cdots, \qquad t = (x - a)^{1/n},$$

when t is a suitably chosen integral or fractional power of $x-a$. This series (1),

* Numbers in brackets refer to references at the end of the paper.

† For the definitions of a few fundamental algebraic terms (group, field, and the like) one may refer to the glossary in Albert **[1]**.

the Puiseux expansion, is usually treated as a convergent series defining an element or "cycle" of the algebraic function y of x, but Lefschetz emphasizes the purely formal character of this expansion. The central fact is that the Puiseux series (1), if substituted in $f(x, y)$, yields an identity $f(x, y) \equiv 0$ in t. Conversely, any series (1) which formally satisfies the equation $f=0$ is the Puiseux series corresponding to some branch of the algebraic curve, or, alternatively, determines a corresponding point P on the Riemann surface of the algebraic function $y(x)$. Any rational function $z=R(x, y)$ in the field $K=k(x, y)$ becomes a power series in t after substitution of the series for y and that for x, $x=a+t^n$;

$$z = t^\nu(a_0 + a_1 t + a_2 t^2 + \cdots), \tag{2}$$

where the integer ν may be positive, negative, or zero, and the a_i are in the coefficient field k of all complex numbers. The set of all such possible power series forms a field $k\{t\}$, because such series can be added, divided, and multiplied by the usual formal procedures. The point P determined by (1) on the Riemann surface can be said to yield a one-to-one map (2) of the functions z of the field K onto a subset of the power series field $k\{t\}$, and this map is an *isomorphism* because any rational relation which holds between several functions of the field must hold between their corresponding power series.

This formal treatment by power series will apply now to algebraic curves even when the field k of constants is not the classical complex number field, but any field k which is algebraically closed, in the sense that every polynomial equation over k has a root in k; that is, for any field k in which the fundamental theorem of algebra holds. Lefschetz pointed out that on this basis most of the classical treatment of algebraic functions of one and two variables, as presented, for instance, by Picard or Weierstrass, can be developed as pure algebra. The abelian integrals and their classification can be managed algebraically in terms of abelian differentials $g(x, y)dx$. Moreover, the genus of an algebraic curve, an important invariant often defined topologically as the number of holes in the Riemann surface (a pretzel), can be defined algebraically, in terms of these integrals, as the maximum number of linearly independent differentials of the "first kind." One or two theorems of an essentially topological character cannot be generalized, but Lefschetz conjectured that the Riemann-Roch theorem could be treated algebraically not only by the usual arithmetic proof [31], but even by one of the classical geometric proofs [21]. This formal series treatment is of utility in other parts of analysis, notably in the treatment of algebraic functions of Dirichlet series [26].

The importance of the application of algebra to geometry was emphasized by Zariski's new proof that the singularities of an algebraic surface can be eliminated by suitable birational transformations. Previously the only sound proof of this important theorem had been one due to Walker [32], using analytic functions. If the algebraic surface S is given in three-space by a single homogeneous algebraic equation $f(x_0, x_1, x_2, x_3)=0$, then the singular points are those points of the surface at which certain partial derivatives simultaneously vanish.

The theorem requires that the surface S be represented in n-space with coördinates $y_0, y_1, \cdots, y_n$ by a non-singular surface S' which is a birational transform of the original surface S in the sense that the y's are rational functions of the x's and conversely. One difficulty in eliminating singular points arises because a transformation carefully constructed to eliminate one singular point may explode some other singular point into a whole singular curve. Zariski treats this difficulty by repeatedly using an "integral closure" process which gets rid of singular curves. In terms of the non-homogeneous coördinates $\xi_1 = x_1/x_0$, $\xi_2 = x_2/x_0$, $\xi_3 = x_3/x_0$, a rational function $\eta = \eta(\xi_1, \xi_2, \xi_3)$ is called *integral* if it satisfies a polynomial equation,

$$\eta^m + a_1(\xi_1, \xi_2, \xi_3)\eta^{m-1} + \cdots + a_m(\xi_1, \xi_2, \xi_3) = 0,$$

with first coefficient 1 and the other coefficients polynomials in the ξ_i. The non-homogeneous integral closure process is a birational transformation replacing the coördinates ξ_1, ξ_2, ξ_3 by integral functions $\eta_1, \eta_2, \cdots, \eta_n$ chosen so that every integral function of the ξ_i is a polynomial in these new coördinates η_j. After this process has removed singular curves, the nature of one of the remaining isolated singular points P depends on the sorts of curves (or "branches") passing through P on the surface. Such a branch can be represented algebraically by a certain corresponding "valuation." The reduction of the singularity is effected by applying to suitable branches at P a "uniformization lemma," which expresses the coördinates as holomorphic functions (integral power series) of two parameters u and v—where these parameters can be chosen as rational functions of the original coördinates (for such valuations, *cf.* Zariski [36]).

The study of valuations occurs not only in algebraic geometry but also in algebraic number theory and other arithmetical questions. If the power series expansion (2) of an algebraic function z at a point P begins with a non-vanishing term a_0t^ν, then the order or *value* $V(z)$ at P may be defined to be the exponent ν of that term.

$$V(z) = \nu \quad \text{if} \quad z = a_0t^\nu + a_1t^{\nu+1} + \cdots, \qquad (a_0 \neq 0). \tag{3}$$

When z has a zero at $t=0$, $V(z)$ is the order of this zero; when z has a pole at $t=0$, $V(z)$ is the negative of the order of the pole. This valuation for a sum or product of two functions z and w can be shown to have the properties

$$V(zw) = V(z) + V(w), \qquad V(z + w) \geqq \min\,(V(z), V(w)). \tag{4}$$

Any real-valued function $V(z)$ defined in a field K and having these two properties is known as a *valuation* of that field. Any such V can also be converted into an "absolute value" $\|z\| = e^{-V(z)}$, with corresponding properties

$$\|zw\| = \|z\| \cdot \|w\|, \qquad \|z + w\| \leqq \max\,(\|z\|, \|w\|).$$

The second of these properties is even stronger than the usual triangle axiom $|z+w| \leqq |z| + |w|$ for complex numbers. Thus $\|z\|$ behaves like the absolute value of z, and limits with respect to it can be defined in the usual way. Especially important are the fields K' which are *complete* with respect to an absolute value, in the sense that every sequence $a_1, a_2, \cdots$ which is a Cauchy sequence has a limit a in the field K'. For instance, the field $K'=k\{t\}$ of all formal power series (2) with $\|z\|=\exp(-V(z))$ given by (3) is a typical complete field.

Arithmetically any prime number p determines a valuation V_p of the integers, if $V_p(n)$ is defined as the exponent of the highest power of p dividing n,

$$V_p(p^\nu b) = \nu, \qquad n = p^\nu b, \qquad b \text{ prime to } p. \tag{5}$$

The field R of rational numbers n/m with this "p-adic" valuation $V_p(n/m) = V_p(n) - V_p(m)$ can be embedded in a larger *p-adic number field* R_p complete with respect to V_p. The structure of this field R_p is determined by the behavior of the residues (mod p), which themselves form a field, the Galois field containing p elements $0, 1, 2, \cdots, p-1$. Similarly the residues (mod t) of the power series (3) form a field, the field k of coefficients. Such *residue-class fields* occur for other fields K with the valuations V. Any complete feld K whose valuation function V takes on only integral values is essentially determined by its residue-class field.

Just as in the Galois theory, the structure of such a field and the form of its subfields depends upon the possible "symmetries" of the field. A symmetry of a field F is technically known as an automorphism: a map of the field upon itself which preserves rational relations. In other words, an *automorphism* S of F is a one-to-one correspondence $x \longleftrightarrow x^S$ of the field F to itself such that sums and products correspond to sums and products:

$$(x + y)^S = x^S + y^S; \qquad (xy)^S = x^S y^S. \tag{6}$$

The successive application of two automorphisms S and T yields a new automorphism $x \longleftrightarrow (x^S)^T$ called the product ST. Under this product the automorphisms of F form a group. For certain complete fields like the p-adic fields this group G has been investigated by MacLane, who finds certain subgroups of G analogous to Hilbert's "inertial" and "ramification" groups for prime ideals (*cf.* also MacLane [**24**]).

2. The lattice representation of the structure of groups. Recent algebraic investigations have shown that the structure of a group depends vitally upon the number and arrangement of its subgroups. An instance in point is the Jordan-Hölder theorem, which asserts that certain chains of relatively normal subgroups of a group must always have the same length. Two subgroups H and K of a given group can be combined in two ways, to yield the *intersection* $H \cap K$ of the two subgroups and the *union* $H \cup K$, which is the smallest subgroup containing both H and K. Relative to these two operations the subgroups are said to form a *lattice* or *structure*. A lattice can also be defined abstractly in terms of the associative and other laws satisfied by intersection and union [**25, 6**].

Since the lattice of subgroups represents many of the properties of a group, one comes inevitably to the question: when is the nature of the group G completely determined by its lattice of subgroups? Since G is an abstract group, its subgroup lattice L will be exactly like the subgroup lattice of any group G' *isomorphic* to G. To say that G is isomorphic to G' means that there is a one-to-one correspondence $T: A \longleftrightarrow A^T$ between the elements A of the group G and the elements $A' = A^T$ of the group G' such that products correspond to products:

$$\text{(7)} \qquad \text{If } A \longleftrightarrow A^T, \quad B \longleftrightarrow B^T, \quad \text{then } AB \longleftrightarrow A^T B^T.$$

Under this isomorphism T each subgroup H of G goes into a subset H^τ of G' composed of all images A^T of elements A of H. This correspondence $H \longleftrightarrow H^\tau$ is a one-to-one correspondence between the subgroups of G and those of G' such that unions and intersections are preserved:

$$\text{(8)} \qquad (H \cap K)^\tau = H^\tau \cap K^\tau, \qquad (H \cup K)^\tau = H^\tau \cup K^\tau.$$

Conversely, a one-to-one correspondence $H \longleftrightarrow H^\tau$ between the subgroups H of G and the subgroups H' of another group G' is called a *subgroup-isomorphism* of G to G' if property (8) holds. Hence the natural question: Are two subgroup-isomorphic groups G and G' necessarily (elementwise) isomorphic? An affirmative answer would mean that the structure of a group is actually determined by the lattice of its subgroups. The answer is not always affirmative, but Baer has an answer in the case of abelian groups. If the group is one of certain listed types, which are all "small" groups in the sense that they have relatively few subgroups, the answer is no; for other abelian groups, the answer is yes: subgroup-isomorphic groups are in fact isomorphic, and in many cases the only subgroup isomorphisms are those generated in the above fashion by ordinary isomorphisms.

3. Generalized quaternions. Systems of elements subject to the rational operations of addition and multiplication but not satisfying the commutative law for multiplication were first discovered in the guise of quaternions during the last century. A real quaternion algebra Q is the set of all linear combinations Y of four basal elements $1, i, j, ij$,

$$\text{(9)} \qquad Y = y_1 + y_2 i + y_3 j + y_4 ij,$$

where the components $y_1, \cdots, y_4$ are real numbers. The rational operations on such elements are *Scalar multiplication*: Y is multiplied by a real number b by multiplying each component y_i by that number b; *Addition*: two elements Y and Z are added by adding corresponding components; *Multiplication*: the product of two elements Y and Z is found by using the usual distributive and associative laws and the following table for the products of the four basal elements:

$$\text{(10)} \qquad i^2 = -1, \qquad j^2 = -1, \qquad ij = -ji.$$

The system Q is called a linear associative algebra because it is a set of elements closed with respect to these three operations, and because these operations satisfy the usual algebraic laws (excluding the law $YZ=ZY$). This particular system Q is an algebra *over* the field of real numbers because the components are elements from that field.

Wedderburn first introduced the consideration of such linear algebras over fields other than the real number field. Over the field of rational numbers, for instance, one has quaternion algebras $Q(\alpha, \beta)$ consisting of all elements Y of (9) with rational components and a multiplication table

$$i^2 = \alpha, \qquad j^2 = \beta, \qquad ij = -ji, \tag{11}$$

where α and β are fixed rational integers. Aside from the fact that the same algebra might be represented with different basal units i' and j' using different constants α' and β', there are infinitely many essentially different quaternion algebras $Q(\alpha,\beta)$, two algebras being the same only if a certain "fundamental number" determined by α and β is the same.* The importance and variety of the algebras possible over the field of rational numbers and other fields was recognized by Dickson, whose researches have paved the way for the far reaching theory of algebras over many types of fields.

The structure of an algebra depends upon the form of its subfields and subalgebras. The quaternion algebra $Q(\alpha, \beta)$ contains the set $R(\sqrt{\alpha})$ of all elements $x=y_1+y_2i$. This set is a field; it is obtained from the field R of rational numbers by adjoining i with $i^2=\alpha$, for $R(\sqrt{\alpha})$ consists of all rational functions of i with coefficients in R. Each element $x=y_1+y_2i$ of the field $R(\sqrt{\alpha})$ has a conjugate $\bar{x}=y_1-y_2i$, and the correspondence $x\longleftrightarrow\bar{x}$ is a one-to-one correspondence of the field $R(\sqrt{\alpha})$ to itself. Since $\overline{x_1x_2}=\bar{x}_1\bar{x}_2$, $\overline{x_1+x_2}=\bar{x}_1+\bar{x}_2$, this correspondence preserves sums and products and hence is an automorphism of the field in the sense of (6). In terms of this subfield $R(\sqrt{\alpha})$ the general quaternion (9) can be written as

$$Y = (y_1 + y_2i) + j(y_3 - y_4i) = x_1 + jx_2 \tag{12}$$

where $x_1=y_1+y_2i$ and $x_2=y_3-y_4i$ are elements in the field $R(\sqrt{\alpha})$. The multiplication table (11) implies that

$$x_1j = (y_1 + y_2i)j = y_1j - y_2ji = j(y_1 - y_2i) = j\bar{x}_1,$$

and hence we can write a new table in terms of conjugates as

$$j^2 = \beta, \qquad xj = j\bar{x}, \quad \text{for } x \text{ in } R(\sqrt{\alpha}). \tag{13}$$

This automorphism formulation of the quaternions has latent potentialities for generalization.

4. Arithmetics of quaternions. Arithmetic can be considered in an algebra if one can select in the algebra a suitable set J of integral numbers which, like the ordinary integers, form a *ring*: that is, a subset J of the algebra closed under

* In terms of this fundamental number a certain simplified canonical form of the table (11) is possible, as shown in Albert [2] and Latimer [19].

addition, subtraction, and multiplication. In the quadratic algebraic number field $K=R(\sqrt{\alpha})$ every element $x=y_1+y_2\sqrt{\alpha}$ satisfies a quadratic equation

$$[t-(y_1+y_2\sqrt{\alpha})][t-(y_1-y_2\sqrt{\alpha})]=t^2-2y_1t+(y_1^2-y_2^2\alpha)=0$$

with rational coefficients. The number x is called an *integer* if this equation has integral coefficients.* The set of all integers then forms a ring. For quaternions one could attempt the same definition, for the quaternion Y of (12) satisfies a rational equation whose roots are $Y=x_1+jx_2$ and its conjugate $\overline{Y}=\bar{x}_1-jx_2$,

$$(14) \qquad (t-Y)(t-\overline{Y})=t^2-2y_1t+N(Y)=0.$$

Here the constant term $N(Y)=Y\overline{Y}$ is the so-called *norm* of Y,

$$(15) \qquad N(Y)=Y\overline{Y}=(x_1+jx_2)(\bar{x}_1-jx_2)=x_1\bar{x}_1-\beta x_2x_2.$$

If the previous definition of an integer is now applied, so that Y is called integral if this equation (14) has rational integers as coefficients, then the set of integers unfortunately may no longer be a ring because there can be two integers whose sum is not integral.

As a substitute one may consider the set $\mathfrak{g}$ of all quaternions $Y=x_1+jx_2$ for which the numbers x_1 and x_2 of the field $R(\sqrt{\alpha})$ are integers. This set is too much dependent upon the particular choice of j, but has most of the requisite properties:

(C): $\mathfrak{g}$ is a ring;

(R'): every element of $\mathfrak{g}$ satisfies a polynomial equation (for instance the equation (14)) which has a first coefficient 1 and the remaining coefficients rational integers;

(U'): $\mathfrak{g}$ contains all the rational integers of R and contains just as many linearly independent elements over R as does the whole quaternion algebra $Q(\alpha,\beta)$.

This last property is immediate, for $\mathfrak{g}$ contains the four basal elements $1, i, j$, and ij which are linearly independent by the construction of the algebra. A subset $\mathfrak{g}$ of an algebra having these three properties (C), (R'), and (U') is called an *order* of the algebra. Dickson recognized that a set of integers in more general algebras is more suitably defined as a *maximal order*† $\mathfrak{g}$; that is, an order which is contained in no larger order of the algebra. Every order is then contained in at least one maximal order; in particular, the ring $\mathfrak{g}$ defined above for the quaternions is not usually itself a maximal order, but can be extended so as to become one.

The arithmetic properties of a maximal order $\mathfrak{O}$ again depend upon a suitable type of subsystem of the order: the *ideals* of the order. A subset $\mathfrak{a}$ of $\mathfrak{O}$ is a *left ideal* of $\mathfrak{O}$ if the difference of two elements of $\mathfrak{a}$ is again an element of $\mathfrak{a}$, and if the product ba is in $\mathfrak{a}$ for any b in $\mathfrak{O}$ and a in $\mathfrak{a}$. One also requires that an ideal

* Zariski's treatment of the singularities of surfaces involves essentially the same notion of an integer (§1).

† This terminology is not that used by Dickson himself. See Dickson [9].

contain at least one rational integer. In particular for any element a_0 in $\mathfrak{O}$ the set (a_0) of all elements ba_0 for b in $\mathfrak{O}$ is an ideal. This ideal is the *principal ideal* generated by a_0. Similarly, left and right ideals can be considered in any ring.

5. Quadratic forms and quaternions. Quaternion algebras reflect many properties of quadratic and hermitian forms, because the norm of a quaternion is itself such a form. The norm $N(Y)$ as calculated in (15) is a simple hermitian form in the variables x_1 and x_2. Directly in terms of the original components $y_1, \cdots, y_4$ of (9) the norm $N(\bar{Y}) = Y\bar{Y}$ with $\bar{Y} = y_1 - y_2 i - y_3 j - y_4 ij$, becomes the quadratic form

$$N(Y) = y_1^2 - \alpha y_2^2 - \beta y_3^2 + \alpha\beta y_4^2.$$

This hermitian form interpretation has been investigated in terms of ideals by Latimer. The order $\mathfrak{g}$ of integers used above has a *basis* $1, j$, in the sense that $\mathfrak{g}$ consists of all linear combinations $x_1 + jx_2$ of 1 and j with coefficients x_1 and x_2 integers of the quadratic field $R(\sqrt{\alpha})$. Similarly, any left ideal $\mathfrak{a}$ of this ring has a basis ω_1 and ω_2, so that any element of $\mathfrak{a}$ can be expressed as $x\omega_1 + y\omega_2$ for x and y again integers of $R(\sqrt{\alpha})$. The norm of this element $x\omega_1 + y\omega_2$ is, except for a constant factor, a form

$$ax\bar{x} + b\bar{x}y + \bar{b}x\bar{y} + cy\bar{y}. \tag{16}$$

This form is hermitian because it is equal to its own conjugate—the coefficients a and c are rational integers and the cross product terms have conjugate coefficients b and $\bar{b}$. The determinant $b\bar{b} - ac$ of this form turns out to be the constant β used in the multiplication tables (13) and (11).

An ideal $\mathfrak{a}$ with a given basis ω_1, ω_2, then corresponds to a hermitian form of determinant β over the field $R(\sqrt{\alpha})$, and conversely. If the basis of the ideal is changed, one gets a new form equivalent to the first in the sense that it can be obtained from the first by a linear homogeneous transformation of determinant 1. The possible classes of equivalent forms can further be put into correspondence with certain* classes of ideals—where two ideals belong to the same class when they differ by principal ideal factors (with positive norms). This means that facts about the classes of forms can be translated into facts about classes of ideals. In particular, Latimer has found certain quaternion algebras with just one class of ideals. This statement means that every (regular) ideal is a principal ideal and hence that the ideal structure of the ring is as simple as possible [**18, 20**].

The quaternion algebras can also be used as a starting point for the treatment of properties of quadratic forms in many variables, of the type

$$f(x_1, x_2, \cdots, x_n) = \sum a_{ij}x_i x_j \qquad (i, j = 1, \cdots, n),$$

where the coefficients a_{ij} are in the field R of rational numbers and can be chosen so that $a_{ij} = a_{ji}$. Many questions of the representation of numbers by such forms

* Because the ring $\mathfrak{O}$ is not a maximal order, it is necessary here to consider not all ideals, but only certain "regular" ideals.

lead to the "representation of zero" by numbers in any field F containing the rational field R. The form f is said to *properly represent* 0 in the field F if there are numbers $x_1, x_2, \cdots, x_n$ not all zero in the field F such that $f(x_1, \cdots, x_n) = 0$. Here F might be the field R_0 of all real numbers, which is complete with respect to the ordinary absolute value, or a p-adic number field R_p which contains R and is complete with respect to the p-adic valuations of (5), §1. If f represents 0 in the original field R, then it certainly represents 0 in any one of these larger fields R_0 and R_p. A theorem due to Hasse now states a converse: If the form $f(x_1, \cdots, x_n)$ properly represents 0 in the real number field R_0 and in every p-adic field R_p, then it also properly represents 0 in the field of rationals [**34, 11**]. This type of theorem, which starts from the behavior "locally" for each prime p and derives the behavior in the original field R, has attracted much current interest (see also the class field theory discussed in §11). Artin has now a new and elegant proof for this theorem of Hasse, using an induction on the number of variables combined with a treatment of the four variable case in terms of the norm forms of quaternion algebras. His proof also applies when R is replaced by an algebraic number field.

6. The structure of algebras. We turn now to the general definition of a linear associative algebra over an arbitrary field F. An algebra A consists of all linear combinations

$$a = \alpha_1 u_1 + \cdots + \alpha_n u_n = \sum \alpha_i u_i \tag{17}$$

of n basal elements $u_1, \cdots, u_n$ with coefficients α_i in the field F. The sum of two such elements is given by

$$(\sum \alpha_i u_i) + (\sum \beta_i u_i) = \sum (\alpha_i + \beta_i) u_i \qquad (i = 1, \cdots, n); \tag{18}$$

and the scalar product of an element a by an element β of the field F is determined by

$$\beta(\sum \alpha_i u_i) = \sum (\beta\alpha_i) u_i, \qquad (i = 1, \cdots, n). \tag{19}$$

The product of two elements a and b in the algebra is defined by the formula

$$(\sum \alpha_i u_i)(\sum \beta_j u_j) = \sum (\alpha_i \beta_j) u_i u_j, \qquad (i, j = 1, \cdots, n) \tag{20}$$

which is completed by a table giving the product $u_i u_j$ of any two basal elements as some particular element of the algebra. This multiplication table must be such that the product $a \cdot b$ satisfies the associative law. The number n of basal elements of the algebra A is its *order*; it can be characterized by the statement that the n basal elements are linearly independent, while any $n+1$ elements of A are necessarily linearly dependent.

We shall assume that the algebra A has a unit element 1 with the property that $1 \cdot a = a \cdot 1 = a$ for every a of the algebra—for if there were no such element we could construct it and add it to the algebra. The algebra then contains the elements $1 \cdot \alpha$ which behave like the corresponding α's of the field F and thus can be identified with the elements of F, so that the algebra then contains its coefficient field F.

Certain types of algebras are important. A *division algebra* is an algebra in which every $a \neq 0$ has an inverse a^{-1} such that $a^{-1}a = aa^{-1} = 1$. An algebra is *simple* if it contains no proper subset which is a two-sided ideal (both a left and a right ideal, as at the end of §4). Such algebras exist: any division algebra is necessarily simple in this sense. A simple algebra is *normal* if it is as uncommutative as possible: that is, if b is an element of the algebra such that $bx = xb$ is true for every x, then b is necessarily an element $b = \beta$ of the coefficient field F.

A major problem in the study of linear algebras is the construction of all algebras out of fields or out of simpler algebras in systematic fashion. The object of such investigations is then a more concrete representation for algebras than that given by the general definition. Matrices yield algebras; for the set of all $m \times m$ square matrices with elements in a field F is an algebra M_m (a so-called *total matrix algebra*). The m^2 matrices with 1 in one position and 0 elsewhere can be considered as its $n = m^2$ basal elements $u_1, \cdots, u_n$. This matrix construction can be generalized if we consider the set of all $m \times m$ matrices with the usual matrix multiplication, but with elements themselves taken from a given normal division algebra D over the field F. The algebras which can be obtained in this fashion are all the normal simple algebras; they are the important type because all other algebras can be decomposed into such normal simple ones by certain structure theorems of Wedderburn [**33**, ch. 10].

New algebras can be formed by combinations of old ones. If A and B are two algebras over a field F one can construct a new algebra, the direct product $A \times B$, which contains both the given algebras A and B. Specifically, $A \times B$ consists of all sums

$$c = b_1u_1 + \cdots + b_nu_n \tag{21}$$

formed from the basal elements u_i of A in the same fashion as in the element (17) of A, except that the coefficients b_i are now elements of B. Multiplication is determined by the multiplication in A and B and by the rule that $u_ib = bu_i$. To be specific, any 2×2 matrix with elements b_{ij} in a division algebra B can be written in the form (21) as

$$\begin{pmatrix} b_{11} & b_{12} \\ b_{21} & b_{22} \end{pmatrix} = b_{11}\begin{pmatrix} 1 & 0 \\ 0 & 0 \end{pmatrix} + b_{12}\begin{pmatrix} 0 & 1 \\ 0 & 0 \end{pmatrix} + b_{21}\begin{pmatrix} 0 & 0 \\ 1 & 0 \end{pmatrix} + b_{22}\begin{pmatrix} 0 & 0 \\ 0 & 1 \end{pmatrix}.$$

Hence the normal simple algebra of all such matrices is simply the direct product $M_2 \times D$ of the division algebra D by the 2×2 total matrix algebra!

7. Cyclic algebras. Since the matrix algebras are readily written down, the construction of normal simple algebras becomes essentially the construction of division algebras. Here explicit constructions are possible after the manner of quaternion algebras. The quaternion algebras have a multiplication table (13) depending on a certain quadratic subfield $R(\sqrt{\alpha})$ and on an automorphism

$x \longleftrightarrow \bar{x}$ of this quadratic subfield. There are similar algebraic extensions $K = F(\theta)$ of other fields F, obtained by adjoining to F a root θ of an irreducible polynomial equation $f(x) = 0$ whose coefficients are in F. The degree m of this irreducible equation is called the *degree* of the field K over F. The nature of this extension depends on the structure of the group G of those automorphisms $x \longleftrightarrow x^S$ of K (*cf.* the definition of (6)) which leaves every element of F fixed. The simplest case arises when there is one such automorphism S with $S^m = I$, I the identity automorphism, so that the order m of S in the group G is exactly the degree of the field. Every automorphism of G is then necessarily one of the powers of S, so that the automorphism group G is the cyclic group $G = [I, S, S^2, \cdots, S^{m-1}]$. A field K with such a group over F is a *cyclic* field and is often denoted by Z. Such fields certainly exist—for let F be the field $R(\omega)$ obtained by adjoining to the rationals a p-th root ω of unity, and let $Z = F(\theta)$ be generated by the adjunction to F of a p-th root $\theta = \sqrt[p]{\alpha}$ not already in F. Then $\omega\theta$ is also a p-th root of α, and the correspondence S defined by

$$\theta^S = \omega\theta, \qquad [b(\theta)]^S = b(\omega\theta),$$

for any polynomial $b(\theta)$ in the field $Z = F(\theta)$, carries a root θ of $t^p - \alpha = 0$ into another root $\omega\theta$ of the same equation and therefore can be shown to be an automorphism. Its powers are

$$S^i\colon \quad \theta^{S^i} = \omega^i\theta, \qquad S^p = I,$$

so that the group and field are actually cyclic.

From any such cyclic field $Z = F(\theta)$ with its automorphism S of order m, one can construct corresponding "cyclic" algebras over F. The algebra is to consist of all elements

$$a = z_0 + z_1 y + z_2 y^2 + \cdots + z_{m-1} y^{m-1}, \tag{22}$$

where the coefficients z_i are now elements of the cyclic field Z and the multiplication rules are given in terms of a fixed element γ of F by the table

$$y^m = \gamma, \qquad zy = yz^S, \quad \text{for } z \text{ in } Z. \tag{23}$$

The symbol (Z, S, γ) designates the algebra so obtained. The quaternions of (12) and (13) are the special case when Z is the quadratic field $R(\sqrt{\alpha})$, $m = 2$, $y = j$, and $\gamma = \beta$. Such a cyclic algebra need not always be a division algebra, but the important fact is that any normal division algebra over a field F of algebraic numbers is necessarily a cyclic algebra [**8**, ch. 7].

The structure of such cyclic algebras is now a matter of central interest. The direct product of two cyclic algebras of the same degree over a field F can be itself represented as another cyclic algebra multiplied directly by a suitable total matrix algebra. A cyclic algebra can even be combined with itself in this fashion to form a new cyclic algebra! The properties of this "direct power" of an algebra are fundamental. Albert [**3**] has shown that these properties may be obtained elegantly by systematically using another simple type of algebra, called the *cyclicsemi-fields*. Such a semi-field is a commutative algebra consisting of all $t\times t$ diagonal matrices whose elements are taken from a given cyclic field Z. Such a diagonal matrix, which has zeros everywhere but on the main diagonal, may be represented by simply writing down the diagonal elements $\{z_1, \cdots, z_t\}$. The automorphism S of the field then generates an automorphism of these matrices which may be obtained by permutating the diagonal elements cyclicly and then applying S to the last element

$$\{z_1, \cdots, z_t\} \longleftrightarrow \{z_2, \cdots, z_t, z_1^S\}.$$

Relative to this correspondence the semi-fields behave very much like ordinary cyclic fields.

An outstanding problem concerns the nature of normal division algebras and normal simple algebras over general fields F—where there might be division algebras which are not cyclic algebras. The cyclic algebras $A=(Z, S, \gamma)$ (*cf*. (23)) are characterized by the fact that they contain a cyclic subfield Z which is maximal (contained in no larger subfield). Any normal division algebra over F still contains a maximal subfield K—but this subfield may have too few automorphisms to be normal* or may not have a cyclic automorphism group. The multiplication table in the algebra may still be described by a more complicated multiplication table depending on certain automorphisms after the fashion of (23). The structure of such algebras has been analyzed by Brauer [7] who makes considerable use of two tools: Firstly, any algebra A over F can be extended to a larger algebra A_K over a given field K containing F—one need only consider the element (17) of the algebra with the same basal units, but now with coefficients α_i taken from the larger field K. Secondly, any algebra A can be considered as an algebra of certain matrices with components in F, for any element a of the algebra has certain products with the basal elements,

$$au_i = \alpha_{i1}u_1 + \cdots + \alpha_{in}u_n. \tag{24}$$

The correspondence between the element a and the matrix $a^*=(\alpha_{ij})$ "represents" A as a matrix algebra. By simplifying these matrices one obtains reductions of this representation to certain irreducible representations over F or K; their construction by Brauer yields information about the algebra.

8. Arithmetic of cyclic algebras. The arithmetic study of a general algebra can be reduced to the study of the normal simple algebras. Here again the basic necessity is the definition of a set of integers—and such sets are defined as maximal orders in exactly the fashion discussed above (§4) for the quaternion algebras. But the definition of such sets and the description of their properties is not the only problem. How can these maximal orders be explicitly found in the case of particular algebras, say for cyclic algebras? How many such orders are there? Some answers have been found for the special case when the cyclic algebra is a quaternion algebra, and Hull has extended the method to any cyclic algebra

* A field K over F is *normal* if it is obtained by adjoining to F *all* roots of an equation; a normal field of degree n over F always has n automorphisms over F.

generated over the field of rational numbers by a cyclic field Z of odd prime degree. Such a cyclic algebra might be represented in several different ways by the generation (Z, S, γ) of (22) and (23), but there is a certain canonical generation which is found by considering the algebra not over the rational number field, but over the larger p-adic number fields constructed in §1, from the valuation (5) belonging to a rational prime p. In this generation one might naturally define an integer to be an element (22) of the algebra in which the coefficients z_i are integers of the algebraic number field Z. This set of integers is an order, but not a maximal order, and Hull [**12**] finds that it can be embedded in exactly n distinct maximal orders which can be explicitly represented in terms of the solutions of certain congruences. All other maximal orders can be obtained from any one of these. This method of finding formulas for maximal orders is a fruitful one, for it has been extended by Perlis [**27**] to the case of cyclic algebras over the rationals where the degree of the cyclic field is not a prime, but a power of a prime.

In any maximal order $\mathfrak{O}$ the study of arithmetic questions of divisibility leads inevitably to the study of the divisors of 1. Such numbers, called *units* of $\mathfrak{O}$, are simply the elements $a \neq 0$ in $\mathfrak{O}$ which have reciprocals $1/a$ also in the order $\mathfrak{O}$. These units form a group under multiplication. Recently [**10**] advances have been made in studying the totality of units and the structure of the group which they form; the typical theorem gives an expression of the units in terms of a finite set of "fundamental" units. Hull has successfully applied these methods to the orders of quaternion algebras, where the group G can be represented geometrically as a certain Fuchsian group.

The ideal theory in a maximal order of a normal simple algebra (we recall that such algebras include the cyclic algebras) has been developed extensively. MacDuffee [**23**] has shown that many of the requisite computations with ideals can be carried out effectively by the use of matrices. This use of matrices depends directly upon the representation of an algebra in terms of matrices, discussed above in (24). Specifically, an ideal $\mathfrak{a}$ in a maximal order $\mathfrak{O}$ of such an algebra over the rational number field always has a basis $\omega_1, \cdots, \omega_n$ such that the ideal consists of all linear combinations

$$b = \beta_1\omega_1 + \cdots + \beta_n\omega_n, \tag{25}$$

for rational integral coefficients β_i. The elements of this basis of the ideal can be represented in terms of the basal elements of the algebra in the form

$$\omega_i = g_{i1}u_1 + g_{i2}u_2 + \cdots + g_{in}u_n, \qquad i = 1, \cdots, n. \tag{26}$$

The matrix of these coefficients (g_{ij}) is said to correspond to the given ideal. MacDuffee has then determined when two matrices correspond to different bases of the same ideal and when a matrix corresponds to an ideal. Greatest common divisor computations on the ideals can then be done elegantly in terms of computations of the greatest common left divisors of the corresponding matrices. Furthermore, the question of the equivalence of two ideals (in the sense that ideals are equivalent when they differ by factors which are principal ideals, *cf.* §4), which was discussed above in connection with hermitian forms, can be treated in terms of matrices. Thus the calculation of the number of classes of equivalent ideals becomes tangible.

9. Canonical forms for matrices. A matrix $A=(a_{ij})$ can always be considered as a linear transformation $y_i=\sum a_{ij}x_j$ $(i=1, \cdots, n)$ which carries a vector $(x_1, \cdots, x_n)$ with n components into another vector $(y_1, \cdots, y_n)$. A change of the coördinate system to which these vectors are referred changes the matrix of the transformation to a new matrix.

$$B = TAT^{-1} \qquad (T \text{ a non-singular matrix}). \tag{27}$$

Such a B is called *similar* to A. Elementary divisors are ordinarily used to reduce A to a canonical form under such similarity transformations. Ingraham has considered this problem in the more general case when the matrix A is one whose elements are taken not in a field but in a division algebra. By using systematically the properties of the space which A transforms he has been able to carry through the study of similarity in this case [**14, 15**]. Because the elements of the matrix are taken from an algebra without a commutative law, the polynomials in a matrix, which are important for the theory, must be redefined in a suitable manner. More generally, the equation (27) can be written in the form $BT=TA$, and this equation can be studied even when T is singular, to determine the maximal possible rank for a matrix T satisfying the equation. Analogous methods work for other matrix equations [**13**].

When the elements of a matrix A are complex numbers, the matrix A is called *hermitian* if $A^*=A$, *unitary* if $AA^*=I$, where $A^*=(\bar{a}_{ji})$ is the conjugate transpose of the matrix A. The reductions of such matrices to canonical form by suitable similarity transformations are important classical problems. These problems can now be treated, as in Williamson [**35**], as a special type of a general reduction of certain normal matrices. A matrix is called *normal* if $AA^*=A^*A$. Both unitary and hermitian matrices have this property. It can be shown that a matrix A is normal in this sense if and only if A can be expressed as a polynomial $f(A^*)$ in its conjugate transpose matrix A^*. This leads to a more general definition: A matrix B is *normal* relative to a non-singular hermitian matrix H if $BH=Hf(B^*)$ for some polynomial f. This includes the previous case with $H=I$. The canonical forms for such relatively normal matrices have been successfully treated by Williamson, using similarity transformations like (27) subject to the natural side condition that the given hermitian matrix H be unchanged by T in the sense that $THT^*=H$.

10. Lie algebras. In the study of the so-called commutators of a group of continuous transformations certain non-associative algebras arise. These algebras differ from the linear associative algebras considered above essentially in the replacement of the associative law of multiplication by two other laws:†

† These laws are not unfamiliar. They are satisfied by the vector product of two vectors. For complete definition of terms, see Jacobson [**16, 17**].

$$(28)\qquad [x, y] = -[y, x], \qquad [x[y, z]] + [y[z, x] + [z[x, y]] = 0.$$

Here, in accord with usage, the product of two elements in a Lie algebra is denoted by $[x, y]$. Such a Lie algebra may readily be obtained from an associative algebra A by the simple device of defining the multiplication of the Lie algebra in terms of the given associative multiplication by the equation

$$[x, y] = xy - yx,$$

so that $[x, y]$ is the "commutator" of x and y. The fact that the so-defined commutator actually satisfies the conditions (28) can be readily verified. By using a suitable automorphism S of the given associative algebra A it is possible to define still other Lie algebras and even to describe their automorphism groups. By systematic use of such construction Jacobson has reduced the problem of determining all "simple" Lie algebras over ordinary fields to questions in underlying associative algebras. These questions can be solved explicitly when the algebra is one over the field of real numbers, but here, as in Cartan's classical theory of Lie groups, certain troublesome exceptional algebras can arise.

11. Local class field theory. An extraordinary combination of the study of fields with valuations, algebraic extensions of fields, automorphism groups of fields and normal simple algebras over fields has arisen in the class field theory. This theory aims at an explicit description of the fields K which are abelian extensions of an algebraic number field k. An *abelian* extension of k is a normal extension of k (*cf.* footnote in §7) such that the group of all automorphisms (6) of K which leave k fixed is *abelian*. The class field theory describes these extensions in terms of automorphism groups by a device which represents the automorphism group in terms of an isomorphic group derived from the multiplicative group in the field k [**4**]. Furthermore, the character of these abelian extensions proves to be intimately connected to the variety and the properties of the normal simple algebras and cyclic algebras possible over k. For many purposes this study of fields and algebras over the rational number field k can be simplified by treating each prime number in k separately; one then considers the extensions not of k but of the p-adic field k_p which is complete with respect to the p-adic valuation (5) defined by this prime p. Schilling has shown in several papers [**28, 29, 30**] that most of the theorems of local class field theory can also be obtained over fields k complete with respect to more general valuations. The p-adic valuations (5) are *discrete* in the sense that the value $V(a)$ of every element a of the field is an integer, but the abelian extensions turn out to have much the same structure if one considers valuations in which $V(a)$ runs over all rational numbers or even includes some irrationalities.

12. What is algebra? To summarize this survey we shall essay here a description of the general tendency of these investigations. Algebra concerns itself with the postulational description of certain systems of elements in which some or all of the four rational operations are possible: fields, linear algebras, Lie algebras, groups. The abstract or postulational development of these systems must then be supplemented by an investigation of their "structure." Under "structure" we include:

(a) the number and interrelations of the subsystems of a given system, either subsystems just like the whole system (lattice of subgroups), or subsystems with especially characteristic properties (sets of integers, maximal orders, ideals, subfields of an algebra, *etc.*);

(b) the group of automorphisms of a system, and connections between the subgroups of this group and the subsystems of the given system (Galois theory, class field theory);

(c) the construction of all systems of specific types out of simpler systems of the same or other types (the construction of cyclic algebras and matrix algebras, the reduction of a given surface to a birationally equivalent surface without singularities, construction of Lie algebras);

(d) alternatively, the description of given systems as subsystems of larger systems (complete fields, power series fields);

(e) criteria or invariants to determine when two explicitly but differently constructed systems are abstractly the same or *isomorphic* (the canonical generation of a cyclic algebra; the genus as an invariant defined by the differentials of a function field).

With this explanation we venture the characterization:

Algebra tends to the study of the explicit structure of postulationally defined systems closed with respect to one or more rational operations. This summary does not account well for the use of topological operations in algebra, of which the valuations form but one example. Furthermore, the reduction of matrices to canonical forms is only indirectly an instance of the criteria intended in (e). As with many hyper-generalizations, our statements fit the facts only when the facts are first slightly distorted!

References

1. A. A. Albert, Modern Higher Algebra, University of Chicago Press, 1937.

2. A. A. Albert, Integral domains of rational generalized quaternion algebras, Bulletin American Mathematical Society, vol. 40, 1934, pp. 164–176.

3. A. A. Albert, On cyclic algebras, Annals of Mathematics, vol. 39, 1938.

4. E. Artin, Beweis des allgemeinen Reziprozitätsgesetzes, Abhandlungen aus dem Mathematischen Seminar, Hamburg, vol. 5, 1927, pp. 353–363.

5. R. Baer, Abelian groups without elements of finite order, Duke Mathematical Journal, vol. 3, 1937, pp. 68–122.

6. G. Birkhoff, On the combinations of subalgebras, Proceedings Cambridge Philosophical Society, vol. 29, 1933, pp. 441–464.

7. R. Brauer, Über die Konstruktion der Schiefkörper, die von endlichem Rang in bezug auf ein gegebenes Zentrum sind, Journal für die Mathematik, vol. 168, 1932, pp. 44–64.

8. M. Deuring, Algebren, Ergebnisse der Mathematik und ihrer Grenzgebiete, vol. 4, part 1, 1935.

9. L. E. Dickson, Algebren und ihre Zahlentheorie, Zürich, 1927.

10. M. Eichler, Neuere Ergebnisse der Theorie der einfachen Algebren, Jahresbericht der Deutschen Mathematiker-vereinigung, vol. 47, 1937, pp. 198–220.

11. H. Hasse, Darstellbarkeit von Zahlen durch quadratische Formen in einem beliebigen algebraischen Zahlkörper, Journal für die Mathematik, vol. 153, 1924, p. 113.

12. R. Hull, Maximal orders in rational cyclic algebras of odd prime degree, Transactions American Mathematical Society, vol. 38, 1935, pp. 515–530.

13. M. H. Ingraham, On certain equations in matrices whose elements belong to a division algebra, Bulletin American Mathematical Society, vol. 44, 1938, pp. 117–124.

14. M. H. Ingraham and M. C. Wolf, Relative linear sets and similarity of matrices whose elements belong to a division algebra, Transactions American Mathematical Society, vol. 42, 1937, pp. 16–31.

15. N. Jacobson, Pseudo-linear transformations, Annals of Mathematics, vol. 38, 1937, pp. 484–507.

16. N. Jacobson, Rational methods on the theory of Lie algebras, Annals of Mathematics, vol. 36, 1935, pp. 875–881.

17. N. Jacobson, Abstract derivation and Lie algebras, Transactions American Mathematical Society, vol. 42, 1937, pp. 206–224.

18. C. G. Latimer, On ideals in generalized algebras and hermitian forms, Transactions American Mathematical Society, vol. 38, 1935, pp. 436–446.

19. C. G. Latimer, On the fundamental number of a rational generalized quaternion algebra, Duke Mathematical Journal, vol. 1, 1935, pp. 433–435.

20. C. G. Latimer, On the class number of a quaternion algebra with a negative fundamental number, Transactions American Mathematical Society, vol. 40, 1936, pp. 318–323.

21. S. Lefschetz, Lectures in algebraic geometry (Mimeographed), Princeton University, 1937.

22. C. C. MacDuffee, Ideals in linear algebras, Bulletin American Mathematical Society, vol. 37, 1931, pp. 841–853.

23. C. C. MacDuffee, Matrices with elements in a principal ideal ring, Bulletin American Mathematical Society, vol. 39, 1933, pp. 564–584.

24. S. MacLane, The uniqueness of the power series representation of certain fields with valuations, Annals of Mathematics, vol. 39, 1938, pp. 370–382.

25. O. Ore, On the foundations of abstract algebra, Annals of Mathematics, vol. 37, 1936, pp. 265–292.

26. A. Ostrowski, Algebraische Funktionen von Dirichletschen Reihen, Mathematische Zeitschrift, vol. 37, 1933, pp. 98–133.

27. S. Perlis, Maximal orders in rational cyclic algebras of composite degree, Bulletin American Mathematical Society, vol. 44, 1938, p. 356.

28. O. F. G. Schilling, The structure of local class field theory, American Journal of Mathematics, vol. 60, 1938, pp. 75–100.

29. O. F. G. Schilling, Class fields of infinite degree over p-adic number fields, Annals of Mathematics, vol. 38, 1937, pp. 469–476.

30. O. F. G. Schilling, A generalization of local class field theory, American Journal of Mathematics, vol. 60, 1938, pp. 667–705.

31. F. K. Schmidt, Zur arithmetischen Theorie der algebraischen Funktionen, I, Mathematische Annalen, vol. 41, 1936, pp. 415–438.

32. R. J. Walker, Reduction of the singularities of an algebraic surface, Annals of Mathematics, vol. 36, 1935, pp. 336–365.

33. J. H. M. Wedderburn, Lectures on matrices, Colloquium Publications of American Mathematical Society, vol. 17, 1934.

34. E. Witt, Theorie der quadratischen Formen in beliebigen Körpern, Journal für die Mathematik, vol. 176, 1937, pp. 31–44.

35. J. Williamson, Matrices normal with respect to a hermitian matrix, American Journal of Mathematics, vol. 60, 1938, pp. 355–373.

36. O. Zariski, Polynomial ideals defined by infinitely near base points, American Journal of Mathematics, vol. 60, 1938, pp. 151–204.

MODULAR FIELDS*

SAUNDERS MAC LANE, Harvard University

1. Introduction. The general theory of modular fields, though elementary in its presuppositions, offers an instructive cross-section of modern algebraic methods. These fields exhibit the generality of subject-matter inherent in abstract algebra, and at the same time illustrate the intimate connection between algebraic and arithmetic problems.

Modular fields arise first in number theory in the consideration of congruences with a prime modulus p. For integers a and b the ordinary definition states that

$$a \equiv b \pmod{p} \qquad \text{means that} \qquad p \text{ divides } (a - b).$$

Any integer a on division by p yields a quotient q and a remainder r,

$$a = qp + r, \qquad 0 \leqq r < p;$$

hence $a \equiv r \pmod{p}$, where the remainder r is one of the integers

$$(1) \qquad F_p: \ 0, 1, 2, \cdots, p - 2, p - 1.$$

Any integer is congruent to one of those in this set of p numbers.

With these numbers alone one can still carry out algebraic operations, provided one adds and multiplies these numbers in the ordinary fashion, and then reduces the answer by congruence to one of the numbers (1). For example, if $p=5$, the product $2\cdot 3=6$ should really be $2\cdot 3\equiv 6-5=1$. In this fashion one can make multiplication and addition tables for $p=5$, as shown. It is strange

+	0	1	2	3	4
0	0	1	2	3	4
1	1	2	3	4	0
2	2	3	4	0	1
3	3	4	0	1	2
4	4	0	1	2	3

·	0	1	2	3	4
0	0	0	0	0	0
1	0	1	2	3	4
2	0	2	4	1	3
3	0	3	1	4	2
4	0	4	3	2	1

that this idea has not appeared more† in texts on number theory, for the idea

* An address delivered before the Mathematical Association of America at Columbus, Ohio, December 30, 1939.

† *Cf.* remarks in Weiss [**26**].

is an essentially simple one. One can introduce it by the intuitively natural algebra of the words "even" and "odd," as

$$\text{even} \cdot \text{even} = \text{even}, \qquad \text{even} \cdot \text{odd} = \text{even}, \qquad \text{odd} \cdot \text{odd} = \text{odd},$$

$$\text{even} + \text{even} = \text{even}, \qquad \text{even} + \text{odd} = \text{odd}, \qquad \text{odd} + \text{odd} = \text{even}.$$

This is just the algebra of integers modulo $p=2$.

A congruence modulo p has all the properties of an equation; congruences can be added and multiplied term by term, and the relation of congruence is reflexive, symmetric, and transitive. If the modulus p is fixed, one might just as well dub congruence "equality." Every integer is then "equal" to one of the p symbols, $0, \cdots, p-1$, and the sums and products of these symbols, so identified, give exactly the algebra of the integers modulo p, as described above.

If one objects to rebaptizing "congruence" by fiat, one may adopt the more sophisticated procedure* of replacing each remainder r modulo p by the *class* r_p of all integers $r, r+p, r+2p, \cdots$ congruent to it. Such "congruence classes" are then added and multiplied according to the rules

$$r_p + s_p = (r + s)_p, \qquad r_p \cdot s_p = (rs)_p. \tag{2}$$

Furthermore the congruence classes r_p and s_p will be equal (*i.e.*, will contain the same elements) if and only if the integers r and s are congruent, so the desired "equality" has now been properly introduced. In any event the *integers modulo p* form a finite set of objects (1) satisfying all rules of algebra.

The presence of such arithmetic objects, which are certainly not ordinary numbers but which still obey ordinary algebra, is the reason why modern algebra is abstract. To separately discuss the algebra of numbers, then the algebra of congruence classes, then the algebra of functions, and so on would be most inefficient. Instead, theorems are better proved for any (abstractly conceived) system of objects whatever to which the basic rules of algebra apply.

These laws of algebra for a set F of objects, such as the integers modulo p, are codified as follows: For a and b in F there is uniquely defined a *sum* $a+b$ and a *product* $a \cdot b$. This product is *commutative* $[ab=ba]$ and *associative* $[a(bc)=(ab)c]$, as is also the sum. The *distributive* law $a(b+c)=ab+ac$ holds for all a, b, and c. The set F contains a zero 0 and a unit 1, with the characteristic properties

$$a + 0 = a = 0 + a, \qquad 1 \cdot a = a = a \cdot 1,$$

respectively. Finally, *subtraction* and *division* are possible, which is to say that the equations $a+x=0$ and $b \cdot y=1$ have solutions x and y in F, except when $b=0$. Any set F of elements with all these properties is called a *field*. One may say that a field is any system of elements within which addition, subtraction, multiplication, and division (excluding division by zero) can be carried out in the usual fashion.

---

* *Cf.* Albert [**1**, p. 7]; van der Waerden [**27**, p. 13]; or Mac Lane [**17**, Chapter I].

Well known fields are: (a) the set of all rational numbers; (b) the set of all real numbers; (c) the set of all complex numbers. The field (1) composed of the integers modulo p is often called the *Galois field* $GF[p]$. A *modular field* is any field containing such a $GF[p]$.

These fields $GF[p]$ are not the only finite fields. One may construct larger fields by simply adjoining to a $GF[p]$ the roots of certain algebraic equations. The process resembles the construction of the complex numbers from the field R of real numbers. Here one adjoins to R a symbol i representing a root of the equation $x^2+1=0$; the field C of all complex numbers $a+bi$ then contains everything which can be expressed rationally in terms of i and real numbers. The fact that C is generated over R by adjoining i is symbolized by $C=R(i)$. Note in particular that the polynomial x^2+1 used to generate this extension is *irreducible* over R, because it cannot be factored into polynomials of smaller degree with coefficients in R.

In similar vein consider the polynomial $f(x)=x^2+x+1$ over the field F_2 with two elements (the integers modulo 2). Neither $f(1)$ nor $f(0)$ is zero, so this polynomial $f(x)$ has no roots in F_2, hence has no linear factors, hence is irreducible over F_2. Invent a symbol u to denote a root of $f(x)=0$, so that

$$u^2 + u + 1 = 0, \qquad u^2 = -u - 1 = u + 1.$$

(Recall that $-1=+1$, modulo 2.) All higher powers of u can thereby be successively reduced to linear expressions in u. Reciprocals can be similarly reduced, so that the field generated by u contains all told just four linear expressions: 0, 1, u, $u+1$. These combine under addition and multiplication

$+$	0	1	u	$u+1$
0	0	1	u	$u+1$
1	1	0	$u+1$	u
u	u	$u+1$	0	1
$u+1$	$u+1$	u	1	0

$\cdot$	0	1	u	$u+1$
0	0	0	0	0
1	0	1	u	$u+1$
u	0	u	$u+1$	1
$u+1$	0	$u+1$	1	u

as shown in the tables. The process of obtaining this field by adjoining to the original F_2 a root u of x^2+x+1 is known as *algebraic extension* of F_2, and the resulting field $F_2(u)$ is called a Galois field of 4 elements.

For each prime p and each integral exponent n one may analogously extend the field of integers modulo p to a field consisting of exactly* p^n elements. As E. H. Moore first showed, *any* two fields with p^n elements each are algebraically indistinguishable (isomorphic). The arithmetic origin of all these finite fields is the study of algebraic integers. If $\mathfrak{p}$ is a prime ideal in a field K of algebraic numbers, then the congruences modulo this ideal behave as do ordinary congruences, and yield like them a finite field with p^n elements, where p^n is the so-called "norm" of the ideal $\mathfrak{p}$. The properties of the resulting finite fields play

* See detailed discussion of finite fields in van der Waerden [**27**, §31]; or Albert [**1**, p. 166].

an essential rôle in the class field theory and in the study of rational division algebras (Albert [2, ch. 9]).

2. Characteristics. The integers modulo p have one peculiar property. The unit 1, added p times to itself, yields $p \equiv 0 \pmod{p}$ as answer; hence

$$1 + 1 + \cdots + 1 = 0, \qquad (p \text{ summands}). \tag{3}$$

On multiplying this equation by any integer a, one has

$$a + a + \cdots + a = 0, \qquad (p \text{ summands}), \tag{4}$$

in the Galois field F_p. Any field F, all of whose elements a have the property (4), is called a field of *characteristic* p, or a *modular field*. It can be shown* that any non-modular field has an infinite characteristic, in the sense that $a \neq 0$ entails $a + a + \cdots + a \neq 0$, for any number of summands. Any finite field of p^n elements essentially contains the integers modulo p, hence satisfies (3) and therefore (4). Thus any finite field is modular.

Watch the effect of (4) on the binomial expansion,

$$(a + b)^p = a^p + pa^{p-1}b + (p(p-1)/2)a^{p-2}b^2 + \cdots + pab^{p-1} + b^p.$$

According to the genesis of this expansion, the term $pa^{p-1}b$ second on the right really represents a sum of p products $a^{p-1}b + a^{p-1}b + \cdots + a^{p-1}b$. In a field of characteristic p this sum is zero. The other intermediate terms of the binomial expansion suffer the same fate, for each binomial coefficient $p(p-1)/2, \cdots, p$ is a multiple of the characteristic p. One has left only

$$(a + b)^p = a^p + b^p, \qquad (a, b \text{ in } F \text{ of characteristic } p). \tag{5}$$

As S. C. Kleene has remarked, a knowledge of the case $p = 2$ of this equation would corrupt freshman students of algebra!

The pth power of a product is always a product of pth powers, so the rules

$$(a \pm b)^p = a^p \pm b^p, \qquad (ab)^p = a^p b^p, \qquad (a/b)^p = a^p/b^p \tag{6}$$

hold in any field of characteristic p. These rules state that the process of raising to a pth power leaves the operations of addition, division, *etc.*, unchanged. This process yields a correspondence

$$a \longleftrightarrow a^p, \qquad (\text{from } F \text{ to } F^p), \tag{7}$$

which carries the field F into the field F^p composed of all pth powers from F. The correspondence is one-to-one, for the equality of two pth powers $a^p = b^p$ would entail $0 = b^p - a^p = (b - a)^p$, and hence $b = a$. To summarize, the correspondence $a \longleftrightarrow a^p$ is an isomorphism, where an *isomorphism* between two fields is defined to be any one-to-one correspondence which preserves sums and products.

Repeated application of the rules in (6) shows that the pth power of any rational expression can be computed by applying the exponent p to each term or

* *Cf.* Albert [1, p. 30]; Mac Lane [17, §21]; van der Waerden [27, §25].

factor in the expression. In particular,

$$(1+1+\cdots+1)^p = 1^p + 1^p + \cdots + 1^p = 1+1+\cdots+1 \tag{8}$$

holds in the field of integers modulo p. If we use m summands here, this is $m^p = m$. In terms of congruences this is $m^p \equiv m \pmod{p}$, which is the little Fermat Theorem!

3. Algebraic and transcendental extensions. Our major concern is the structure of the general modular field, finite or infinite. In the analogous case of fields of numbers it is customary to distinguish the algebraic numbers, such as $\sqrt{3}$, which satisfy some polynomial equation with rational coefficients, from the transcendental numbers (e, π), which satisfy no such equation. In general, let a given field F be contained in any larger field K. An element u of K is *algebraic* over F if u is a root of a polynomial

$$f(x) = a_n x^n + a_{n-1}x^{n-1} + a_{n-2}x^{n-2} + \cdots + a_1 x + a_0 \tag{9}$$

with coefficients a_i in F. If this equation $f(x)=0$ for u be chosen with a degree n as small as possible, the polynomial $f(x)$ is *irreducible* over F. For, a reducible $f(x)$ would have factors $f(x)=f_1(x)f_2(x)$ with coefficients in F, and u would satisfy one of the equations $f_1(x)=0$, $f_2(x)=0$, of degree smaller than n. An element u in K not algebraic over F is called *transcendental*; for u transcendental, $f(u)=0$ implies that all the coefficients in $f(x)$ are zero.

Important is not the element u in K by itself, but the field $F(u)$ which it generates. The field consists of all rational combinations of u with coefficients in F, and is called a *simple extension* of F, "algebraic" or "transcendental" according as u is algebraic or transcendental over F. This dichotomy is the root of one of the basic results found by Steinitz in his pioneering investigations of fields (Steinitz [**23**]): *Any modular field can be obtained by successive transcendental and algebraic extensions of a field (isomorphic to the field) of integers modulo p.*

Such extensions can be used not only to build up a given field K from a subfield F, but also to manufacture new fields from old. Given a polynomial $f(x)$ irreducible over a field F, one can concoct a symbol u for a root of this polynomial and construct therewith an algebraic extension $F(u)$ generated by the root u. In point of fact, $F(u)$ consists of elements expressible as polynomials $b_0+b_1u+\cdots+b_{n-1}u^{n-1}$, with coefficients in F and of degree less than the degree n of the given $f(x)$.

Alternatively, a variable t over a modular field gives rise to rational functions

$$\frac{g(t)}{h(t)} = \frac{b_0 + b_1 t + \cdots + b_r t^r}{c_0 + c_1 t + \cdots + c_m t^m}, \qquad (c_i,\ b_j \text{ in } F, \text{ not all } c_i = 0). \tag{10}$$

Under the usual rules for adding and multiplying such expressions, the totality of these rational functions is a field $F(t)$ which is a simple transcendental extension of F. If F is a finite field, the resulting field $F(t)$ is the simplest instance of an infinite modular field.

4. Inseparable equations. Over the transcendental extension $F(t)$ there are in turn algebraic extensions, such as that generated by a root of the polynomial $f(x)=x^p-t$. This $f(x)$ is irreducible over $F(t)$, for if it could be factored, the denominators in t could be eliminated, and we could write $x^p-t=g(x, t)h(x, t)$, with factors which are polynomials in x and t. Since the product of these two polynomials is linear in t, one of them must be linear in t, while the other cannot involve t at all! This is absurd unless one of the factors is a constant; hence $f(x)$ is indeed irreducible.

But trouble arises with the introduction of a root u for this equation $x^p-t=0$. Since this u is a pth root of t, we have a factorization

$$x^p - t = x^p - u^p = (x - u)^p, \tag{11}$$

according to the rule (6) for the pth power of a difference. This means that u is a p-fold root of x^p-t, so this irreducible polynomial has all its roots equal, and t has only one pth root.

This differs drastically from the usual situation with ordinary complex nth roots, for an irreducible polynomial $f(x)$ with *rational* coefficients can never have a multiple root. Let us trace the proof of this fact. If $f(x)$ has a complex number r as m-fold root, then $f(x)=(x-r)^m g(x)$, with $m>1$. The derivative is

$$f'(x) = (x - r)^{m-1}[mg(x) + (x - r)g'(x)]. \tag{12}$$

Since $m>1$, this insures that $f(x)$ and $f'(x)$ have a common factor $(x-r)^{m-1}$, not a constant. But the highest common factor of $f(x)$ and $f'(x)$ can be found by the euclidean algorithm, using only rational operations. This highest common factor then has rational coefficients, and its degree is at most that of $f'(x)$. It must divide $f(x)$, counter to the assumed irreducibility of that polynomial.

Can this contradiction be deduced for a polynomial $f(x)$, irreducible not over the rationals but over some modular field, and having a multiple root r in a larger field? The derivative $f'(x)$ of calculus is no longer available, but for any polynomial $f(x)$ as in (9) a "formal" derivative can still be defined as

$$f'(x) = na_n x^{n-1} + (n - 1)a_{n-1}x^{n-2} + (n - 2)a_{n-2}x^{n-3} + \cdots + a_1. \tag{13}$$

Here the coefficient ia_i of the term x^{i-1} is to denote the sum

$$ia_i = a_i + a_i + \cdots + a_i, \qquad (i \text{ summands}). \tag{14}$$

Apply this derivative to the troublesome polynomial x^p-t of (11). We find

$$(x^p - t)' = px^{p-1} = x^{p-1} + \cdots + x^{p-1} = 0, \qquad (p \text{ summands}).$$

No wonder that an argument on the H. C. F. of x^p-t and 0 runs aground! Looking back, one sees that the argument following (12) about multiple roots will work, except in such cases when $f'(x)$ vanishes.

When do all coefficients ia_i of $f'(x)$ vanish? In a modular field $ia_i=0$ means either that a_i itself is zero, or that the number i of summands, in (14), is a multiple of the characteristic p. A coefficient a_i can thus differ from zero only for terms $a_i x^i$ with exponent $i\equiv 0 \pmod p$. The vanishing of $f'(x)$ means therefore that $f(x)$ can involve x only as powers of x^p, so that $f(x)$ has the form

$$(15) \qquad g(x) = b_m x^{mp} + b_{m-1}x^{(m-1)p} + \cdots + b_1 x^p + b_0.$$

An irreducible polynomial $g(x)$ of this form must always have p-fold roots. Such a polynomial is called *inseparable* (its roots cannot be "separated" into distinct roots). Many properties of ordinary equations fail for inseparable equations.

An element u algebraic over a modular field F is called *separable* over F if the irreducible equation for u is separable (*i.e.*, has no multiple roots). Of the inseparable algebraic elements the simplest examples are pth roots which satisfy inseparable equations $x^p = a$. Consider an arbitrary inseparable element u, root of an inseparable polynomial (15) of degree mp. This polynomial involves only pth powers of its variable, so u^p is a root of an equation

$$(16) \qquad h(y) = b_m y^m + b_{m-1}y^{m-1} + \cdots + b_1 y + b_0,$$

of smaller degree m. The adjunction of the root u to our field F can then be effected in two stages

$$F \to F(u^p) \to F(u^p, \sqrt[p]{u^p}) = F(u).$$

The element u^p first adjoined may still belong to an inseparable equation $h(y) = 0$; in that event the process can be reapplied to get u^{p^2} satisfying an equation of still smaller degree. The adjunction of an inseparable algebraic element to a modular field can be accomplished by adjoining successive pth roots of a suitable separable algebraic element (Steinitz [23]). This reduction of algebraic extensions to separable extensions followed by extensions by pth roots, indicates that the novel properties are concerned chiefly with the latter type of extension.*

5. Perfect fields. There are no inseparable algebraic elements over the field of integers modulo p, for this field already contains the pth roots of all of its elements—indeed, the Fermat Theorem, $a^p = a$, asserts that every element is its own pth root. A *perfect* field F of characteristic p is a field in which each element a has a pth root. Over such a field each pth root equation $x^p = a$ is reducible, as $x^p - a = (x - \sqrt[p]{a})^p$. More generally *any inseparable polynomial $g(x)$ involving only pth powers of x must be reducible over a perfect field.* For, each coefficient b_i of the polynomial $g(x)$ in (15) has in F a pth root $b_i^{1/p}$; according to the simple behavior of pth powers this gives a factorization

$$g(x) = (b_m^{1/p} x^m + b_{m-1}^{1/p} x^{m-1} + \cdots + b_1^{1/p} x + b_0^{1/p})^p.$$

Every finite field F is perfect, hence has no inseparable algebraic extensions. To prove this, recall the correspondence $a \longleftrightarrow a^p$ of (7), which is a one-to-one

* Technically, the least power $q = p^e$ such that u^q is separable over F is known as the *exponent* of u over F. The *degree* of u over F is the degree of its irreducible equation, while the degree of u^q is known as the *reduced degree* of u.

correspondence between *all* elements of F and those elements a^p which are pth powers. Since there are but a finite number of elements in F, there must be the same number of pth powers. This means that every element is a pth power.

A simple transcendental extension $F(t)$ of a modular field can never be perfect. To verify this we need only produce an element with no pth root in the field. The variable t itself is such an element, for if t had as pth root some rational function $g(t)/h(t)$ in the field, t would equal $[g(t)/h(t)]^p$, a pth power which can be calculated by the rule (6). In the notation of (10), the result is

$$(c_0^p + c_1^p t^p + \cdots + c_m^p t^{mp})t = b_0^p + b_1^p t^p + \cdots + b_r^p t^{rp},$$

an identity which clearly cannot hold good. For similar reasons a multiple transcendental extension $F(t_1, t_2, \cdots, t_n)$, consisting of all rational functions of n independent variables t_i, cannot be a perfect field.

6. Galois theory. To what extent can one generalize to modular fields the ordinary properties of fields of rational and algebraic numbers? A major topic is the Galois theory, which analyzes the solvability of a polynomial equation $f(x)=0$ over a field F. The roots $r_1, \cdots, r_n$ of this equation generate over F a *root field*

$$(17)\quad K = F(r_1, r_2, \cdots, r_n), \qquad \text{where} \qquad f(x) = (x - r_1)(x - r_2) \cdots (x - r_n);$$

the Galois Theory studies K in terms of its group of automorphisms, each of which is an isomorphism of the field K with itself, induced by a permutation of the roots r_i. Should these roots all be equal, the only such permutation is the identity, and the theory breaks down. Only if one assumes that the roots are all distinct, *i.e.*, that $f(x)$ is separable, does the standard theory of root fields hold* over a modular F.

This straightforward generalization does not suffice for irreducible *in*separable polynomials. The first process to fail is the construction of a "Galois resolvent," which is an equation with a root u in K such that all the roots r_i can be rationally expressed in terms of this single quantity u. In terms of fields, this means that the multiple algebraic extension $K=F(r_1, \cdots, r_n)$ can be represented as a simple extension $F(u)$. Over an imperfect field F there may be multiple algebraic extensions which cannot be so represented. Consider for instance the rational function field,

$$(18)\qquad F_0 = P(t_1, t_2), \qquad P \text{ perfect},$$

in two independent variables t_1 and t_2. An adjunction of pth roots will yield an extended field

$$(19)\qquad K_0 = F_0(u_1, u_2); \qquad u_1^p = t_1, \qquad u_2^p = t_2,$$

which consists of all elements expressible as polynomials

* *Cf.* Albert [1, ch. VIII]; van der Waerden [27, ch. 7]; Mac Lane [17, §68].

$$w = \sum_{i,j} a_{ij} u_1^i u_2^j = h(u_1, u_2), \qquad (i, j = 0, \cdots, p-1), \tag{20}$$

with coefficients a_{ij} in F_0. This field K_0 is not a simple extension $K_0 = F_0(w)$ for any w. For, if there were a generator w, then by the rule for pth powers,

$$w^p = \sum_{i,j} a_{ij}^p u_1^{ip} u_2^{jp} = \sum_{i,j} a_{ij}^p t_1^i t_2^j$$

is in F_0, so w is a pth root of an element of F_0. That such a single pth root could generate the field K_0 containing two independent pth roots u_1 and u_2 is unreasonable. This hunch can be substantiated by an argument on the degree* of the extension K_0 of F_0.

If a multiple extension does not have one generator, what is then the *minimum* number of generators? Miriam Becker [6] has recently found the answer. Over the particular field $P(t_1, t_2)$ of (18) it appears that *any* multiple algebraic extension can be expressed by two generators, just as in the case of the special extension K_0 of (19). The underlying reason is the presence of just two independent pth roots, $\sqrt[p]{t_1}$ and $\sqrt[p]{t_2}$, not in the field $P(t_1, t_2)$; the pth root of any other rational function $g(t_1, t_2)$ in the field can be expressed by the rule (6) in terms of these two pth roots, together with pth roots of coefficients which already lie in the perfect base field P.

Over any modular field F one calls the r pth roots $a_1^{1/p}, a_2^{1/p}, \cdots, a_r^{1/p}$ *p-independent* if no one of them can be rationally expressed in terms of F and the others. Becker proves that *any multiple algebraic extension of an imperfect field F can be generated by m elements, where m is the maximum number of independent pth roots over F*. If $m = 0$, F is perfect: if $m = 1$, any multiple algebraic extension is simple, as shown by Steinitz.

7. Derivatives. The solution of an ordinary equation $f(x) = 0$ by radicals (if possible) proceeds in successive stages which correspond to successive fields lying between the coefficient field F and the root field K. For a separable equation the whole array of possible intermediate fields is finite—but not so for some inseparable extensions. Between the fields F_0 and K_0 of (19) lie infinitely many distinct fields $F_0((t_1 + t_2^m)^{1/p})$, with $m = 1, p+1, 2p+1, \cdots$. For a separable equation the fields intermediate between K and F can be put into one-to-one correspondence with the sub-groups of the Galois group of automorphisms of K over F. This certainly fails for an inseparable extension like (19), for in that case the Galois group of K_0 over F_0 consists of the identity alone and so has no proper sub-groups to correspond to intermediate fields. Specifically, the Galois group consists of all isomorphisms of K_0 with itself which leave fixed each element in the base field F_0; but an isomorphism leaving fixed the elements t_1 and t_2 of F_0

* This degree is the maximum number of elements of K_0 "linearly independent" over F_0. This maximum is p^2, for any w is linearly dependent on the p^2 elements $u_1^i u_2^j$ of (20). For a simple extension $F_0(w)$ the degree would be only p. Hence $F_0(w)$ cannot equal K_0.

must likewise leave fixed their *unique* pth roots u_1 and u_2 and hence must leave all elements of K_0 fixed.

For this description of intermediate fields by the Galois group Jacobson has found a substitute, in the special case of extensions K obtained by adjoining any number of pth roots to a modular field F, as

$$K = F(a_1^{1/p}, a_2^{1/p}, \cdots, a_n^{1/p}), \qquad \text{each } a_i \text{ in } F. \tag{21}$$

By a piece of poetic justice, his solution depends on exploiting the very formal derivatives whose misbehavior (*cf.* §4) is at the root of inseparability. For example, in the field K_0 of (19) one has two "derivative" operators D_1 and D_2, defined for the arbitrary element $w=h(u_1, u_2)$ of (20) by

$$h(u_1, u_2)D_1 = \partial h(u_1, u_2)/\partial u_1, \qquad h(u_1, u_2)D_2 = \partial h(u_1, u_2)/\partial u_2. \tag{22}$$

This time the properties of pth powers are fortunate, for $u_1^pD_1=pu_1^{p-1}=0$, as it ought to be, for $u_1^p=t_1$ is in the base field and so should have derivative 0 according to the definition (22). These derivatives can be used to characterize subfields of K_0; for example, the sub-field $F_0(u_1)$ consists of everything annihilated by the operator D_2 (*i.e.*, of all w with $wD_2=0$).

In general, Jacobson considers [12] all *formal differentiation operators D* which map K into itself by a correspondence $w\rightarrow wD$ which carries elements of F into zero and which obeys the usual formal rules for differentiation:

$$(v + w)D = vD + wD, \qquad (vw)\, D = v(wD) + (vD)w.$$

From any two such operators D_1 and D_2 one may construct new differentiations $D_1 \pm D_2$, D_1^p, and D_1c, for c in F. Furthermore, the commutator $[D_1, D_2]=D_1D_2-D_2D_1$ is again a formal differentiation. This commutator satisfies the identity

$$[[D_1, D_2], D_3] + [[D_2, D_3], D_1] + [[D_3, D_1], D_2] = 0,$$

which is one of the essential postulates for a Lie algebra. The set $\mathfrak{L}$ of all differentiations is in fact a Lie algebra over the base field F. This algebra acts as a substitute for the Galois group of a field K of type (21), in the sense that *there is a one-to-one correspondence between the fields intermediate between K and F and the restricted Lie sub-algebras of the algebra $\mathfrak{L}$ of all formal differentiations of K over F.* For this purpose a *restricted* sub-algebra of $\mathfrak{L}$ is a sub-set $\mathfrak{L}'$ of $\mathfrak{L}$ which is itself a Lie algebra and which is restricted to contain D^p for each D of $\mathfrak{L}'$.

8. Algebraic geometry. A skew curve can be represented as the intersection of two surfaces, which may often be taken as cylinders

$$f(x, y) = 0, \qquad g(x, z) = 0 \tag{23}$$

with axes parallel to the z and y coördinate axes, respectively. If f and g are polynomials, the intersection of these cylinders is an algebraic curve. Alternatively, x may be viewed as a quantity transcendental over the field C of complex numbers; the polynomial equations then make the quantities y and z algebraic over the field $C(x)$ of rational functions of x. All told they give a field $C(x, y, z)$

generated by "algebraic functions" y and z of x. This field is the algebraic invariant of the curve (23). The ordinary analytic theory of these algebraic function fields can be developed, without using the geometry of the Riemann surface, if the base field C of complex numbers is replaced by a perfect modul field P or even by an imperfect one.*

In an n-dimensional euclidean space an r-dimensional algebraic manifold can be described as the set of points common to $n-r$ suitable algebraic hypersurfaces. These hypersurfaces may be taken, as in (23), in the form of "cylinders"

$$(24)\quad f_1(y_1, \cdots, y_r, y_{r+1}) = f_2(y_1, \cdots, y_r, y_{r+2}) = \cdots = f_{n-r}(y_1, \cdots, y_r, y_n) = 0,$$

where each f_i is an irreducible polynomial actually containing y_{r+i}. As coefficients in (24) we use not complex numbers but elements from a perfect modular field P. If this field P is finite, this means that we are considering a manifold in some finite affine (or projective) geometry, consisting of a finite number of "points" specified by coördinates in P. Algebraically, the symbols $y_1, \cdots, y_n$ related by (24) generate a field $K=P(y_1, \cdots, y_r, y_{r+1}, \cdots, y_n)$, consisting of all rational functions of these quantities, subject only to the rules of algebra and the special conditions (24). This field is obtained from the base field P by r successive simple extensions by the transcendentals $y_1, \cdots, y_r$, followed by $n-r$ successive algebraic extensions by the roots $y_{r+1}, \cdots, y_n$ of the polynomial equations (24). In a sense, the geometry of the manifold depends on the structure of this field.

What of the presence of inseparable equations in the definition (24) of such a manifold? Suppose, for instance, that the equation $f_1=0$ is inseparable in y_{r+1}, so that this variable appears only as a pth power. Certainly this could not simultaneously be the case for all the variables $y_1, \cdots, y_r, y_{r+1}$ in f_1, for in that event we could extract the pth root of every term in the equation $f_1=0$, thus making $f_1=(g_1)^p$, counter to the assumed irreducibility of f_1 over the perfect field P. Suppose then that y_1 is one of the variables which does not appear in $f_1(y_1, \cdots, y_r, y_{r+1})$ only as a pth power. The equation $f_1(y_1, \cdots, y_{r+1})$, which originally defined y_{r+1} inseparably over the field $P(y_1, \cdots, y_r)$, can be turned about and viewed as a definition of y_1 as a quantity *separable* and algebraic over the field $P(y_2, \cdots, y_r, y_{r+1})$, generated by the r independent transcendentals $y_2, \cdots, y_{r+1}$. A further juggling of the independent variables can then be applied to any subsequent equations of (24) which may be inseparable. Hence the result: *If a field $K=P(y_1, \cdots, y_n)$ is obtained from a perfect field P by adjoining a finite number of elements $y_1, \cdots, y_n$, one can find for K a generation $K=P(t_1, \cdots, t_r; u_1, \cdots, u_{n-r})$ involving r simple transcendental extensions by variables t_i, followed by $n-r$ separable algebraic extensions.* Whenever independent transcendents t_i in K have this property, that every element in K is *separable*

* *Cf.* general discussion of these abstract algebraic functions in Mac Lane-Nilson [**19**] or Schilling [**21**]. Especially interesting is the introduction of a Riemann Zeta function when P is finite (Hasse [**9**]), the peculiar behavior of the Weierstrass points whenever P is modular (Schmidt [**22**]), and the generalizations of Abelian functions (Schilling [**20**]).

and algebraic over $P(t_1, \cdots, t_r)$, we say that the $t_1, \cdots, t_r$ form a *separating transcendence basis* for K over P.

This construction of separating transcendence bases was discovered independently for different purposes: by the author, in connection with Albert's theory of pure forms (Albert [4]); by van der Waerden [28], for a new proof of the theorem that two distinct irreducible algebraic manifolds M_r and M_{n-r} in projective n-space intersect in a finite number of points, and, moreover, that the "number" of points, properly counted, is the product of the degrees of M_r and M_{n-r}.

9. Preservation of independence. The troubles of inseparable equations can be avoided whenever we find a separating transcendence basis for the field under consideration. Unfortunately this cannot always be done. Suppose, for instance, that the base field is the field $F_0 = P(t_1, t_2)$ of all rational functions of two transcendents t_1 and t_2 over a perfect field P, and construct a larger field L by adjoining first a new transcendent z and then an algebraic element u, with

$$u^p = t_1 + t_2 z^p, \qquad L = F_0(z, u). \tag{25}$$

Since the pth root u is inseparable over $F_0(z)$, this z is surely not a *separating* transcendence basis for L over F_0. The order of adjunction might have been inverted, adding u first as a transcendent to L and then z, but the equation (25) indicates that z would then be a pth root. The same trouble would always arise: one can prove that L has over F_0 *no* separating transcendence basis.* The same troublesome example arises in Krull's general ideal theory [13].

To find the reason for this absence of separability one must look at the possible independent pth roots in the base field F_0. In §6 we saw that the pth roots $\sqrt[p]{t_1}$ and $\sqrt[p]{t_2}$ were p-independent there, because neither can be expressed in terms of F_0 and the other. These pth roots are no longer p-independent in the top field L, for the defining equation (25) of that field gives an expression $\sqrt[p]{t_1} = u - z\sqrt[p]{t_2}$. This suggests that we restrict attention to those extensions L over F which *preserve p-independence*, in the sense that any set of p-independent pth roots over F remains p-independent over L. The relevance of this concept is indicated by the following alternative description: *a field L preserves p-independence over F if and only if the adjunction to F of any finite set of elements $y_1, \cdots, y_n$ from L yields a field $F(y_1, \cdots, y_n)$ which has over F a separating transcendence basis.*

This concept also makes it possible to find explicit conditions that given extensions have separating transcendence bases (Mac Lane [16]). One simply stated result is this: *If a field K has a finite separating transcendence basis over a sub-field M, then any field L between K and M also has a finite separating transcendence basis over M.* In other words, one can find a set S of independent tran-

* Even though, according to the Theorem of §8, L has over the *original* perfect field P a separating transcendence basis consisting of u, z, and t_1.

scendents in L, such that every element of L satisfies over $M(S)$ an algebraic irreducible equation without multiple roots.

10. General field towers. What can be said of the structure of arbitrarily complicated modular fields? The fields $P(y_1, \cdots, y_n)$ associated with algebraic manifolds had separating transcendence bases over a perfect field P. Does every modular field have a separating transcendence basis T over a suitable perfect sub-field?

The answer is no. A simple counterexample may be built from the extension $P(t)$ of a finite field P by a transcendental t. We saw in §5 that $P(t)$ is imperfect because t has in it no pth root. If we try to embed $P(t)$ in a larger field P' which will be perfect, we must have in P' a pth root $t^{1/p}$ and hence the whole rational function field $P(t^{1/p})$ generated by this root. In this field $t^{1/p}$ has no pth root, so we add $t^{p^{-2}}$, and so on, till we have the "tower"

$$P(t) \subset P(t^{p^{-1}}) \subset P(t^{p^{-2}}) \subset P(t^{p^{-3}}) \subset \cdots . \tag{26}$$

The field enveloping everything in this tower may be called $P(t^{p^{-\infty}})$; it consists of all elements lying in any one of the fields (26). Furthermore this sum field $P(t^{p^{-\infty}})$ is perfect, for an element in any one of the fields of (26) does have a pth root in the next field of the tower.

This perfect field $P(t^{p^{-\infty}})$ can have over P no separating transcendence basis. Any such basis would consist of a single transcendent t', which must lie in some one of the fields $P(t^{p^{-e}})$ of the tower (26). The generating element $t^{p^{-(e+1)}}$ of the next field is then a quantity inseparable over $P(t')$, so t' cannot have been the desired separating basis.

The tower (26) as written shows $P(t^{p^{-\infty}})$ generated by a transcendental extension followed by successive (inseparable) extractions of pth roots. Nevertheless each field of this tower, considered by itself, is a simple transcendental extension of P by $t^{p^{-e}}$. The whole field is thereby approximated by a tower of fields, each of which has a separating transcendence basis over the base field P, and each of which consists of pth powers of elements in the next field. F. K. Schmidt has shown that any perfect field P' has a similar "separating tower" over any one of its perfect sub-fields. He also stated without proof an analogous tower theorem for an imperfect field, but it was later shown by examples* that this general theorem could not hold. Recently F. K. Schmidt and the author have jointly [18] found a modified tower theorem: *If a modular field K is generated from a perfect sub-field P by a denumerable number of elements, then there is a sub-field L with a separating transcendence basis over P and a tower of fields $L \subset M_0 \subset M_1 \subset \cdots$ which collectively exhaust K, such that each M_i has over L a separating transcendence basis and is generated over L by pth powers from M_{i+1}.* The non-denumerable cases can then be broken down into a transfinite sequence

* *Cf.* Mac Lane [15]. Curiously enough, these examples involve a use of the modular law of lattice theory!

of denumerable steps, each of which "preserves p-independence" in the sense discussed in §9.

The separability of these field towers is essential to get polynomials with distinct roots, in order to apply an implicit function theorem.* This is used in the proof of the structure theorem for p-adic fields (*cf.* Hasse-Schmidt [10]). These p-adic fields are fields topologically complete with respect to a suitable norm (or "absolute value"), obtained by extending the norm for the p-adic numbers of Hensel.† These p-adic fields are not themselves modular fields, but they determine a congruence relation $a \equiv b \pmod{p}$ from which modular fields can be obtained by the standard arithmetic device.

11. Troublesome examples. The extent of our ignorance of general modular fields can be forcibly illustrated by various startling examples. The field $P(t^{p^{-\infty}})$ used to illustrate §10 was still manageable, for though it had no separating transcendence basis, it at least was itself perfect. But can there be an imperfect field K which has no separating transcendence basis over some perfect sub-field P? There is indeed such a K, for which P may even be chosen as the maximum perfect sub-field. Over a finite field P choose a countable set of indeterminates $t_1, t_2, \cdots$, and then introduce additional algebraic elements in accord with the inseparable relations

$$y_1^p = t_1 + t_2 t_3^p, \qquad y_2^p = t_2 + t_3 t_4^p, \qquad y_3^p = t_3 + t_4 t_5^p, \cdots. \tag{27}$$

Our example is the field $K = P(t_1, t_2, \cdots; y_1, y_2, \cdots)$. Since the y's are pth roots, the t's clearly cannot form a separating transcendence basis. One might try to invert the equations (27) to define everything in terms of the basis $t_1, t_2, y_1, y_2, y_3, \cdots$, but that still leaves the pth roots such as $t_3^p = (y_1^p - t_1)/t_2$. It can be shown that no method of picking a transcendence basis for K over P will yield a basis which is separating, and this example is but a taste of the trouble possible (*cf.* [15], [16]).

12. p-Algebras. The relevance of the study of inseparable extensions to other algebraic questions is clearly illustrated by the p-algebras, which are defined‡ as linear algebras over a field F of characteristic p which have as degree some power of the characteristic. The theory of these algebras, which culminates in the theorem that every such algebra is "similar" to a cyclic algebra, depends essentially on the construction of inseparable fields contained in the algebra (in technical parlance, every p-algebra has a purely inseparable splitting field). To illustrate this, choose as the base field the field $P(t)$ of all rational functions of t with coefficients in a perfect field P. Introduce a pth root u, with $u^p = t$, and a

* The so-called Hensel-Rychlik theorem; *cf.* Albert [1] or Mac Lane-Nilson [19, §11].

† See the description in C. C. MacDuffee [14].

‡ *Cf.* Albert [2, ch. 7]; and also Jacobson [11], Teichmüller [25].

quantity v with $v^p = v + t$. The set of all sums

$$w = \sum_{i,j} a_{ij} u^i v^j, \qquad (i = 0, \cdots, p-1; j = 0, \cdots, p-1; a_{ij} \text{ in } F),$$

then forms a linear algebra of degree p over F, if one uses the multiplication table

$$u^p = t, \qquad v^p = v + t, \qquad vu = u(v + 1).$$

The essential point for the theory is that this algebra contains both the inseparable extension $F(u)$ and the cyclic separable extension $F(v)$ of the base field F.

There are many further ways in which modular fields can arise in other algebraic investigations. We mention here only the use of fields of characteristic 2 in discussing Boolean algebras (Stone [**24**]), the theory of matrices over a modular field (Albert [**5**]), the definition of modular fields by special polynomials (Carlitz [**7**]), and the quasi-algebraic closure of finite fields (Chevalley [**8**]).

13. Summary. Modular fields include finite fields, Galois extensions of fields, algebraic function fields, and fields for algebraic manifolds, as well as for more bizarre types. The study of such fields is suggested by their origin in arithmetic questions about congruences, p-adic numbers, and ideal theory. On the other hand, an independent survey of their structure is indicated by the program of abstract algebra: first the development of the abstract concept ("field") in order to cover the variegated known examples, then the derivation of general theorems touching this concept, and lastly a classification of the types of systems which fall under the concept. We have seen that the straightforward generalization of the known properties of number fields is but one phase of our structure theory. There is also the investigation of characteristic new phenomena, of inseparability, of p-independence and the like, which distinguish the modular fields from the non-modular. The presence of curious examples of fields, which must at present still be given individual treatment, indicates that the present situation abounds in new questions, and that abstract algebra can very well give rise to concrete conundrums.

Bibliography

Albert, A. A., [1] Modern Higher Algebra, Chicago, 1937. [2] Structure of Algebras, American Mathematical Society Colloquium Publications, vol. XXIV, New York, 1939. [3] p-Algebras over a field generated by one indeterminate, Bulletin of the American Mathematical Society, vol. 43, 1937, pp. 733–736. [4] Quadratic null forms over a function field, Annals of Mathematics, vol. 39, 1938, pp. 494–505. [5] Symmetric and alternate matrices in an arbitrary field, I, Transactions of the American Mathematical Society, vol. 43, 1938, pp. 386–436.

Becker, M. F., and Mac Lane, Saunders, [6] The minimum number of generators for inseparable algebraic extensions, Bulletin of the American Mathematical Society, vol. 46, 1940, pp. 182–186.

Carlitz, L., [7] A class of polynomials, Duke Mathematical Journal, vol. 43, 1938, pp. 167–182.

Chevalley, C., [8] Demonstration d'une hypothèse de M. Artin, Abhandlungen aus dem mathematischen Seminar der Universität Hamburg, vol. 11, 1936, pp. 73–75.

Hasse, H., [9] Theorie der Kongruenzzetafunktionen, Sitzungsbericht der Preussischen Akademie der Wissenschaften, Berlin, 1934, pp. 250–255.

Hasse, H., and Schmidt, F. K., [10] Die Struktur diskret bewerteter Körper, Journal für die reine und angewandte Mathematik, vol. 170, 1934, pp. 4–63.

Jacobson, N., [11] p-Algebras of exponent p, Bulletin of the American Mathematical Society, vol. 43, 1937, pp. 667–670. [12] Abstract derivation and Lie algebras, Transactions of the American Mathematical Society, vol. 42, 1937, pp. 206–224.

Krull, W., [13] Beiträge zur Arithmetik kommutativer Integritätsbereiche VII, Inseparable Grundkörpererweiterung, Bemerkungen zur Körpertheorie, Mathematische Zeitschrift, vol. 45, 1939, pp. 319–334.

MacDuffee, C. C., [14] The p-adic numbers of Hensel, this MONTHLY, vol. 45, 1938, pp. 500–508.

Mac Lane, Saunders, [15] Steinitz field towers for modular fields, Transactions of the American Mathematical Society, vol. 46, 1939, pp. 23–45. [16] Modular fields *I*, Separating transcendence bases, Duke Mathematical Journal, vol. 5, 1939, pp. 372–393. [17] Notes on Higher Algebra, (planographed) Ann Arbor, 1939. With Schmidt, F. K., [18], Ueber inseparable Körper, forthcoming in Mathematische Zeitschrift. With Nilson, E. N., [19] Algebraic functions, (planographed) Ann Arbor, 1940.

Schilling, O. F. G., [20] Foundations of an abstract theory of Abelian functions, American Journal of Mathematics, vol. 61, 1939, pp. 59–80. [21] Modern Aspects of the Theory of Algebraic Functions, Mimeographed, Chicago, 1938.

Schmidt, F. K., [22] Zur arithmetischen Theorie der algebraischen Funktionen II, Allgemeine Theorie der Weierstrasspunkte, Mathematische Zeitschrift, vol. 45, 1939, pp. 75–97.

Steinitz, E., [23] Algebraische Theorie der Körper, Journal für die reine und angewandte Mathematik, vol. 137, 1910, pp. 167–308; also edited by R. Baer and H. Hasse, Berlin, 1930.

Stone, M. H., [24] The theory of representations for Boolean algebras, Transactions of the American Mathematical Society, vol. 40, 1936, pp. 37–111.

Teichmüller, O., [25] p-Algebren, Deutsche Mathematik, vol. 1, 1936, pp. 362–388.

Weiss, Marie J., [26] Algebra for the undergraduate, this MONTHLY, vol. 46, 1939, pp. 635–642.

van der Waerden, B. L., [27] Moderne Algebra, vol. I, First edition, Berlin, 1930 (also second edition, 1938). [28] Zur algebraischen Geometrie XIV, Schnittpunktszahlen von algebraischen Mannigfaltigkeiten, Mathematische Annalen, vol. 115, 1938, pp. 619–644.

7

ROBERT H. CAMERON

Robert Cameron was born in Brooklyn on May 17, 1908. He received his A.B. from Cornell in 1929, his A.M. in 1930, and his Ph.D. in 1932. He was instructor at Cornell from 1929–33, a National Research Fellow at Brown University and at the Institute for Advanced Study in 1933–35. He was instructor of mathematics at M.I.T., 1935–37, assistant professor, 1937–43, associate professor, 1943–45. In 1945 he moved to the University of Minnesota as professor and was chairman of the State College Mathematics Department, 1957–63. He retired in 1974. He was a member of the panel of the Office of Research and Development on the group project on applied mathematics at New York University during the War, 1944–45. He is a member of the American Association for the Advancement of Science, the American Mathematical Society, and the Mathematical Association of America.

His principal interests in mathematics lie in the area of almost periodic functions and transformations, integration in function spaces, and non-linear integral equations.

SOME INTRODUCTORY EXERCISES IN THE MANIPULATION OF FOURIER TRANSFORMS*

ROBERT H. CAMERON, Massachusetts Institute of Technology

1. Introduction. In this paper we will not be so much interested in the intrinsic properties of Fourier transforms themselves as in what we can do with them. "What formal manipulations can we carry on and what problems can we solve by using the Fourier transforms as one of our tools?" will be the questions we try to answer. It might therefore be well at the outset before even telling what a Fourier transform is, to give a few samples of problems it can solve for us. Perhaps if the problems interest the reader and their formal solutions intrigue or mystify him, he will be willing to read further and find out how those queer looking solutions were obtained.

Let us take as one sample the non-linear integral equation

$$\int_{-\infty}^{\infty} f(x-t)f(t)\,dt+2f(x)=g(x); \tag{1.1}$$

in which $g(x)$ is a given function of the real variable x, and $f(x)$ is the unknown function that we wish to find. To make the problem even more specific, let us assume that the given function is

$$g(x)=\frac{4x^2+10}{\pi(x^4+5x^2+4)}.$$

Now I think the reader will agree that this is a problem to which no ordinary formal methods of approach apply. Linear integral equations are bad enough, but this is not even linear since $f(x-t)$ and $f(t)$ are multiplied together. Yet it is possible by means of Fourier transforms to write down a formal answer to this problem, and then by accurate analysis justify the formal process under certain conditions. In order (we hope) to whet the reader's curiosity, we shall write out the formal solution to (1.1) immediately, withholding the explanation of how it was obtained until a later part of the paper. Here it is:

$$f(x)=\frac{1}{2\pi}\int_{-\infty}^{\infty} e^{-iux}\left\{\left[1+\int_{-\infty}^{\infty} e^{iu\xi}g(\xi)\,d\xi\right]^{\frac{1}{2}}-1\right\}du.$$

If we use the specific function

$$g(x)=\frac{4x^2+10}{\pi(x^4+5x^2+4)}$$

and substitute in the above formula, we find (after carrying out the indicated operations with the aid of a table of definite integrals) that

$$f(x)=\frac{1}{\pi(x^2+1)}.$$

* This is the fifth article in a series of expository articles solicited by the Editors.

Moreover we can readily verify by substituting this function in the original equation (1.1) that it is actually a solution of the equation.

Perhaps the reader might be interested at the start to see a few other equations whose solutions will be found by means of Fourier transforms. In many cases the solutions have to be left in the form of definite integrals, since these integrals cannot be evaluated finitely in the terms of elementary functions. However, even being able to express the answer in such a form is better than not being able to express it at all. For instance, we shall see that the differential equation

$$\frac{d^2Y}{dx^2}+\frac{dY}{dx}+xY=0 \tag{1.2}$$

has as its general solution

$$Y=A\int_0^{\infty}e^{-t^2/2}\cos\left(tx-(t^3/3)\right)dt$$
$$+B\int_0^{\infty}\left[e^{-t^2/2}\sin\left(tx-(t^3/3)\right)-e^{(t^2/2)-(t^3/3)-tx}\right]dt; \tag{1.3}$$

that the integral equation

$$\rho(x)+\int_0^{\infty}\rho(x-t)e^{-t}\,dt=\frac{1}{x^2+1} \tag{1.4}$$

has the bounded solution

$$\rho(x)=\int_0^{\infty}\frac{(2+u^2)\cos ux-u\sin ux}{4+u^2}e^{-u}\,du$$

and that the difference-differential equation

$$\frac{d}{dx}\left[f(x)\right]+f(x)+f(x+1)=\frac{1}{x^2+1} \tag{1.5}$$

has the bounded solution

$$f(x)=\int_0^{\infty}\frac{\cos xs+\cos(xs-s)+s\sin xs}{2+2\cos s+2s\sin s+s^2}e^{-s}\,ds.$$

2. The formal definition of a Fourier transform. The average paper on Fourier transforms or their applications is apt to present a rather forbidding aspect to the casual reader. One is assumed to have rather extensive knowledge of the Lebesgue integral and its properties; and in particular one is assumed to be very much at home in the spaces L_1, L_2, and L_p. Moreover it is usually taken for granted that the reader is well acquainted with the whole (very extensive) literature of Fourier Transforms, and that he is able to fit that particular paper right into the appropriate notch. Worst of all the theorems themselves are apt to merely deepen the mystery of the subject and completely discourage the reader; for in a large number of cases the hypotheses and conclusions seem to be entirely haphazard, having no relation whatsoever to anything else in mathematics, or even to each other.

Guessing the behavior of the stock market five years in advance seems a small matter in comparison to guessing what conclusions will go with what hypotheses in the theory of Fourier transforms.

Now obviously a great deal goes on under the surface; for mathematicians do not go around making altogether haphazard guesses, and then pulling out of a hat chains of logic which prove these guesses to be correct. This underlying creative thinking is almost certain to be obscured in the mass of detailed work necessary to prove the theorems rigorously; and if the author does not give a preliminary sketch or final summary of his work, his most important ideas may be lost to the reader who is not a specialist. All this is unfortunate, because many men who might advantageously use Fourier transforms as a tool in their work are discouraged and prevented from doing so.

In order to avoid this difficulty and show the simplicity of the underlying ideas of the subject, we shall in this introductory paper lay aside all ideas of mathematical rigor. We shall not be explicit as to the type of integrals used or the sense in which infinite integrals are to be interpreted. These things are to exist in some reasonable sense; and just what is "reasonable" belongs to a later phase of the subject. In this spirit we make the following definition.

DEFINITION. *If $f(x)$ is a given function, and $i=\sqrt{-1}$, the function*

$$F(x)=\frac{1}{\sqrt{2\pi}}\int_{-\infty}^{\infty} e^{isx}f(s)\,ds \tag{2.1}$$

is called the Fourier transform of $f(x)$; and we shall denote this relationship in the following way:

$$f(x)\rightrightarrows F(x).$$

This definition will seem more concrete if we actually apply it to a particular function and find its Fourier transform. Let us apply it to

$$f(x)=\exp(-a^2x^2)=e^{-a^2x^2},$$

(where a is a positive number). Replacing x by the variable of integration s, we have $f(s)=\exp(-a^2s^2)$; and substituting in (2.1), we have

$$F(x)=\frac{1}{\sqrt{2\pi}}\int_{-\infty}^{\infty} e^{isx}e^{-a^2s^2}\,ds=\frac{1}{\sqrt{2\pi}}\int_{-\infty}^{\infty} e^{-(x^2/4a^2)-(as-(ix/2a))^2}\,ds. \tag{2.2}$$

Let $t=as-(ix)/(2a)$, and substitute in (2.2) then we have*

$$F(x)=\frac{1}{\sqrt{2\pi}}\int_{-\infty}^{\infty} e^{-(x^2/4a^2)-t^2}\,dt/a$$

$$=\frac{e^{-(x^2/4a^2)}}{a\sqrt{2\pi}}\int_{-\infty}^{\infty} e^{-t^2}\,dt=\frac{e^{-(x^2/4a^2)}}{a\sqrt{2}},$$

---

* Since x goes from $-\infty$ to $+\infty$ along the real axis, t really goes from

$$-\infty-(ix)/(2a) \quad \text{to} \quad +\infty-(ix)/(2a)$$

along a line parallel to the real axis and $x/(2a)$ units below it. But it is easy to see that this line can be moved up to the real axis without changing the value of the integral.

since the probability integral

$$\int_{-\infty}^{\infty} \exp(-t^2)\,dt = \sqrt{\pi}$$

(see for instance B. O. Pierce's table of integrals or Wood's *Advanced Calculus*). We have thus obtained the result:

$$e^{-a^2x^2} \rightrightarrows \frac{1}{a\sqrt{2}} e^{-(x^2/4a^2)},$$

and in particular, when $a = 1/\sqrt{2}$,

$$e^{-\frac{1}{2}x^2} \rightrightarrows e^{-\frac{1}{2}x^2}.$$

Thus we see that $\exp(-\frac{1}{2}x^2)$ is its own Fourier transform; and we anticipate that it will play an important role in the theory for this reason.

3. Formal properties of Fourier transforms. The concept we have just defined is a useful one because of its many useful formal properties. Some of these are:

(a) *It is linear.* This means that if

$$f(x) \rightrightarrows F(x) \quad \text{and} \quad g(x) \rightrightarrows G(x),$$

then

$$f(x) + g(x) \rightrightarrows F(x) + G(x)$$

and

$$cf(x) \rightrightarrows cF(x)$$

(where c is any real or complex constant). These facts are self evident when we write them out:

$$\int_{-\infty}^{\infty} e^{isx}[f(s) + g(s)]\,ds = \int_{-\infty}^{\infty} e^{isx} f(s)\,ds + \int_{-\infty}^{\infty} e^{isx} g(s)\,ds$$

$$\int_{-\infty}^{\infty} e^{isx}[cf(s)]\,ds = c\int_{-\infty}^{\infty} e^{isx} f(s)\,ds.$$

(b) *It replaces multiplication by ix by differentiation.* Symbolically

$$ixf(x) \rightrightarrows \frac{d}{dx} F(x);$$

or written out,

$$\frac{d}{dx}\int_{-\infty}^{\infty} e^{isx} f(s)\,ds = \int_{-\infty}^{\infty} e^{isx}[isf(s)]\,ds.$$

This formula can be verified by merely carrying out the indicated differentiation; and it holds whenever differentiation under the integral sign is permissable. We shall of course not worry about such a detail at present, but will operate with this formula as though it were universally true, and check up after all formal operations are completed.

(c) *It replaces differentiation by multiplication by* $-ix$. This property is practically the same as the preceding, except that the differentiation is applied to the original function and the multiplication is applied to the transform. Symbolically

$$\frac{d}{dx} f(x) \rightrightarrows -ixF(x); \tag{3.1}$$

or written out

$$-ix \int_{-\infty}^{\infty} e^{isx} f(s)\, ds = \int_{-\infty}^{\infty} e^{isx} \left[\frac{d}{ds} f(s) \right] ds. \tag{3.2}$$

To see this formal relationship, integrate the right hand member by parts. We have

$$\int_{-\infty}^{\infty} e^{isx} f'(s)\, ds = \left[(e^{isx}) f(s) \right]_{s=-\infty}^{s=+\infty} - \int_{-\infty}^{\infty} ixe^{isx} f(s)\, ds. \tag{3.3}$$

Now if $f(s) \to 0$ as $s \to \pm\infty$, the expression in brackets drops out, and the integral which remains equals the left member of (3.2). The condition that $f(s) \to 0$ as $s \to \pm\infty$ will usually hold for the functions with which we deal; and we will regard (3.1) as a formal identity for practical manipulative purposes.

4. A second order differential equation. Before going any further with our study of the properties of Fourier transforms, we shall see how the second example mentioned in the introduction can be partially solved by the use of properties (a), (b), (c) alone. Let us suppose that the solution $Y(x)$ of the differential equation

$$\frac{d^2Y}{dx^2} + \frac{dY}{dx} + xY = 0$$

is the Fourier transform of a function $y(x)$; and let us see what differential equation $y(x)$ must satisfy. Then since $y(x) \rightrightarrows Y(x)$, we have by (b):

$$ixy(x) \rightrightarrows \frac{d}{dx} Y(x),$$

and

$$(ix)^2 y(x) \rightrightarrows \frac{d^2}{dx^2} Y(x).$$

Also by (c)

$$\frac{d}{dx} y(x) \rightrightarrows -ix\, Y(x),$$

or

$$\frac{1}{-i} \frac{d}{dx} y(x) \rightrightarrows xY(x);$$

so that we obtain finally by using (a),

$$-x^2 y(x) + ixy(x) + i \frac{d}{dx} y(x) \rightrightarrows \frac{d^2Y}{dx^2} + \frac{dY}{dx} + xY.$$

But the transform of zero is zero, so we expect $y(x)$ to satisfy the equation

$$-x^2 y + ixy + i \frac{dy}{dx} = 0$$

or

$$\frac{dy}{dx} = -(ix^2 + x) y.$$

Moreover this equation is of the first order and the variables are separable, so we may write

$$\int \frac{dy}{y} = -\int (ix^2+x)\,dx$$

and

$$\log y = -\left(i\frac{x^3}{3} + \frac{x^2}{2}\right) + \log c$$

and

$$y(x) = ce^{-i(x^3/3)-(x^2/2)}.$$

Thus we have solved the transformed equation and found $y(x)$; and since $y(x) \rightleftarrows Y(x)$, we have

$$Y(x) = \frac{c}{\sqrt{2\pi}} \int_{-\infty}^{\infty} e^{isx} e^{-i(s^3/3)-(s^2/2)}\,ds.$$

We can express this answer in terms of real quantities by using the fact that $\exp(iu) = \cos u + i \sin u$, and we obtain on putting

$$A = c\sqrt{2/\pi}\,,$$

$$\begin{aligned}
Y(x) &= \frac{A}{2}\int_{-\infty}^{\infty} e^{i(sx-(s^3/3))} e^{-(s^2/2)}\,ds \\
&= \frac{A}{2}\int_{-\infty}^{\infty} \cos\big(sx-(s^3/3)\big) e^{-(s^2/2)}\,ds \\
&\quad + i\frac{A}{2}\int_{-\infty}^{\infty} \sin\big(sx-(s^3/3)\big) e^{-(s^2/2)}\,ds \\
&= A\int_{0}^{\infty} \cos\big(sx-(s^3/3)\big) e^{-(s^2/2)}\,ds. \qquad (4.1)
\end{aligned}$$

In the last step, the sine integral vanishes since its positive and negative parts cancel; and the cosine integral from $-\infty$ to $+\infty$ is twice its value from 0 to $+\infty$ since the cosine is an even function.

The reader will note that we have obtained only one part of the solution (1.3) and may wonder why. The answer is that the second part is not the Fourier transform of any function; so when we assumed that $Y(x)$ was the transform of $y(x)$, we ruled out the second part. The missing part can be obtained by modifying our definition of a Fourier transform and will be discussed later in section 12. If the coefficient of dY/dx in (1.2) had been negative (say -1) instead of positive, we would have formally obtained

$$Y(x) = A\int_{0}^{\infty} \cos\big(sx-(s^3/3)\big) e^{s^2/2}\,ds;$$

but the exponential now becomes infinite as $s \to \pm\infty$ and the integral diverges. Thus in this case we would get neither part of the solution by the present unmodified method. However, the modification of the method given in section 12 would give both solutions in this case.

It still remains to verify that (4.1) is actually a solution of (1.2), for we have just seen that the solution need not be a Fourier transform at all. We obtained the answer by purely formal manipulation of the properties (a), (b), (c); and we have already noted that these properties depend on certain extra conditions which are not necessarily satisfied in every case. Thus we must verify first that the integral (4.1) converges and second that it satisfies the equation (1.2). We therefore note that the factor $\exp(-s^2/2)$ goes to zero so rapidly as $s\to\infty$ that the integral converges and permits all necessary manipulations; and by direct substitution in (1.2) we find that it satisfies the equation.

5. A non-homogeneous differential equation. If we analyze the methods used in partially solving (1.2), we find that the steps are these:

(1) We let $y(x)\rightrightarrows Y(x)$ and see what equation $y(x)$ must satisfy when $Y(x)$ satisfies a given equation. We call this equation for $y(x)$ the transformed equation.

(2) We solve the transformed equation for $y(x)$.

(3) We calculate $Y(x)$ by taking the transform of $y(x)$.

Any one of these steps may be impossible to carry out; but consider the second particularly. If the transformed equation is no simpler than the original equation, the method is useless. Now since (ix) factors go into derivatives and vice versa, the order of the new equation must equal the highest degree of the coefficients of the original equation and vice versa; so the method improves the situation only when the given differential equation has polynomial coefficients of lower degree than the order of the equation. But first order linear differential equations are the only ones we can formally solve for coefficients which are general functions of x, and it thus appears that our method is likely to be useful only when the coefficients of the given equation are linear functions of x. However, such equations

$$(a_0+b_0x)\frac{d^nY}{dx^n}+(a_1+b_1x)\frac{d^{n-1}Y}{dx^{n-1}}+\cdots+(a_n+b_nx)y=0 \tag{5.1}$$

form an important class; and our method does formally apply to this type of equation.

We might next enquire whether we could solve (5.1) if the right hand side were a function of x instead of zero. For instance, apply the method to

$$\frac{d^3Y}{dx^3}-3xY=x^3e^{-\frac{1}{2}x^2}.$$

Letting $y(x)\rightrightarrows Y(x)$, we have

$$(ix)^3y(x)\rightrightarrows\frac{d^3}{dx^3}Y(x)$$

$$i\frac{d}{dx}y(x)\rightrightarrows xY(x) \quad \text{and}$$

$$-ix^3y(x)-3i\frac{d}{dx}y(x)\rightrightarrows\frac{d^3Y(x)}{dx^3}-3x\,Y(x). \tag{5.2}$$

But the right hand side equals $x^3\exp(-\frac{1}{2}x^2)$, so we must find what function has this as its transform in order to know what the left side equals. Thus we see that it is necessary in working with Fourier transforms to be able to work backwards and forwards. We must not only know how to get the transform of a function, but also how to find the function corresponding to a given transform.

In the present case, this causes no difficulty, for we know that

$$e^{-\frac{1}{2}x^2} \rightrightarrows e^{-\frac{1}{2}x^2},$$

so that

$$\frac{d^3}{dx^3} e^{-\frac{1}{2}x^2} \rightrightarrows (-ix)^3 e^{-\frac{1}{2}x^3}; \quad \text{or}$$

$$i^3(-x^3+3x)e^{-\frac{1}{2}x^2} \rightrightarrows x^3 e^{-\frac{1}{2}x^2}. \tag{5.3}$$

Since the right members of (5.2) and (5.3) are equal, we shall assume that the left members are also equal (the validity of such an assumption will be discussed later). Thus we have

$$-ix^3 y(x) - 3i\frac{d}{dx}y(x) = i(x^3-3x)e^{-\frac{1}{2}x^2}, \quad \text{or} \tag{5.4}$$

$$\frac{d}{dx}y(x) + \frac{x^3}{3}y(x) = -\tfrac{1}{3}(x^3-3x)e^{-\frac{1}{2}x^2}. \tag{5.5}$$

Being a first order linear differential equation, this has the integrating factor

$$e^{\int (x^3/3)\,dx} = e^{x^4/12};$$

and multiplying (5.5) by this, we have

$$e^{x^4/12}\left\{\frac{dy}{dx} + \frac{x^3}{3}y\right\} = -\tfrac{1}{3}(x^3-3x)e^{(x^4/12)-(x^2/2)},$$

or

$$\frac{d}{dx}\left[ye^{x^4/12}\right] = -\frac{d}{dx}\left[e^{(x^4/12)-(x^2/2)}\right].$$

Integrating, we have

$$ye^{x^4/12} = -e^{(x^4/12)-(x^2/2)} + c;$$

or

$$y(x) = -e^{-\frac{1}{2}x^2} + ce^{-(x^4/12)}.$$

Finally, since $y(x) \rightrightarrows Y(x)$, we obtain

$$Y(x) = -e^{-\frac{1}{2}x^2} + \frac{c}{\sqrt{2\pi}}\int_{-\infty}^{\infty} e^{isx}e^{-(s^4/12)}\,ds.$$

The complex integral can be reduced to real form as in the preceding problem; and other parts of the solution involving other arbitrary constants can be found by the method of section 12.

6. Fourier's theorem. We have seen in the last problem that it is likely to be necessary not only to calculate Fourier transforms, but also their inverses. We need to know how to find the original function when its Fourier transform is given. Fortunately, this problem has a very simple formal solution; though the underlying theory is far from simple. This leads us to our fourth property of the Fourier transformation:

(d) *When repeated, it reproduces the original function with the sign of the independent variable changed.* Stated symbolically, this says that if

$$f(x) \rightrightarrows F(x),$$

then

$$F(x) \rightrightarrows f(-x);$$

or

$$f(x) \rightrightarrows F(x) \rightrightarrows f(-x).$$

Written out, this says that if

$$\frac{1}{\sqrt{2\pi}} \int_{-\infty}^{\infty} e^{isx} f(s)\, ds = F(x),$$

then (formally)

$$\frac{1}{\sqrt{2\pi}} \int_{-\infty}^{\infty} e^{-isx} F(s)\, ds = f(x). \tag{6.1}$$

Combining the two integrals and replacing x by s and s by t in the first, the statement is that

$$\frac{1}{2\pi} \int_{-\infty}^{\infty} e^{-isx} \left[\int_{-\infty}^{\infty} e^{its} f(t)\, dt \right] ds = f(x)$$

or

$$\frac{1}{2\pi} \int_{-\infty}^{\infty} \int_{-\infty}^{\infty} e^{is(t-x)} f(t)\, dt\, ds = f(x).$$

This can be (and originally was) stated in terms of real numbers. Thus, if we put $\exp(iu) = \cos u + i \sin u$, we find that sine cancels out by symmetry while the cosine doubles up, and we have

$$\frac{1}{2\pi} \int_{0}^{\infty} \int_{-\infty}^{\infty} \cos[s(t-x)] f(t)\, dt\, ds = f(x).$$

This is Fourier's theorem, which holds for a wide class of functions, though not by any means for all functions. We shall not however try to prove it in this form; but shall go back to Fourier transforms and our symbolic notation.

If we wish to indicate that $f(x) \rightrightarrows F(x)$ and $F(x) \rightrightarrows \mathfrak{F}(x)$, we will write

$$f(x) \rightrightarrows F(x) \rightrightarrows \mathfrak{F}(x).$$

However, if we are not interested in $F(x)$ and merely wish to indicate that $\mathfrak{F}(x)$ is the double transform of $f(x)$, we shall omit the $F(x)$ and merely write

$$f(x)\rightrightarrows\rightrightarrows\mathfrak{F}(x).$$

Thus property (d) says that

$$f(x)\rightrightarrows\rightrightarrows f(-x);$$

and we shall begin by verifying this for some simple functions.

We have already found that

$$e^{-\frac{1}{2}x^2}\rightrightarrows e^{-\frac{1}{2}x^2};$$

and of course if we apply the transformation again, we still get the same function; so

$$e^{-\frac{1}{2}x^2}\rightrightarrows\rightrightarrows e^{-\frac{1}{2}x^2}.$$

But $\exp(-\frac{1}{2}x^2)=\exp\left[-\frac{1}{2}(-x)^2\right]$, so $\exp(-\frac{1}{2}x^2)$ has the specified property (d). Let us also verify that $x^n\exp(-\frac{1}{2}x^2)$ has this property when n is a positive integer. Applying (b) n times, we have

$$(ix)^n e^{-\frac{1}{2}x^2}\rightrightarrows\frac{d^n}{dx^n}\left(e^{-\frac{1}{2}x^2}\right),$$

and applying (c) n times, we obtain

$$\frac{d^n}{dx^n}\left(e^{-\frac{1}{2}x^2}\right)\rightrightarrows(-ix)^n e^{-\frac{1}{2}x^2}.$$

Thus

$$(ix)^n e^{-\frac{1}{2}x^2}\rightrightarrows\rightrightarrows(-ix)^n e^{-\frac{1}{2}x^2};$$

or

$$x^n e^{-\frac{1}{2}x^2}\rightrightarrows\rightrightarrows(-x)^n e^{-\frac{1}{2}(-x)^2};$$

and $x^n\exp(-\frac{1}{2}x^2)$ has property (d) when n is a positive integer. But since the Fourier transformation is linear, sums of functions of this type must also have the property (d) and it follows that if $P(x)$ is any polynomial, the product

$$P(x)e^{-\frac{1}{2}x^2}=\left(a_0x^n+a_1x^{n-1}+\cdots+a_n\right)e^{-\frac{1}{2}x^2} \tag{6.2}$$

has the same property. By approximating other functions by functions of the form (6.2), it is possible to show that a large class of these other functions have property (d). In his proof of Plancherel's theorem, Wiener applies limiting processes to sequences of functions of the form (6.2) and thus shows that (d) holds for the important and extensive class of functions known as *the class* L_2. A function $f(x)$ is said to belong to the class L_2 if $f(x)$ is Lebesgue integrable between every pair of finite limits a and b and $\left[f(x)\right]^2$ is absolutely integrable from $-\infty$ to $+\infty$. Of course this includes all functions $f(x)$ which are Riemann integrable on all finite intervals and for which

$$\int_{-\infty}^{+\infty}[f(x)]^2dx$$

converges absolutely. In particular, it includes all continuous functions $f(x)$ which approach zero at $\pm\infty$ as fast or faster than $1/x$ does. Thus the functions

$$\frac{1}{\sqrt{x^2+1}},\quad e^{-x^4},\quad \frac{\sin x}{x^2+1},\quad \text{etc.} \tag{6.3}$$

belong to L_2 and so have property (d). Another class of functions having the property (d) is the class L_1 which consists of functions that are absolutely integrable from $-\infty$ to $+\infty$ in the Lebesgue sense. Such functions may have more violent discontinuities than those of L_2, but they have to approach zero at $\pm\infty$ somewhat faster. Thus,

$$\frac{1}{x^{2/3}\sqrt{x^2+1}},\quad e^{-x^4},\quad \frac{\sin x}{x^2+1},\quad \text{etc.} \tag{6.4}$$

belong to L_1 and so have the property (d). However, functions of L_1 and L_2 do not have identical properties with regard to their Fourier transforms, particularly in regard to the way the definition (2.1) is to be interpreted. Moreover the transform of a function of L_2 is again a function of L_2, while the transforms of functions of L_1 need not belong to either class. The first function of (6.3) does not belong to L_1 because it goes to zero too slowly at $\pm\infty$; while the first function of (6.4) does not belong to L_2 because its discontinuity at zero is too violent; (it approaches ∞ too fast).

Returning again to formal considerations, we find that the property (d) is very useful because it enables us to calculate many new definite integrals. For instance, if we denote the transform of* $\exp(-|x|)$ by $F(x)$, we have

$$\begin{aligned}\sqrt{2\pi}\,F(x)&=\int_{-\infty}^{\infty}e^{isx}e^{-|s|}\,ds=\int_{0}^{\infty}e^{isx}e^{-s}\,ds+\int_{-\infty}^{0}e^{isx}e^{s}\,ds\\&=\left[\frac{e^{s(ix-1)}}{ix-1}\right]_0^{\infty}+\left[\frac{e^{s}(ix+1)}{ix+1}\right]_{-\infty}^{0}\\&=-\frac{1}{ix-1}+\frac{1}{ix+1}=\frac{-2}{(ix-1)(ix+1)}=\frac{2}{x^2+1};\end{aligned}$$

so

$$e^{-|x|}\overset{\rightarrow}{\rightarrow}\sqrt{\frac{2}{\pi}}\cdot\frac{1}{x^2+1}.$$

Now applying rule (d), we reverse the order and have

$$\sqrt{\frac{2}{\pi}}\;\frac{1}{x^2+1}\overset{\rightarrow}{\rightarrow}e^{-|-x|}; \tag{6.5}$$

* The symbol $|x|$ means the absolute value (numerical value) of x, and thus $|x|=x$ when x is positive, and $|x|=-x$ when x is negative. It is never negative, and $|x|=|-x|$ for all values of x.

and this when written out gives us the new integration formula

$$\int_{-\infty}^{\infty} e^{isx} \frac{ds}{s^2+1} = \pi e^{-|x|}.$$

The correctness of this formula may be checked by means of a table of definite integrals.

7. List of the formal properties of Fourier transforms. To facilitate further formal calculations, it seems worth while to collect the various properties of Fourier transforms into one list; and we include in this list both the formulas already obtained and those which will be obtained later:

If c is a real or complex constant, τ a real constant, and

$$f(x) \rightrightarrows F(x) \quad \text{and} \quad g(x) \rightrightarrows G(x),$$

then it follows that

(a) $$f(x)+g(x) \rightrightarrows F(x)+G(x)$$

$$cf(x) \rightrightarrows cF(x)$$

(b) $$ixf(x) \rightrightarrows \frac{d}{dx} F(x)$$

(c) $$\frac{d}{dx} f(x) \rightrightarrows -ix\,F(x)$$

(d) $$F(x) \rightrightarrows f(-x)$$

(e) If $F(x) = G(x)$, then $f(x) = g(x)$

(f) $$e^{i\tau x} f(x) \rightrightarrows F(x+\tau)$$

(g) $$f(x+\tau) \rightrightarrows e^{-i\tau x} F(x)$$

(h) $$\frac{1}{\sqrt{2\pi}} \int_{-\infty}^{\infty} f(x-u)\,g(u)\,du \rightrightarrows F(x)G(x)$$

(i) $$f(x)\,g(x) \rightrightarrows \frac{1}{\sqrt{2\pi}} \int_{-\infty}^{\infty} F(x-u)G(u)\,du.$$

8. Uniqueness. From property (d) there follows immediately another important property of the Fourier transformation, namely:

(e) *It is a one-to-one transformation.* This means that to each one single function $f(x)$ there corresponds only one transform $F(x)$; and conversely that each one transform $F(x)$ is the transform of only one function $f(x)$. A function cannot have two transforms (as we see from the definition, which is not multiple valued); and a single transform cannot belong to two distinct functions. The latter fact is deeper, and says that if $F(x)=G(x)$, then $f(x)=g(x)$; or in terms of integrals, if

$$\int_{-\infty}^{\infty} e^{isx}f(s)\,ds=\int_{-\infty}^{\infty} e^{isx}g(s)\,ds \quad \text{for all real } x,$$

then

$$f(x)=g(x) \quad \text{for all* real } x.$$

This theorem holds for all classes of functions for which the property (d) holds, as the following argument shows. For if $f(x)\rightrightarrows F(x)$ and $g(x)\rightrightarrows G(x)$ then $F(x)\rightrightarrows f(-x)$ and $G(x)\rightrightarrows g(-x)$; so if $F(x)\equiv G(x)$, then $f(-x)\equiv g(-x)$, and $f(x)$ and $g(x)$ are identical.

This uniqueness property has actually been used before we formally stated it. Thus, in section 5 we drew the conclusion (5.4) by noting that the right numbers of (5.2) and (5.3) are equal and assuming that that implied that the left members were equal. This amounts to assuming that (e) holds.

9. Translation properties and difference equations. Another type of equation that Fourier transforms help to solve is the type known as difference equations, in which different values of the independent variable occur in the same function. Thus,

$$f(x)+f(x+1)+f(x+2)=e^{x}$$

is called a difference equation; while (1.5) is called a difference-differential equation because the derivatives of $f(x)$ occurs as well as $f(x)$ and $f(x+1)$. The reason Fourier Transforms can be applied in certain difference equations is that the Fourier transformation has the following translation properties:

(f) *It replaces a multiplication by* $\exp(i\tau x)$ by a translation of τ units to the left, and

(g) *Similarly, it replaces a translation of τ units to the left by multiplication by* $\exp(-i\tau x)$.

If we write these statements out symbolically they read

$$e^{i\tau x}f(x)\rightrightarrows F(x+\tau)$$

$$f(x+\tau)\rightrightarrows e^{-i\tau x}F(x),$$

and in terms of integrals they read

$$\frac{1}{\sqrt{2\pi}}\int_{-\infty}^{\infty} e^{is(x+\tau)}f(s)\,ds=\frac{1}{\sqrt{2\pi}}\int_{-\infty}^{\infty} e^{isx}\left[e^{i\tau s}f(s)\right]ds$$

---

* Actually the conclusion is true for all x except a set of Lebesgue measure zero. Thus, if $f(x)$ and $g(x)$ were equal with the exception of one single value of x where they differed, the integrals would still be equal for all x. But sets of Lebesgue measure zero are negligible for all of the calculations in which we are interested, and we therefore consider $f(x)$ and $g(x)$ as being equivalent.

and

$$e^{-i\tau x}\frac{1}{\sqrt{2\pi}}\int_{-\infty}^{\infty}e^{isx}f(s)\,ds=\frac{1}{\sqrt{2\pi}}\int_{-\infty}^{\infty}e^{isx}f(s+\tau)\,ds.$$

Written in this form, the truth of the first statement is self-evident and the second is seen to be correct as soon as we put the factor $\exp(-i\tau x)$ under the integral sign in the first member and replace s by $s'-\tau$ in the second member. The statement (g) may also be obtained from (f) by the use of (d).

Now to solve (1.5), let $f(x)\rightrightarrows F(x)$, and transform the equation

$$\frac{d}{dx}f(x)+f(x)+f(x+1)=\frac{1}{x^2+1}$$

by applying the formulas listed in section 7. Thus we have

$$\frac{d}{dx}f(x)\rightrightarrows -ix\,F(x)$$

$$f(x+1)\rightrightarrows e^{-ix}F(x)$$

and

$$\frac{d}{dx}f(x)+f(x)+f(x+1)\rightrightarrows -ixF(x)+F(x)+e^{-ix}F(x).$$

But by (6.5),

$$\frac{1}{x^2+1}\rightrightarrows\sqrt{\frac{\pi}{2}}\;e^{-|x|},$$

so the transformed equation is

$$(-ix+1+e^{-ix})F(x)=\sqrt{\frac{\pi}{2}}\;e^{-|x|}.$$

Solving for $F(x)$, we find

$$F(x)=\sqrt{\frac{\pi}{2}}\;\frac{e^{-|x|}}{(-ix+1+e^{-ix})};$$

and since by (d)

$$F(x)\rightrightarrows f(-x),$$

it follows that

$$f(-x)=\tfrac{1}{2}\int_{-\infty}^{\infty}e^{isx}\frac{e^{-|s|}}{(-is+1+e^{-is})}\,ds;$$

so

$$f(x)=\tfrac{1}{2}\int_{0}^{\infty}\frac{e^{-isx}e^{-s}\,ds}{1+e^{-is}-is}+\tfrac{1}{2}\int_{-\infty}^{0}\frac{e^{-isx}e^{s}\,ds}{1+e^{-is}-is}.$$

We next substitute $-s$ for s in the second integral and write the complex exponentials in terms of trigonometric functions:

$$f(x)=\tfrac{1}{2}\int_0^{\infty}\frac{(\cos sx-i\sin sx)e^{-s}\,ds}{1+\cos s-i\sin s-is}$$

$$+\tfrac{1}{2}\int_0^{\infty}\frac{(\cos sx+i\sin sx)e^{-s}\,ds}{1+\cos s+i\sin s+is}.$$

Finally, we put the two fractions together under one integral sign, adding the fractions in the usual way by first reducing to common denominator. All the imaginary terms now drop out, and we obtain the final answer

$$f(x)=\int_0^{\infty}\frac{[\cos sx+\cos(sx-s)+s\sin sx]}{2+2\cos s+2s\sin s+s^2}e^{-s}\,ds.$$

Of course we need to check this answer; for we have obtained it by purely formal manipulations based on very shaky logical foundations; and we have not attempted to justify each step by seeing that the functions are of the type to which the formulas apply. However, questionable methods of arriving at an answer do not invalidate the correctness of an answer if it actually satisfies the required equation; and it is easy to verify that the integral we have just obtained does converge and does satisfy the required equation. We of course do not claim that this is the only solution.

10. Certain integral equations. The integral equations that Fourier transforms help us to solve are those in which the unknown function occurs in what is called a *convolution* or *faltung* or *resultant*. The convolution $h(x)$ of two functions $f(x)$ and $g(x)$ may* be defined to be

$$h(x)=\frac{1}{\sqrt{2\pi}}\int_{-\infty}^{\infty}f(x-s)\,g(s)\,ds;$$

so that a convolution is a doubly infinite integral of the product of the two functions, the variables $x-s$ and s being substituted in the functions in place of x. Of course any other letter would do in place of s, and it does not matter in which function we substitute the s. For if $x-s=t$, we have

$$\int_{-\infty}^{\infty}f(x-s)\,g(s)\,ds=\int_{-\infty}^{\infty}f(t)\,g(x-t)\,dt;$$

since the change of sign in the differential nullifies the change of sign due to the necessary interchange of limits after substitution.

Integral equations involving convolutions frequently arise in physics and other branches of applied mathematics, and it is therefore important to know how to solve them. The reason that we can solve them is that the Fourier transformation changes convolutions into ordinary products. The transformation has properties (h) and (i):

* The constant $1/\sqrt{(2\pi)}$ is only included in the definition for convenience. The term *convolution* does not necessarily include this constant.

(h) *It replaces convolutions by products.*

(i) *It replaces products by convolutions.* Written symbolically, these statements are

$$\frac{1}{\sqrt{2\pi}}\int_{-\infty}^{\infty} f(x-u)\,g(u)\,du \rightrightarrows F(x)G(x)$$

and

$$f(x)\,g(x) \rightrightarrows \frac{1}{\sqrt{2\pi}}\int_{-\infty}^{\infty} F(x-u)G(u)\,du;$$

and written out in detail, they are

$$\frac{1}{2\pi}\int_{-\infty}^{\infty} e^{isx}\left[\int_{-\infty}^{\infty} f(s-u)\,g(u)\,du\right]ds = \frac{1}{2\pi}\left[\int_{-\infty}^{\infty} e^{itx}f(t)\,dt\right]\cdot\left[\int_{-\infty}^{\infty} e^{iux}g(u)\,du\right]$$

and

$$\frac{1}{\sqrt{2\pi}}\int_{-\infty}^{\infty} e^{isx}f(s)\,g(s)\,ds$$
$$=\frac{1}{(2\pi)^{3/2}}\int_{-\infty}^{\infty}\left\{\left[\int_{-\infty}^{\infty} e^{is(x-u)}f(s)\,ds\right]\cdot\left[\int_{-\infty}^{\infty} e^{itu}g(t)\,dt\right]\right\}du.$$

The first of these statements has a simple formal proof based on interchanging order of integration and replacing $s-u$ by t:

$$\begin{aligned}\int_{-\infty}^{\infty} e^{isx}\int_{-\infty}^{\infty} f(s-u)\,g(u)\,du\,ds &= \int_{-\infty}^{\infty}\int_{-\infty}^{\infty} e^{isx}f(s-u)\,g(u)\,du\,ds\\ &= \int_{-\infty}^{\infty}\int_{-\infty}^{\infty} e^{ix(s-u)}f(s-u)e^{ixu}g(u)\,ds\,du.\\ &= \int_{-\infty}^{\infty}\int_{-\infty}^{\infty} e^{ixt}f(t)e^{ixu}g(u)\,dt\,du\\ &= \int_{-\infty}^{\infty} e^{ixu}g(u)\left[\int_{-\infty}^{\infty} e^{ixt}f(t)\,dt\right]du\\ &= \left[\int_{-\infty}^{\infty} e^{ixt}f(t)\,dt\right]\cdot\left[\int_{-\infty}^{\infty} e^{ixu}g(u)\,du\right].\end{aligned}$$

The second statement may be obtained by combining the first statement with (d).

It is now possible to solve the first problem mentioned in the introduction:

$$\int_{-\infty}^{\infty} f(x-t)f(t)\,dt + 2f(x) = g(x);$$

for we let

$$f(x) \rightrightarrows F(x) \quad\text{and}\quad g(x) \rightrightarrows G(x),$$

and have

$$\frac{1}{\sqrt{2\pi}}\int_{-\infty}^{\infty} f(x-t)f(t)\,dt \rightrightarrows F(x)\cdot F(x)$$

so that

$$\int_{-\infty}^{\infty} f(x-t)f(t)\,dt+2f(x) \rightrightarrows \sqrt{2\pi}\,\left[F(x)\right]^2+2F(x)$$

and

$$\sqrt{2\pi}\,\left[F(x)\right]^2+2F(x)=G(x).$$

But this transformed equation is an ordinary quadratic equation in $F(x)$, and we can solve it by the quadratic formula, obtaining

$$\begin{aligned} F(x)&=\frac{-2\pm\sqrt{4+4\sqrt{2\pi}\,G(x)}}{2\sqrt{2\pi}} \\ &=\frac{1}{\sqrt{2\pi}}\left(-1\pm\sqrt{1+\sqrt{2\pi}\,G(x)}\right) \\ &=\frac{1}{\sqrt{2\pi}}\left[-1\pm\sqrt{1+\int_{-\infty}^{\infty} e^{isx}g(s)\,ds}\right]. \end{aligned}$$

Transforming back by the inverse Fourier transformation (6.1), we have

$$f(x)=\frac{1}{2\pi}\int_{-\infty}^{\infty} e^{-itx}\left[-1\pm\sqrt{1+\int_{-\infty}^{\infty} e^{ist}g(s)\,ds}\right]dt.$$

Though this is formally two solutions, the formal statement is very deceptive in this regard. It may, as a matter of fact, represent infinitely many different solutions, because we may choose the signs one way for some values of t and the other way for other values of t. However, there is one important case in which there cannot be more than one solution. This is the case in which $g(x)$ is of class L_1 and its Fourier transform is nowhere equal to $-1/\sqrt{(2\pi)}$ and we are seeking for solutions $f(x)$ of class L_1. In this case it can be shown that thre is not more than one solution of class L_1. Moreover there is a simple rule for determining whether there is one or no solution. This consists of tracing out the values taken on by the Fourier transform of $g(x)$ as x varies continuously from $-\infty$ to $+\infty$. These values will trace a continuous curve in the complex plane, beginning and ending at zero. If this curve winds an even number of times around the point $-1/\sqrt{(2\pi)}$, there will be a solution of class L_1; but if it winds an odd number of times around $-1/\sqrt{(2\pi)}$, there will be no solution of class L_1. The reason is roughly that the square root must assume the value $+1$ at both $-\infty$ and $+\infty$ if the outside integral is to converge; and this can only happen if the expression under the radical winds an even number of times around the origin. In particular, if

$$g(x)=\frac{4x^2+10}{\pi(x^4+5x^2+4)},$$

we obtain

$$f(x)=\frac{1}{\pi(x^2+1)}$$

as the unique solution of class L_1.

11. A linear integral equation. While dealing with integral equations it seems worth while to take up the more ordinary case of the linear integral equation. We shall take such a case for our last illustrative example. Let us therefore consider the equation given in (1.4), namely

$$\rho(x)+\int_0^\infty \rho(x-t)e^{-t}\,dt=\frac{1}{x^2+1}.$$

The integral in this equation does not appear to be a convolution because it is only taken from 0 to $+\infty$ instead of from $-\infty$ to $+\infty$. However, we can replace the lower limit by $-\infty$ if we replace $\exp(-t)$ by a function $g(t)$ which equals $\exp(-t)$ whenever t is positive but equals zero when t is negative. For then we have

$$\int_{-\infty}^\infty \rho(x-t)\,g(t)\,dt=\int_{-\infty}^0 \rho(x-t)\,g(t)\,dt+\int_0^\infty \rho(x-t)\,g(t)\,dt$$

$$=\int_{-\infty}^0 \rho(x-t)\cdot 0\cdot dt+\int_0^\infty \rho(x-t)e^{-t}\,dt=\int_0^\infty \rho(x-t)e^{-t}\,dt;$$

and the equation becomes

$$\rho(x)+\int_{-\infty}^\infty \rho(x-t)\,g(t)\,dt=\frac{1}{x^2+1}.$$

Now if $\rho(x)\rightleftarrows\Phi(x)$ and $g(x)\rightleftarrows G(x)$, it follows that

$$\rho(x)+\int_{-\infty}^\infty \rho(x-t)\,g(t)\,dt\rightleftarrows\Phi(x)+\sqrt{2\pi}\,\Phi(x)G(x);$$

and since

$$\frac{1}{x^2+1}\rightleftarrows\sqrt{\frac{\pi}{2}}\;e^{-|x|},$$

we obtain the transformed equation

$$\Phi(x)\left[1+\sqrt{2\pi}\,G(x)\right]=\sqrt{\frac{\pi}{2}}\;e^{-|x|}.$$

But

$$G(x)=\frac{1}{\sqrt{2\pi}}\int_{-\infty}^\infty e^{isx}g(s)\,ds$$

$$=\frac{1}{\sqrt{2\pi}}\int_{-\infty}^0 e^{isx}\cdot o\cdot ds+\frac{1}{\sqrt{2\pi}}\int_0^\infty e^{isx}e^{-s}\,ds$$

$$=\frac{1}{\sqrt{2\pi}}\int_0^\infty e^{s(ix-1)}\,ds=\frac{1}{\sqrt{2\pi}}\,\frac{1}{1-ix}$$

and hence

$$\Phi(x)\left[1+\frac{1}{1-ix}\right]=\sqrt{\frac{\pi}{2}}\; e^{-|x|}$$

and

$$\Phi(x)=\sqrt{\frac{\pi}{2}}\left(\frac{1-ix}{2-ix}\right)e^{-|x|}.$$

Transforming back, we have

$$\rho(x)=\tfrac{1}{2}\int_{-\infty}^{\infty} e^{-isx}\frac{(1-is)}{(2-is)}e^{-|s|}\,ds;$$

and after the usual simplification, we obtain the answer given in the introduction. Substitution shows that this is correct.

12. Integrals allied to the Fourier transform. Let us consider a function $f(x)$ defined on the interval from 0 to $+\infty$; and from this function let us construct four functions, all of which are defined on the whole interval from $-\infty$ to $+\infty$, as follows:

$$f_c(x)=\begin{cases} f(x) & \text{when } x \text{ is positive} \\ f(-x) & \text{" " " negative} \end{cases}$$

$$f_s(x)=\begin{cases} -if(x) & \text{" " " positive} \\ if(-x) & \text{" " " negative} \end{cases}$$

$$f_l(x)=\begin{cases} \sqrt{2\pi}\; f(x) & \text{" " " positive} \\ 0 & \text{" " " negative} \end{cases}$$

$$f_m(x)=\sqrt{2\pi}\; f(e^x) \qquad \text{for all real } x.$$

Then if $F_c(x)$, $F_s(x)$, $F_l(x)$, $F_m(x)$ are the Fourier transforms of $f_c(x)$, $f_s(x)$, $f_l(x)$, $f_m(x)$, we obtain by formal substitution in the definition (2.1) and formal simplification the integrals

$$F_c(x)=\sqrt{\frac{2}{\pi}}\int_0^{\infty} f(s)\cos sx\, ds$$

$$F_s(x)=\sqrt{\frac{2}{\pi}}\int_0^{\infty} f(s)\sin sx\, ds$$

$$F_l(ix)=\int_0^{\infty} f(s)e^{-sx}\, ds$$

$$F_m(ix)=\int_0^{\infty} f(t)t^{x-1}\, dt.$$

These four important integrals are known as the Fourier cosine transform, the Fourier sine transform, the Laplace transform, and the Mellin transform respectively of $f(x)$. They have properties somewhat similar to those of the Fourier transforms, yet differing from them in many important points. For a discussion of these properties, the reader should consult Titchmarsh's *Introduction to the Theory of Fourier Integrals*.

A different type of modification of the Fourier transform is obtained when we deal with functions of a complex variable and use some path of integration other than the real axis. Thus the modified Fourier transform is

$$F(x)=\frac{1}{\sqrt{2\pi}}\int_C e^{isx}f(s)\,ds \tag{12.1}$$

where the contour C is chosen so that as many as possible of the formal properties of Fourier transforms still hold. In particular, C should either be a closed curve or a curve which goes to infinity in some direction at both ends. This is necessary to preserve property (c); for if there are finite end points, the values at these end points will have to be substituted in the UV term of the integration by parts, and an extra term will crop up and spoil property (c). If the contour is infinite, it must of course be chosen so that $f(s)$ approaches zero as we approach infinity.

The contour integral (12.1) satisfies properties (a), (b), and (c); and since these were the only properties used in section 4 in solving (1.2), this integral could have been used there instead of the integral going from $-\infty$ to $+\infty$. But the only place in which the integral itself was used was in the last step, where we obtain $Y(x)$ from $y(x)$. Thus the only difference that the use of (12.1) could make would be that the final integral would be taken over C instead from $-\infty$ to $+\infty$. But such an answer would have just as much formal justification as the one we actually obtained, and this leads us to wonder whether every contour would produce the same answer or would at least produce a solution to the problem. As a matter of fact, every contour along which the integral converges properly does give us a solution to the equation; but not necessarily the same solution; and this enables us to complete the solution of the problem. You remember that we obtained only part of the general solution and mentioned that a method would later be given by which we could obtain the rest. That method is merely to obtain different parts of the solution by varying the contour used, and then to add these parts together to form the general solution. Since the equation with which we are dealing is linear and has its right hand member zero, the sum of two solutions is again a solution, and this process is valid.

In the present case, our solution is to be a multiple of

$$\int_C e^{i(sx-(s^3/3))}e^{-(s^2/2)}\,ds,$$

where C is to be chosen so as to approach infinity in some other way than positively and negatively along the real axis. We wish the integrand to approach zero, and this means that the real part of the exponent

$$i(sx-(s^3/3))-(s^2/2)$$

must approach $-\infty$. Now for numerically large s, the numerically largest term is $-is^3/3$; and this will be real if s is pure imaginary. If $s=it$, then $-is^3/3=-t^3/3$, and it approaches $-\infty$ as s approaches infinity. Thus we can let one end of C go out along the upper part of the imaginary axis; and of course the other end can go out in either of the directions used before. Let us therefore take C as a contour starting at $i\infty$ and coming down the imaginary axis to zero, and then turning right and going out along the real axis to $+\infty$. Using this contour, we have

$$-Bi\int_C e^{i(sx-(s^3/3))}e^{-(s^2/2)}ds$$

$$=-Bi\int_\infty^0 e^{i(itx+(it^3/3))}e^{t^2/2}i\,dt-Bi\int_0^\infty e^{i(sx-(s^3/3))}e^{-(s^2/2)}ds$$

$$=-B\int_0^\infty e^{-(t^3/3)+(t^2/2)-tx}dt$$

$$-Bi\int_0^\infty\left[\cos(sx-(s^3/3))+i\sin(sx-(s^3/3))\right]e^{-(s^2/2)}ds.$$

We may as well drop the cosine term of this solution, as it is just like the part already obtained and is therefore itself a solution and may be included with the first part by a change of the constant A. We therefore have as the second part of our solution

$$B\int_0^\infty\left[e^{-(s^2/2)}\sin(sx-(s^3/3))-e^{-(s^3/3)+(s^2/2)-sx}\right]ds;$$

and the problem is completely solved. The convergence of this second integral and the fact that it is a solution of the differential equation can be directly verified; and it can also be shown that neither part of the solution is identically zero or a constant multiple of the other part.

13. Conclusion and warning. One frequently hears the statement that a little knowledge of medicine is a dangerous thing. A similar statement might well be made in regard to certain branches of mathematics; particularly Fourier transforms. A mere formal knowledge of Fourier transforms will lead the manipulator to all sort of false conclusions. The situation here is much worse than in elementary calculus where lack of mathematical rigor and optimistically formal use of limit theorems *may* lead to false conclusions, but usually do not. Here purely formal work is sure to lead one into difficulties, and rather soon at that. You see, in this work there is really nothing that can be called "formally correct," because the formal rules are not even self consistent unless we put strong restrictions on the functions; and when we begin to state these restrictions there is no half way about it. We have to go all the way and do exact, rigorous mathematics.

As we stated in the introduction, our purpose here is merely to give the reader a general idea of the way a Fourier transform ordinarily behaves when suitably restricted. We hope that the reader may become interested in Fourier transforms as he sees the sort of thing that can be done with them, and that he may be willing to take time to learn the details of their exact behavior after he has had this little non-technical glance at the way they act when the machinery is well oiled with sufficiently powerful hypotheses. If the reader would like to gain a real understanding of the subject, he should study one of the standard works such as Titchmarsh's *Theory of Fourier Integrals*, Carslaw's *Fourier Series and Integrals*, Bochner's *Vorlesungen über Fouriersche Integrale*, Wiener's *The Fourier Integral*, and Paley and Wiener's *Fourier Transforms in the Complex Domain.*

8

PAUL RICHARD HALMOS

Paul Richard Halmos was born in Budapest, Hungary, on March 3, 1916. He came to the United States in 1929 and received his B.S. from the University of Illinois in 1934, his M.S. in 1935, and Ph.D. in 1938 (under the supervision of J. L. Doob). At Illinois he was assistant in mathematics, 1935–36, instructor, 1938–39, and associate, 1942–43. During the period 1939–42 he was a Fellow at the Institute for Advanced Study at Princeton where he came under the influence of John von Neumann. During this period he wrote the preliminary edition of his first book, *Finite Dimensional Vector Spaces*. This book is still in use today in a revised edition, 1958, and has been influential in bringing linear algebra and matrix theory into the undergraduate curriculum of today.

Professor Halmos has had a varied career in many of the great American universities. He was assistant professor at Syracuse University, 1943–46, and then at the University of Chicago, 1946–49. He became associate professor in 1949 and professor in 1956. In 1961 he went to the University of Michigan where he remained until 1968. During the academic year 1968–69 he served as chairman of the Department of Mathematics at the University of Hawaii. He was Distinguished Professor at Indiana University, 1969–76, and is now at the University of California at Santa Barbara. During the war he was with the Radiation Laboratory at the Massachusetts Institute of Technology. He was named Guggenheim Fellow for 1947–48. He has also served as visiting professor at Montevideo, 1951–52, and at Miami University in 1965–66. He is a member of the American Mathematical Society and the Mathematical Association of America. He served as editor of the *Proceedings*, 1956–62, *Mathematical Reviews*, 1964–69, *Mathematical Surveys*, 1973–75, and the *Bulletin* since 1974. He was editor of two book series for Van Nostrand, and is currently editor of three for Springer-Verlag. In addition to winning the Chauvenet Prize in 1947, he won the Lester R. Ford award in 1970.

Professor Halmos' areas of research have been mainly in three parts of modern mathematics: Operator theory in Hilbert spaces, ergodic theory, and algebraic logic. He has written a total of eight books, including *Measure Theory*, 1950; *Introduction to Hilbert Spaces*, 1951; *Lectures on Ergodic Theory*, 1956; *Naive Set Theory*, 1960; *Algebraic Logic*, 1962; *Lectures on Boolean Algebras*, 1963; and a *Hilbert Space Problem Book*, 1967. Each of these books has left its mark, both as a work of mathematics and as an example of the best in exposition.

Professor Halmos has always been the kind of mathematician and expositor that the Chauvenet prize was designed to encourage. From the very beginning he has shown concern for his students. In the preface to the first edition of his *Finite Dimensional Vector Spaces* he states:

> "Addressing the advanced undergraduate or beginning graduate student, I treat linear transformations of finite dimensional vector spaces by the methods of more general theories. My purpose is to emphasize the simple geometric notions common to many parts of mathematics and its applications, and to do this in a language which gives away the trade secrets and tells the student what is in the back of the mind of people proving theorems about integral equations and Banach spaces."

He has consistently held to these principles of making the secrets of mathematical research clear to his audiences, avoiding the abstruse language which all too frequently obscures the true meaning of mathematical ideas. At the 1974 annual meeting of the Association he participated in a symposium on the Problems of Learning to Teach where he reiterated his feeling that students must learn "to ask and to do." Throughout his books he is concerned, not so much with exhibiting his own prowess, as with developing the student's ability to do mathematics. His writing is accessible to anyone who wants to learn, and this, together with his delightful sense of humor, makes it a joy to read. His books, his lectures, and his teaching reflect the very spirit which led Julian Lowell Coolidge to suggest the creation of a prize to stimulate the best in expository mathematics.

THE FOUNDATIONS OF PROBABILITY

P. R. HALMOS, Syracuse University

1. Introduction. Probability is a branch of mathematics. It is not a branch of experimental science nor of armchair philosophy, it is neither physics nor logic. This is not to say that the experimenter and the philosopher should not discuss probability from their points of view. They should, and they do. The situation is analogous to that in geometry. No one denies that the physicist and the philosopher have made valuable contributions to our understanding of the space concept, nor, in spite of this, that geometry is a rigorous part of modern mathematics.

Like Euclidean geometry, and for that matter like most mathematical theories, probability has four aspects: axiomatization, development, coordinatization, and application. We proceed to explain our use of these words.

"Axiomatization" is clear. We all know that the study of geometry begins with a list of undefined terms and a list of postulates. It is important in this connection to remember two facts. First: the selection of the list of terms and postulates is not entirely arbitrary, but is derived only after a thorough examination of our intuitive notions of the subject. Second: the selection of terms and postulates is not uniquely determined. When several different axiomatizations of the same subject exist then only extra mathematical considerations, such as practical convenience or personal prejudice, can lead us to prefer one among the many. The greater part of this paper is devoted to a prepostulational examination of probability. The axiomatic system to which this examination leads is not the only possible approach to probability, but it is the approach which has been adopted by the majority of workers in this field.

By "development" we mean simply the main part of the theory, the definitions and theorems which chiefly occupy the professional mathematician. "Coordinatization" is a general process the most familiar instance of which is the proof of the equivalence of the synthetic and analytic aspects of Euclidean geometry. The isomorphism of a finite group to a group of permutations and the representation of an algebra by matrices are further examples of this process. Properly speaking coordinatization is just one of the theorems belonging to development, but a theorem of such fundamental implications that it effects basic changes in the appearance, methods, and results of the entire theory.

The hardest philosophical problem in geometry as well as in probability is the problem of "application." Do the theorems derived from the postulates reflect any light on the physical world which suggested them, and if so, how and why?

The purpose of this paper is exposition, exposition intended to convince the professional mathematician that probability is mathematics. To this end we shall discuss the four features just enumerated. The paper contains almost no proofs, very few precise definitions and theorems, and many heuristic derivations. Despite however the small number of rigorous statements, they form the foundation on which the remainder is built. For the convenience of the reader they are italicized. If these italicized statements are lifted from their context and read consecutively, they will furnish at least a partial answer to the question "what is probability?"

2. Boolean algebra. The principal undefined term in probability theory is "event." Intuitively speaking an event is one of the possible outcomes of some physical experiment.

To take a rather popular example consider the experiment of rolling an ordinary six-sided die and observing the number $v(=1, 2, 3, 4, 5, \text{ or } 6)$ showing on the top face of the die. "The number v is even"—"it is less than 4"—"it is equal to 6"—each such statement corresponds to a possible outcome of the experiment. From this point of view there are as many events associated with the experiment as there are combinations of the first six positive integers taken any number at a time. If for the sake of aesthetic completeness and later convenience we consider also the impossible event, "the number v is not equal to any of the first six positive integers," then there are altogether 2^6 admissible events associated with the experiment of the rolling die. For the purpose of studying this example in more detail let us introduce some notation. We write $\{246\}$ for the event "v is even," $\{123\}$ for "v is less than 4," and so on. The impossible event and the certain event $(=\{123456\})$ deserve special names: we reserve for them the symbols o and e respectively.

Everyday language concerning events uses such phrases as these: "two events a and b are incompatible or mutually exclusive," "the event a is the opposite of the event b or complementary to b," "the event a consists of the simultaneous occurrence of b and c," "the event a consists of the occurrence of at least one of the two events b and c." Such phrases suggest that there are relations between events and ways of making new events out of old that should certainly be a part of their mathematical theory.

The notion of complementary event is probably closest to the surface. If a is an event we denote the complementary event by a': an experiment one of whose outcomes is a will be said to result in a' if and only if it does not result in a. Thus if $a=\{246\}$ then $a'=\{135\}$. We may also introduce combinations of events suggested by the logical concepts of "and" and "or." With any two events a and b we associate their "join" $a\cup b$ (also called union or sum and often denoted by $a+b$), and their "meet" $a\cap b$ (or intersection or product, often denoted by ab). Here $a\cup b$ occurs if and only if at least one of the two events a or b occurs, while $a\cap b$ occurs if and only if both a and b occur. Thus if $a=\{246\}$ and $b=\{123\}$ then $a\cup b=\{12346\}$ and $a\cap b=\{2\}$.

The operations a', $a\cup b$, and $a\cap b$ satisfy some simple algebraic laws. It is clear for example that both the expressions $a\cup b$ and $a\cap b$ are independent of the order of the terms (commutative law), and that neither of the expressions $a\cup b\cup c$ and $a\cap b\cap c$ depends on the order in which the two indicated operations are performed (associative law). These facts are intuitively obvious from the verbal definition of the operations and are easily verified in any finite case such as the rolling die. There are many other similar identities satisfied by these methods of combining events: the following is a list of the most important ones.

$$o' = e \qquad (a')' = a \qquad e' = o$$

$$(a \cap b)' = a' \cup b' \qquad (a \cup b)' = a' \cap b'$$

$$a \cap a' = o \qquad a \cup a' = e$$

$$o \cap a = o \qquad o \cup a = a$$

$$e \cap a = a \qquad e \cup a = e$$

$$a \cap b = b \cap a \qquad a \cup b = b \cup a$$

$$(a \cap b) \cap c = a \cap (b \cap c) \qquad (a \cup b) \cup c = a \cup (b \cup c)$$

$$a \cap (b \cup c) = (a \cap b) \cup (a \cap c) \qquad a \cup (b \cap c) = (a \cup b) \cap (a \cup c)$$

A system B of elements $o, a, b, \cdots, e$ in which operations a', $a \cup b$, and $a \cap b$ are defined in such a way that each of the above list of identities is satisfied is called a "Boolean algebra." For the traditional theory of probability, concerned with simple gambling games such as the rolling die, in which the total number of possible events is finite, the above heuristic reduction of events to elements of a Boolean algebra is adequate. For situations arising in modern theory and practice, and even for the more complicated gambling games, it is necessary to make an additional assumption. This assumption, in descriptive terms, is that the operations $\cup$ and $\cap$, assumed defined for two elements and immediately extended by mathematical induction to any finite number, should make sense also for an infinite sequence. In other words it is desirable to have an interpretation for symbols such as $a_1 \cup a_2 \cup \cdots$ and $a_1 \cap a_2 \cap \cdots$. In order to phrase precisely this assumption of infinite operations it is necessary to use a few simple facts from the theory of Boolean algebras.

If a and b are any two elements of the Boolean algebra B which satisfy the relation $a \cup b = b$ (or the equivalent relation $a \cap b = a$) we shall write $a \subset b$ and say that "a is smaller than b" or "a is contained in b" or "a implies b." The intuitive interpretation of this relation is as follows: the event a implies the event b, or is contained in the event b, if the occurrence of a is a sub-case of the occurrence of b. Thus in the example of the die $\{123\} \subset \{1234\}$ and "$v=2$" $\subset$ "v is even." The technical significance of the relation $\subset$ is that the operations $\cup$ and $\cap$ may be defined in terms of it. For example $a \cup b$ is the smallest of all elements which contain both a and b. In more detail: given a and b, consider all c's for which both $a \subset c$ and $b \subset c$. The assertion concerning $a \cup b$ is two fold: first, $a \cup b$ is an admissible c, and second, for any admissible c we have $a \cup b \subset c$. As an example consider $a = \{12\}$ and $b = \{24\}$. The elements $\{1234\}$, $\{1246\}$, $\{12456\}$, $\cdots$ all have the property of containing both a and b. However the element $\{124\}$, which also has that property, is smaller than any other such element, and it is in fact true that $\{12\} \cup \{24\} = \{124\}$.

Motivated by the relation between $\cup$ and $\subset$ we now proceed as follows. Let B be a Boolean algebra. If for every infinite sequence $a_1, a_2, \cdots$ of elements of

B there exists among the elements containing all the a_n a smallest one, say a, we say that B is a σ-algebra and we write $a = a_1 \cup a_2 \cup a_3 \cup \cdots$. Not every Boolean algebra is a σ-algebra; the assumption that B is one (the hypothesis of countable additivity) is an essential restriction.

Perhaps an example, though a somewhat artificial one, might illustrate the need for the added assumption. Suppose that a player determines to roll a die repeatedly until the first time that the number showing on top is 6. Let a_n be the event that the first 6 appears only on the nth roll. The event $a = a_1 \cup a_2 \cup a_3 \cup \cdots$ occurs if and only if the game ends in a finite number of rolls. The occurrence of the opposite event a' is at least logically (even if not practically) conceivable and it seems reasonable to want to include a discussion of it in a general theory of probability. Numerous examples of this kind together with some rather deep lying technical reasons justify therefore the following statement.

The mathematical theory of probability consists of the study of Boolean σ-algebras.

This is not to say that all Boolean σ-algebras are within the domain of probability theory. In general statements concerning such algebras and the relations between their elements are merely qualitative: probability theory differs from the general theory in that it studies also the quantitative aspects of Boolean algebras. In the next section we shall describe and motivate the introduction of numerical probabilities.

3. Measure algebra. When we ask "what is the probability of a certain event?" we expect the answer to be a number, a number associated with the event. In other words probability is a numerically valued function P of events a, that is of elements of a Boolean σ-algebra B, $P = P(a)$. On intuitive and practical grounds we demand that the number $P(a)$ should give information about the occurrence habits of the event a. If in a large number of repetitions of the experiment which may result in the event a we observe that a actually occurs only a quarter of the time (the remaining three quarters of the experiments resulting therefore in a') we may attempt to summarize this fact by saying that $P(a) = 1/4$. Even this very rough first approximation to what is desired yields some suggestive clues concerning the nature of the function P.

If, to begin with, $P(a)$ is to represent the proportion of times that a is expected to occur, then $P(a)$ must be a positive real number, in fact a number in the unit interval $0 \leqq P(a) \leqq 1$. The extreme value 0 has a special significance. Since the impossible event o will never occur, it is clear that we must write $P(o) = 0$. Conversely however if an event a refuses ever to occur, we are tempted to declare its occurrence impossible and thus from the relation $P(a) = 0$ to deduce $a = o$. The other extreme value of $P(a)$ has of course a similar interpretation: $P(a) = 1$ if and only if $a = e$.

The relation between proportion and probability has further consequences. Suppose that a and b are mutually exclusive events—say $a = \{1\}$ and $b = \{246\}$ in the example of the die. (In the algebraic theory mutually exclusive events correspond to "disjoint" elements of the Boolean algebra B, that is to elements

a and b for which $a \cap b = o$.) In this case the proportion of times that the join $a \cup b (= \{1246\}$ for the example) occurs is clearly the sum of the proportions associated with a and b separately. If an ace shows up one-sixth of the time and an even number half the time, then the proportion of times in which the top face is either an ace or an even number is $\frac{1}{6}+\frac{1}{2}$. It follows therefore that the function P cannot be completely arbitrary—it is necessary to subject it to the condition of additivity, that is to require that if $a \cap b = o$ then $P(a \cup b)$ should be equal to $P(a)+P(b)$.

We are now separated from the final definition of probability theory only by a seemingly petty (but in fact very important) technicality. If $P(a)$ is an additive function of the sort just described on a Boolean σ-algebra B, and if $a_1, a_2, \cdots, a_n$ is any finite set of pairwise disjoint elements of B (this means that for $i \neq j$, $a_i \cap a_j = o$) then it's easy to prove by mathematical induction that $P(a_1 \cup a_2 \cup \cdots \cup a_n) = P(a_1)+P(a_2)+ \cdots +P(a_n)$. If however $a_1, a_2, a_3, \cdots$ is an infinite sequence of pairwise disjoint elements then it may or may not be true that $P(a_1 \cup a_2 \cup a_3 \cup \cdots) = P(a_1)+P(a_2)+P(a_3)+ \cdots$. The general condition of countable (that is, finite or enumerably infinite) additivity is a further restriction on the probability measure P—a restriction without which modern probability theory could not function. It is a tenable point of view that our intuition demands infinite additivity just as much as finite additivity. At least however infinite additivity does not contradict any of our intuitive ideas and the theory built on it is sufficiently far developed to assert that the assumption is justified by its success. We shall therefore adopt this assumption as our final postulate.

Numerical probability is a measure function, that is a finite, nonnegative, and countably additive function P of elements in a Boolean σ-algebra B, such that if the null and unit elements of B are o and e respectively then $P(a)=0$ is equivalent to $a=o$ and $P(a)=1$ is equivalent to $a=e$.

In the next section we shall discuss a general method of constructing examples of probability measures.

4. Measure space. Let $\omega_j (j=1, \cdots, 6)$ be the point on the real axis whose directed distance from the origin is j, and let Ω be the set whose elements are these six points. Consider the system B^* of all subsets of Ω. (The empty set o and the full set $e=\Omega$ are counted as belonging to B^*.) With any element a of B^* (that is, with any subset of Ω) we may associate the complementary element (set) consisting of exactly those points ω_j which do not belong to a. Similarly with any two subsets a and b of Ω we may associate their union (the set of points belonging to either a or b or both), and their intersection (the set of points belonging simultaneously to a and b). It is easy to verify that under the operations of complementation (a'), formation of unions ($a \cup b$), and formation of intersections ($a \cap b$), the system B^* forms a Boolean algebra, in fact, though somewhat vacuously, a σ-algebra. Suppose moreover that for each $j=1, \cdots, 6$, p_j is a positive number such that $p_1+ \cdots +p_6=1$. Then we may define $P(a)$ for any subset a of Ω, to be the sum of those p_j whose ω_j belongs to a. Thus if

$a = \{135\}$ then $P(a) = p_1 + p_3 + p_5$; if $a = o$ then $P(a) = 0$. The function P and the algebra B^* satisfy all the assumptions of probability theory and the reader has doubtless recognized that this B^* and P were implicit in our earlier discussion of the rolling die. It is often customary on philosophical and practical grounds to discuss only the case $p_1 = \cdots = p_6 = \frac{1}{6}$. We shall say a word about this special case later; for the moment it is sufficient to point out that any other choice of the p_j furnishes an equally acceptable probability structure and does in fact constitute the mathematical theory of some carefully loaded die.

The above example of a Boolean algebra can be generalized: we attempt next to obtain a similar but more geometrical example. For this purpose we again choose a set Ω, but, instead of a finite set, we choose a set with infinitely many points, in fact all the points of a continuum. To be specific let us choose for Ω the points ω of a square of unit area in the Cartesian plane. In analogy with the preceding example we consider the system B^* of all subsets of Ω and define complement, union, and intersection as before. Once more B^* is a Boolean σ-algebra; it is not however the one on which we shall base our probability theory. (It can be shown that it is not possible to define a probability measure P with the desired properties on B^*.) We shall instead consider a certain subsystem (sub-algebra) of B^*, constructed as follows:

We begin with the system R of all rectangles contained in Ω (where for the sake of definiteness we consider closed rectangles, that is sets consisting of the interior plus the perimeter of a rectangle). The system R is not closed under the Boolean operations: in general not even a finite (let alone a countably infinite) union or intersection of rectangles is itself a rectangle, and similarly the complement of a rectangle isn't one. We have therefore to enlarge the system R to a system R' including all complements and countable unions and intersections of elements of R. It turns out that even this is not enough: R' is still not a Boolean algebra, and the extension process has to be continued. If however the extension process is continued sufficiently (and this happens to mean transfinitely) often, we reach eventually a Boolean σ-algebra B of subsets of Ω. (The algebra B is important in analysis: sets of B are called the Borel sets of the square.)

We face next the task of defining P. For those familiar with the theory of Lebesgue measure it will suffice to say that we define $P(a)$, for each a in B, to be the Lebesgue measure of the set a. It is not difficult to get an intuitive idea of how P is defined. If a is a rectangle (that is an element of R) we define $P(a)$ to be the area of a. If a is an element of R' we proceed to determine $P(a)$ in accordance with the requirement of countable additivity. Thus for example if b is the complement of a rectangle a, we write $P(b) = 1 - P(a)$, and if b is the union of a finite or infinite sequence of disjoint rectangles $a_1, a_2, \cdots$ we define $P(b) = P(a_1) + P(a_2) + \cdots$. By repeating this extension process ad transfinitum we succeed eventually in defining $P(a)$ for every a in B.

There is an objection to the construction just described. If the set a consists of a single point then it is intuitively obvious (and follows easily from the rigorous definition of P) that $P(a)$ (= the area of a) is zero. More generally if a

consists of any finite or enumerably infinite set of points we still have $P(a)=0$, and it is even possible (if for example a is a line segment) to have $P(a)=0$ for sets a containing uncountably many points. This definitely contradicts our explicitly formulated axiom that $P(a)=0$ should happen if and only if $a=o$. The customary way to get around this difficulty is by redefining the notion of equality that occurs in the equation $a=o$. It is proposed that we agree to consider as identical two subsets of Ω whose difference has probability zero. (In technical language, we consider, instead of the sets a, equivalence classes of sets modulo the class of sets of probability zero.) Through this agreement we are committed in particular to identifying any set of probability zero with the empty set o, and it follows therefore that in the reduced algebra B (that is, the algebra obtained from B by making the suggested identifications) all the axioms of probability are valid.

The long and tortuous process just described is very general. If Ω is any space (such as an interval or a cube) on a certain σ-algebra B of subsets of which a countably additive measure P is defined (such as length or volume), subject only to the restriction that the measure of all Ω is equal to 1, we obtain from B and P a system satisfying all the axioms of probability theory by the process of identification according to sets of measure zero. Thus there are as many probability systems as there are examples of "measure spaces."

The reason for the introduction of measure spaces into a discussion of probability theory is not merely to give examples. It can in fact be shown that the two theories (measure and probability) are coextensive. More precisely:

If B is any Boolean σ-algebra and P a probability measure on B, then there exists a measure space Ω such that the system B is abstractly identical with an algebra of subsets of Ω reduced by identification according to sets of measure zero, and the value of P for any event a is identical with the values of the measure for the corresponding subsets of Ω.

Hence measure is probability and probability is measure and, in virtue of the theorem just stated, the entire classical theory of measure and integration may be and has been carried over and used to give rigorous proofs of probability theorems.

5. Measure vs. probability. Having discussed the extent to which probability and measure are the same, we now dedicate a few words to describing the extent to which they are different. One feature that differentiates the two theories is that in the general theory of measure it is usual to admit the possibility that the measure of the entire space is infinite. This possibility is not admissable in probability theory. As long, however, as the measure of the whole space is finite it is always possible to introduce a scale factor which makes it equal to 1, and hence it is always possible to think of it (even if somewhat artificially) as a "probability space." Thus for example the language and notation of probability may be and have been used in such seemingly widely separated parts of mathematics as ergodic theory, topological groups, and integral geometry.

Even however if the infinite case is ruled out, it is a conspicuous fact that most theorems in which the word measure is used (rather than the word probability) have a very different appearance from the theorems of probability theory. The best way to explain the difference between measure and probability is to liken it to the difference between analytic and synthetic geometry. It isn't stretching a point too far to say that the representation of a probability algebra by a measure space is similar to the introduction of coordinates into geometry. Synthetic and analytic geometry are of course abstractly identical in the sense that any theorem in the one domain may be stated and proved in the language and machinery of the other—may be, but isn't. The theorems in the two fields differ in their intuitive content. It is natural to discuss linear transformations in analytic geometry and the nine point circle in synthetic geometry—and even though the interchange is possible, it isn't desired. The abstract identity of the two fields is however an extremely useful fact, exploited mostly by the synthetic side which often finds it convenient to lean on the analytic crutch. Similarly, probability is measure, and research in the field would be very greatly hampered if we were not permitted to use this analytic crutch—but the notions suggested by probability, the notions which are important and intuitive and natural inside the field, appear sometimes extremely special and artificial in the frame work of general measure theory.

In this section and the preceding ones we have treated axiomatization and coordinatization. We proceed now to development. In the following sections we shall define the basic concepts of probability theory, and discuss in particular those which serve in the sense described above to give to probability its distinguishing flavor.

6. Independent events. In order to motivate the definitions of the concepts to be studied in the sequel we return to the example of the die. For simplicity we make the classical assumption that any two faces are equally likely to turn up and that consequently the probability of any particular face showing is $\frac{1}{6}$. Consider the events $a=\{246\}$ and $b=\{12\}$. The first notion we want to introduce, the notion of conditional probability, can be used to answer such questions as these: "what is the probability of a when b is known to have occurred?" In the case of the example: if we know that v is less than 3, what can we say about the probability that v is even? The adjective "conditional" is clearly called for in the answer to a question of this type: we are evaluating probabilities subject to certain preassigned conditions.

To get a clue to the answer consider first the event $c=\{2\}$ and ask for the conditional probability of a, given that c has already occurred. The intuitive answer is perfectly clear here, and is independent as it happens of any such numerical assumptions as the equal likelihood of the faces. If v is known to be 2 then v is certainly even, and the probability must be 1. What made the answer easy was the fact that c implied a. The general question of conditional probability asks us to evaluate the extent (measured by a numerical probability or proportion) to which the given event b implies the unknown event a. Phrased in this way the question almost suggests its own answer: the extent to which b is contained in a can be measured by the extent to which a and b are likely to occur simultaneously, that is by $P(a \cap b)$. Almost—not quite. The trouble is that $P(a \cap b)$ may be very small for two reasons: one is that not much of b is contained in a, and the other is that there isn't very much of b altogether. In other words it isn't merely the absolute size of $a \cap b$ that matters: it's the relation or proportion of this size to the size of b that's relevant.

We are led therefore to define the conditional probability of a, given that b has occurred, in symbols $P_b(a)$, as the ratio $P(a \cap b)/P(b)$. For $a = \{246\}$ and $c = \{2\}$ this gives the answer we derived earlier, $P_c(a) = 1$; for $a = \{246\}$ and $b = \{12\}$ we get the rather reasonable figure $P_b(a) = \frac{1}{2}$. In other words if it's known that v is either 1 or 2 then v is even or odd (that is equal to 1 or equal to 2) each with probability $\frac{1}{2}$.

Consider now the following two questions: "b happened, what is the chance of a?" and simply "what is the chance of a?." The answers of course are $P_b(a)$ and $P(a)$ respectively. It might happen, and does in the example given above, that the two answers are the same, that in other words knowledge of b contributes nothing to our knowledge of the probability of a. It seems natural in this situation to use the word "independent": the probability distribution of a is independent of the knowledge of b. This motivates the precise definition: two events a and b are independent if $P_b(a) = P(a)$. The definition is transformed into its more usual form and at the same time gains in symmetry if we recall the definition of $P_b(a)$. In symmetric form: a and b are independent in the sense of probability (statistically or stochastically independent) if and only if $P(a \cap b) = P(a)P(b)$.

7. Repeated trials. Suppose next that we wish to make two independent trials of the same experiment—say, for example, to roll an honest die twice in succession. We shall presently exploit the precise definition of independence to clarify the notion of independent trials; first however it's worth while to remark on the intuitive content of the concept. Suppose that in a crude attempt to even things up we resolve on the following procedure: if the first die shows an even number we choose for the second experiment a die on which all the numbers are odd, and vice versa. The two experiments are not independent of each other in this case: whereas the a priori probability of getting an even number with the second die is $\frac{1}{2}$, the conditional probability of getting an even number with the second die, given that the first one showed an odd number, is one. We say that the two experiments are performed independently of each other only if the conditions under which the second experiment is to be performed are unaffected by the outcome of the first experiment.

If an experiment consists of two rolls of a die we don't expect the reported outcome of the experiment to be a number v, but rather a pair of numbers (v_1, v_2). The measure space Ω associated with the two-fold experiment consists not of 6 but of 36 points. (It is convenient to imagine these points laid out along the regular pattern of a 6×6 square.) The problem is to determine how the probability is distributed among these points. For a clue to the answer consider the events $a =$ "$v_1<3$" and $b =$ "$v_2<4$." We have $P(a)=\frac{1}{3}$ and $P(b)=\frac{1}{2}$; hence if we interpret the independence of the trials to mean the independence of any two events such as a and b we should have $P(a\cap b)=\frac{1}{6}$. If in the suggested diagram for the measure space associated with this discussion we encircle the points belonging to $a\cap b$ we get the following figure.

v_2 \ v_1	1	2	3	4	5	6
1	⊙	⊙	·	·	·	·
2	⊙	⊙	·	·	·	·
3	⊙	⊙	·	·	·	·
4	·	·	·	·	·	·
5	·	·	·	·	·	·
6	·	·	·	·	·	·

We see therefore that the formula $P(a\cap b)=P(a)P(b)$ appears analogous to the fact that the area of a rectangle is the product of the lengths of its sides.

We say therefore, if the analytic description of an experiment is given by a measure space Ω with a Boolean σ-algebra B of subsets on which a probability measure P is defined, that the analytic description of the experiment consisting of two independent trials of the given experiment is as follows. The space of points ω is replaced by the space of pairs of points (ω_1, ω_2) (the so called product space $\Omega\times\Omega$), B is replaced by the Boolean σ-algebra generated by the "rectangular" sets of the form $\{\omega_1 \text{ is in } a_1, \omega_2 \text{ is in } a_2\}$ where a_1 and a_2 belong to B, and the probability measure on this space of pairs is determined by the requirement that its value for rectangular sets of the kind described should be given by the product $P(a_1)P(a_2)$. The ideas involved in this procedure are not essentially original nor characteristic of probability theory: they are the same as the ideas involved in defining the area of plane sets in terms of the length of linear sets. There is of course a theorem hidden in this definition—a theorem which asserts that a probability measure satisfying the stated product requirement indeed exists and is in fact uniquely determined by this requirement.

What we can do once, we can do again. Just as two repetitions of an experiment gave rise to ordered pairs (ω_1, ω_2), similarly any finite number of repetitions (say n) give rise to the space of ordered n-tuples $(\omega_1, \omega_2, \cdots, \omega_n)$, with a multiplicatively determined probability measure. The procedure can be extended also to infinity: the analytic model of an infinite sequence of independent repetitions of an experiment is a measure space Ω whose points ω are infinite sequences $\{\omega_1, \omega_2, \omega_3, \cdots\}$. Even if an actually infinite sequence of repetitions of an experiment is practically unthinkable, there is a point in considering the infinite dimensional space Ω. The point is that many probability statements are assertions concerning what happens in the long run—assertions which can be made precise only by carefully formulated theorems concerning limits. Hence even if practice yields only approximations to infinity, it is the infinite sequence space Ω that is the touchstone whereby the mathematical theory of probability can be tested against our intuitive ideas. The first and most important such long run statement is described in the following paragraphs.

Suppose that an experiment is capable of producing an event a with probability p, and suppose that an infinite sequence of independent trials of this experiment is performed. We consider therefore the space of all sequences $\omega = \{\omega_1, \omega_2, \omega_3, \cdots\}$ where for each n, ω_n may or may not belong to a. Once the experiments have been performed so that we are given a particular point ω we may start asking numerical questions. We may ask for example: out of the first n trials of the basic experiment how many resulted in a? This means: out of the first n coordinates $\omega_1, \omega_2, \cdots, \omega_n$ of ω how many belong to a? The answer to this question depends obviously on n and just as essentially on the particular sequence ω—let us denote it by $m_n(\omega)$.

Now what does our intuition say? The usual statement (one which we have already exploited in our heuristic derivation of the notion of probability) is that the ratio of the number of successes to the total number of trials should be approximately equal to the probability of the event being tested. In our notation this seems to mean that for large n the ratio $m_n(\omega)/n$ should be close to the constant $p=P(a)$. The question arises: for which ω's should this be true? Not surely for all of them. For the sequence space Ω contains sequences none of whose coordinates belong to a, and for such a sequence ω, $m_n(\omega)$ is zero for all n. The best that we have a right to demand is that the ω's for which our statement is not true should be equivalent to the empty set of ω's in the sense of probability—that is that their totality should have probability zero. And this is true.

To sum up: we have just derived the statement (not the proof) of the most important special case of the so called strong law of large numbers. In mathematical language the assertion of this law is that as $n\to\infty$, $\lim m_n(\omega)/n$ exists and is equal to $p(=P(a))$ except for a set of ω's of measure zero. In more classical terms: it is almost certain that the "success ratios" converge to the probability of the event being tested.

8. Random variables. In order to gain a more thorough understanding of the law of large numbers and at the same time to introduce the language in which most of the theorems of probability theory are stated, we proceed to discuss the notion of a random variable.

"A random variable is a quantity whose values are determined by chance." What does that mean? The word "quantity" is meant to suggest magnitude—numerical magnitude. Ever since rigor has come to be demanded in mathematical definitions it has been recognized that the word "variable," particularly a variable whose values are "determined" somehow or other, means in precise language a function. Accordingly a random variable is a function: a function whose numerical values are determined by chance. This means in other words that a random variable is a function attached to an experiment—once the experiment has been performed the value of the function is known. The spatial model of probability is extremely well adapted to making this notion still more precise. If the analytic correspondent of an experiment is a measure space Ω then any possible outcome of the experiment is by definition represented by a point ω in this space. Hence a function of outcomes is a function of ω's: a random variable is a real valued function defined on a probability space Ω.

The preceding sentence does not yet constitute our final definition of a random variable. For suppose that $x=x(\omega)$ is a function on the space Ω. We shall call x a random variable only if probability questions concerning the values of x can be answered. An example of such a question is: what is the probability that x is between α and β? In measure theoretic language: what is the measure of the set of those ω's for which the inequality $\alpha \leqq x(\omega) \leqq \beta$ is satisfied? In order for such questions to be answerable it is necessary and sufficient that the sets that occur in them belong to the basic σ-algebra B of Ω. A function $x(\omega)$ for which this is true for every interval (α, β) is called "measurable." Accordingly we make the following definition:

A random variable is a measurable function defined on a measure space with total measure 1.

Instances of random variables can be found even in that part of our discussion which preceded their definition. The quantity v associated with the rolling die is an example, as are also the quantities v_1 and v_2 associated with the two fold repetition of this experiment. To obtain some further examples, consider any fixed event a which may result from an experiment and let the random variable x be the number of times that a actually occurs. If the experiment is performed only once then x has only two possible values: 1 if a occurs and 0 otherwise. More generally if the experiment is repeated n times the random variable x becomes the function $m_n(\omega)$ introduced in the discussion of the law of large numbers.

9. Expectation, variance, and distribution. Let us consider in detail the random variable v associated with an honest die. The possible values of v are the first six positive integers. The arithmetic mean of these values, that is the number $(1+\cdots+6)/6$, is of considerable interest in probability theory. It is called the average, or mean value, or expectation of the random variable v and it is denoted by $E(v)$. If the die is loaded so that the probability p_j associated with j is not necessarily $\frac{1}{6}$ then the arithmetic mean is replaced by a weighted average: in this case $E(v)=1\cdot p_1+\cdots+6\cdot p_6$. It is well known that the analogs of such weighted sums in cases where the number of values of the function (random variable) need not be finite are given by integrals. The kind of integral that enters into probability theory is similar in every detail to the Lebesgue integral and we shall not reproduce its definition here.

If the measurable function $x(\omega)$ is integrable then its expectation $E(x)$ is by definition the value of its integral $\int x(\omega)dP(\omega)$ over the entire domain Ω.

As a useful though extremely special case we mention that if x is a counting variable of the sort mentioned in the preceding paragraph ($x=1$ if a certain even a occurs and $x=0$ otherwise) then $E(x)=P(a)$.

It is obviously of interest to ask not only what is the expected value of a random variable x but also how closely the values of x are clustered about its expected value. The customary measure of clustering of a random variable x is one inspired by the method of least squares and called the "variance" or "dispersion" of x.

The variance of x is the expression $\sigma^2(x)=E(x-\alpha)^2$, where $\alpha=E(x)$.

(The square root of the variance is called the "standard deviation.") In words: take the square of the deviation of x from its expected value α, and use the sum (weighted sum, integral) of these squared deviations as a measure of clustering. Since a sum of squares vanishes only if each term does, the vanishing of the variance indicates that x is identically equal to its expected value (except perhaps for a set of probability zero). In general, the smaller the variance the closer the values of x lie to $E(x)$.

Such numbers as $E(x)$ and $\sigma^2(x)$ yield partial information about the distribution of the values of x. Complete information would mean an answer to every question of the form "what is the probability that x lies in the interval (α, β)?" In order to deal with such questions we introduce the notion of distribution function.

The distribution function $F_x(\lambda)$ of a random variable x is a function of a real variable λ defined for each λ to be the probability that $x<\lambda$.

These functions can be used to answer every probability question concerning random variables; for example the expression $F_x(\beta)-F_x(\alpha)$ represents the probability that x belong to the (half open) interval $\alpha\leqq x<\beta$, and the Stieltjes integrals $\int_{-\infty}^{\infty}\lambda dF_x(\lambda)$ and $\int_{-\infty}^{\infty}\{\lambda-E(x)\}^2dF_x(\lambda)$ represent the expectation and variance of x respectively. Distribution functions are useful because being comparatively simple real functions of real variables they are amenable to treatment by the methods of classical analysis. It is the whole purpose of a large part of probability theory to find the distribution functions of certain random variables.

10. Independent variables. Let us consider next two random variables x and y which are comparable in the sense that they are both represented by measurable functions on the same measure space Ω, so that $x=x(\omega)$ and $y=y(\omega)$. It is easy to see that the function $E(x)$, being defined by an integral, is homogeneous of degree 1 and additive, that is $E(\lambda x)=\lambda E(x)$ for every real constant λ and $E(x+y)=E(x)+E(y)$. Similarly the variance $\sigma^2(x)$ is homogeneous of degree 2, that is $\sigma^2(\lambda x)=\lambda^2\sigma^2(x)$. One way to prove this latter fact is to make use of the following identity connecting σ^2 and E:

$$\sigma^2(x) = E(x)^2 - E^2(x), \tag{1}$$

(where for later convenience we write $E(x)^2$ for $E(x^2)$ and $E^2(x)$ for $\{E(x)\}^2$). This identity in turn follows from the definition of σ^2. Since $\sigma^2(x)=E(x-\alpha)^2$

where $\alpha = E(x)$, we have $\sigma^2(x) = E(x^2 - 2\alpha x + \alpha^2) = E(x^2) - 2\alpha E(x) + \alpha^2$. (We used here the fact that the expected value of a constant is equal to that constant.) The identity (1) follows by substituting for α its value $E(x)$. Letting the formalism guide us we may inquire whether σ^2 is additive, that is whether or not the identity

$$\sigma^2(x+y) = \sigma^2(x) + \sigma^2(y) \tag{2}$$

is valid. The answer in general is no. In order to investigate conditions under which (2) is true we proceed to a brief discussion of some possible relations between pairs of random variables.

Let a and b be two independent events and let x and y be the associated counting random variables (so that x for example is 1 if and only if a occurs and $x=0$ otherwise). The product random variable xy in this case can be equal to 1 if and only if both a and b occur, so that xy is the counting variable of $a \cap b$. Since $E(x) = P(a)$, $E(y) = P(b)$, and similarly $E(xy) = P(a \cap b)$, we have in this special case

$$E(xy) = E(x)E(y). \tag{3}$$

The validity of this formula is sufficiently important in the applications of probability to bear a name of its own: two random variables, not necessarily the counting variables of a pair of independent events, satisfying it are called "uncorrelated." The reason for the terminology is that the coefficient of correlation $r = r(x, y)$ of two random variables x and y is defined by $r = \{E(xy) - E(x)E(y)\}/\sigma^2(x)\sigma^2(y)$; this coefficient vanishes if and only if (3) holds.

It is now easy to state the facts concerning the formula (2): it is valid if and only if (3) is. In other words the variance is additive for a pair of random variables if and only if the expectation is multiplicative, that is if and only if they are uncorrelated. For the proof we merely expand the left member of (2), thus:

$$\begin{aligned}\sigma^2(x+y) &= E(x+y)^2 - E^2(x+y) \\ &= \{E(x)^2 - 2E(xy) + E(y)^2\} - \{E^2(x) - 2E(x)E(y) + E^2(y)\} \\ &= \sigma^2(x) + \sigma^2(y) - 2\{E(xy) - E(x)E(y)\}.\end{aligned}$$

Let us now return to the pair of counting variables x and y associated with two independent events a and b. Because of the independence of a and b, any probability statement concerning y is unaffected by our knowledge or ignorance of the value of x. More precisely, any two events defined by x and y, for example the events "$x=0$" and "$y=1$", are independent. If every two events defined by two random variables x and y respectively, that is any two events defined by inequalities of the form $\alpha \leqq x \leqq \beta$ and $\gamma \leqq y \leqq \delta$, are independent events, no matter what α, β, γ and δ are, we say that x and y are independent random variables. It is not too difficult to generalize what we proved about the special case of counting variables: for independent random variables the expectation, if it exists, is multiplicative and consequently the variance is additive. In still other words: independence implies absence of correlation—a proposition which certainly sounds natural enough.

One word of caution before we leave this brief introduction to the notion of independence for random variables. What we defined was the independence of two random variables. It would be natural to try to define the independence of a finite or infinite sequence of random variables $x_1, x_2, \cdots$, by the requirement that any pair be independent. Natural, but as it happens, not very useful. The correct definition replaces two-term products by many-term products in the following way.

The random variables $x_1, x_2, \cdots$, are independent if the probability of the simultaneous occurrence of any finite number of the events defined by $\alpha_n \leqq x_n \leqq \beta_n$ is the product of the separate probabilities, no matter what real constants the α's and β's are.

It is easy to construct examples to show that this notion is indeed different from the notion of pairwise independence.

11. Law of large numbers. We are now in a position to reformulate and generalize the strong law of large numbers in terms of random variables. Let the sequence space of points $\omega = \{\omega_1, \omega_2, \cdots\}$ be the analytic model of the infinite repetition of an experiment one of whose possible outcomes is the event a. Let a_n be the event "ω_n belongs to a" or equivalently the event "the nth experiment results in a," and let $x_n = x_n(\omega)$ be the counting variable associated with a_n. In this context that means that $x_n(\omega)$ has the value 1 for all those sequences $\omega\{\omega_1, \omega_2, \cdots\}$ for which the nth coordinate ω_n belongs to a, and $x_n(\omega)$ has the value 0 otherwise. What significance has the sum $x_1 + \cdots + x_n$? Since a particular term x_j contributes one unit to this sum if and only if the jth experiment results in a, it is clear that the value of the sum, for any sequence ω, is the number of those coordinates among the first n coordinates of ω which do belong to a. But this is exactly the function we denoted above by $m_n(\omega)$. Hence our version of the law of large numbers is equivalent to the assertion that the averages $(x_1 + \cdots + x_n)/n$ converge (except possibly for a set of ω's of probability zero) to the constant $p = P(a)$. For the generalization of this result that we are about to formulate it is worth while to observe that $p = E(x_n)$ is also equal to the common value of the expectations of the x's.

The sequence of random variables $x_1, x_2, \cdots$ has two important properties which are sufficient to ensure the validity of the law of large numbers. One of these properties is independence. It follows very easily from the fact that the experiments yielding the values of the various x's are independently performed, that the variables $x_1, x_2, \cdots$ are indeed independent. The other essential property of the sequence is usually expressed by the statement that the random variables x_n all have the same distribution. The definition of this concept is as follows.

Two random variables x and y have the same distribution if for every interval (α, β) the probabilities of the two events $\alpha \leqq x \leqq \beta$ and $\alpha \leqq y \leqq \beta$ are equal, or equivalently if the distribution functions $F_x(\lambda)$ and $F_y(\lambda)$ are identical.

In our particular case it is the fact that the probability that ω_n belong to a is the same for all n (namely $P(a)$) that implies that the x_n all have the same distribution. That independence and equidistribution are indeed the crucial hypotheses for the law of large numbers is shown by the following general formulation of that law.

If $x_1, x_2, \cdots$ is a sequence of independent random variables with the same distribution, and if the expectations $E(x_n)$ exist and have the value α (necessarily the same for all n) then the averages $x_1 + \cdots + x_n/n$ converge as $n \to \infty$ (except perhaps on a set of probability zero) to the constant α.

12. Central limit theorem. Sums (such as $x_1 + \cdots + x_n$) of independent random variables with the same distribution occur very often in probability theory. It is of considerable practical importance to investigate the precise distribution of such sums and if possible the limiting behavior of these distributions. We assume concerning the x's that their expectations and variances both exist and write $E(x_j) = \alpha$, $\sigma^2(x_j) = \beta$. It follows from the independence and equidistribution of the x's that $E(x_1 + \cdots + x_n) = n\alpha$ and $\sigma^2(x_1 + \cdots + x_n) = n\beta$. At first sight this seems like a discouraging phenomenon: if both the expectation and the variance become infinite, how can we expect a reasonable asymptotic behavior from the much more delicate distribution function? But the way out of the difficulty is easy: by a translation and a change of scale (different to be sure for each n) it is possible to normalize the sum $x_1 + \cdots + x_n$ so that its expectation is 0 and its variance 1 for every positive integer n. To get the expectation to be 0 we merely subtract its actual value, $n\alpha$, from the sum—the additivity of the expectation ensures the desired result. To get the variance to be 1 we divide by a constant factor. It is important to recall that the variance is homogeneous of degree 2, so that the constant factor will be not $n\beta$ but $\sqrt{n\beta}$. We arrive thus at the normalized sums

$$\frac{x_1 + \cdots + x_n - n\alpha}{\sqrt{n\beta}}$$

and inquire again after the distribution function of this random variable and the limit of such distribution functions. The answer here is known and is embodied in the so called central limit theorem (or Laplace-Liapounoff theorem) stated as follows.

If $x_1, x_2, \cdots$ is a sequence of independent random variables with the same distribution, expectation α, and variance β, then the distribution functions of the modified sums $(x_1 + \cdots + x_n - n\alpha)/\sqrt{n\beta}$ converge as $n \to \infty$ to a fixed distribution function, the same no matter what the original distribution of the x's is. In more detail, the limit as $n \to \infty$ of the probability of the event defined by the inequality

$$\frac{x_1 + \cdots + x_n - n\alpha}{\sqrt{n\beta}} < \lambda$$

exists and is equal to

$$G(\lambda) = \frac{1}{\sqrt{2\pi}} \int_{-\infty}^{\lambda} e^{-u^2/2} du.$$

The distribution function $G(\lambda)$ is called the Gaussian or normal distribution.

With this statement we end our discussion of the development of probability theory and turn to a few remarks connected with the problem of application.

13. Determination of initial probabilities. When the mathematician announces that the probability of an event is a certain number, he is immediately faced with two questions. First the practical man asks what is the practical meaning of a probability statement? How should one act on it? If the mathematician succeeds in answering this question then the philosopher wants to know the reason for the answer. What establishes the connection between mathematical theory and practice? Our remarks in what follows will bear on these very old and very difficult questions only incidentally—they are dedicated mainly to a smaller problem of the theory, but one which frequently worries the layman.

The problem is how the probability of concretely given events is really defined. It is all very well to talk about Boolean algebras and measure theory, but what is the probability that a coin will fall heads up? What the layman realizes and what we now wish to emphasize is that the mathematician has not answered any such questions. He cannot. He can no more say that the probability of obtaining two heads in succession with a coin is $\frac{1}{4}$ than he can say that the volume of a cube is 8. The volume of a cube is given by a formula. If the hypotheses under which the formula applies are verified and if the variables entering into the formula are given specific values then the volume of a cube can be calculated. In exactly the same sense the mathematical theory of probability is a collection of formulae which enable us to calculate certain probabilities assuming that certain other ones are given. If we know that the probability of obtaining heads with a certain coin is $\frac{1}{2}$ and if we know that two successive tosses of the coin were performed independently then we can assert that the probability of getting two heads is $\frac{1}{4}$.

Despite the fact that probability theory shares with all other mathematical theories its inability to state a conclusion without hypotheses, the above answer to the layman's question will probably seem unsatisfactory to many readers. There must be some reason why most people believe that the probability of heads is $\frac{1}{2}$. It is often even proved. The usual proof is based on symmetry arguments, or equivalently on the principle of sufficient reason. (Why should heads have any greater likelihood of appearing than tails?) Do these proofs have any mathematical validity?

The answer is definitely yes. In some cases it is more pleasing to the intuition or more convenient for practice to formulate our hypotheses purely qualitatively. In almost all such cases the hypotheses take the form of invariance—the probabilities entering into the problem are required to be invariant under a certain group of transformations. It often turns out then that an existence and uniqueness theorem is true, that is it can be proved that there exists one and only one probability measure satisfying the stated hypotheses. Theorems of this type are certainly a part, an increasingly important part, of the theory of probability, and as long as their hypotheses are clearly formulated and recognized as hypotheses, the professional mathematician is the last person to sneer at them. Their advantage at the level of elementary pedagogy seems to lie in the fact that the statement "heads and tails are equally likely" is easier to grasp intuitively than the statement "the probability of heads is $\frac{1}{2}$."

We see thus that a mathematical statement on probability has to have certain either explicitly or implicitly given probabilities to begin with. In practice the physicist (or actuary, or anyone else interested in applying the theory) obtains these initial numbers experimentally. If he wants to know what is the probability of a coin falling heads up, he tosses the coin a large number of times and then uses the law of large numbers to assure himself that he may use the obtained frequency ratio as an approximation to the correct value of the probability. Or he may observe that the values of a random variable are obtained as the sum of a large number of independent variables each with a negligible variance and thus be led to introduce the normal distribution. Such approximative procedures are of course common to all parts of applied mathematics.

14. Conclusion. Our exposition is finished. If the reader has been patient enough to read this far he may be curious enough to read farther. Our scanty bibliography will furnish a basis for such reading. For certainly not all probability theory is contained in this paper, nor as yet in any collection of books or papers. There is still much room in the field for the exercise of the analytic ingenuity and abstract generality of both classical and modern mathematics. If this paper will be instrumental in persuading mathematicians that probability is mathematics, and in causing some to look into the subject more deeply than they had previously thought worth while, it will have more than accomplished its purpose.

BIBLIOGRAPHY

G. Birkhoff, Lattice Theory, New York, 1940.

H. Cramér, Random Variables and Probability Distributions, Cambridge, 1937.

A. Khintchine, Asymptotische Gesetze der Wahrscheinlichkeitsrechnung, Berlin, 1933.

A. Kolmogoroff, Grundbegriffe der Wahrscheinlichkeitsrechnung, Berlin, 1933.

S. Saks, Theory of the Integral, Warsaw, 1937.

J. V. Uspensky, Introduction to Mathematical Probability, New York, 1937

APPENDIX TO "THE FOUNDATIONS OF PROBABILITY"

C. R. BLYTH, Queen's University, Kingston, Ontario

The content and language of this essay are still modern and readable 33 years later: these foundations stand unchanged, and so universally used that investigations of possible alternative systems use not the name "probability" but qualified names such as "finitely-additive probability" and "comparative probability". Readers of Halmos's *How to write mathematics* in **[1]** will notice how closely he was following his 1971 advice already in 1944. One point of language is that instead of formally identifying A with the empty set whenever $P(A)=0$, current usage is to simply say "a set of probability 0."

In the years since 1944 there has been an immense development of Probability, which is now a large and flourishing branch of Mathematics, having important interactions with many other branches. No one would now doubt that Probability is Mathematics, although there are still a few troglodytes who think that Statistics isn't: Time is gradually replacing them with people more at home in the second half of the 20th century.

Statistics is a branch of Probability that has had especially rapid development during 1944–1977. It is a mathematical model for the use of experimental evidence in deciding which P function to use as mathematical model for each particular real process. Statistics is inescapably and essentially involved in every use of Probability as a model for a real process. So it is worth while to add to Halmos's essay a short introduction to the foundations of Statistics. And then a little more needs to be said about what it is in the real world that probability is intended to be a model for. (Halmos's essay is not concerned with this question, and touches on it only briefly.)

The foundations of statistics. In deciding on a probability model for a particular real process, we leave the P measure unspecified—except possibly for symmetry and independence restrictions that correspond to apparent symmetry and independence in the real process—and to be *estimated* (not *determined*) from observations. Notice how similar this is to the familiar deterministic procedure of specifying the structure only of a physical law, leaving the values of constants to be determined from observations. We then make some number n of repetitions of the actual real process [Model: independent copies of the same P model] and use observed proportions to estimate the $P(\ )$ values.

Example. For a coin toss we take the completely unspecified model $P(H)=p$ and $P(T)=1-p$, with p an unknown parameter, $0 \leqslant p \leqslant 1$, whose value is to be estimated from actual tosses of the coin. (Of course, Halmos's Ω and B are specified: it's P that isn't.) For a practical example using the same model, consider a cross that produces an offspring of two possible types, H and T. We now make n tosses with the actual coin [Model: independent copies with $P(H)=p$ each time] and count the total number x of heads seen [Model: x is a random variable whose

Binomial (n,p) distribution can be deduced mathematically from the P model just taken] and for p estimate $T(x)$, usually x/n [Model: the random variable $T(x)$, or the function T that generates it, is called an estimator of p]. Next we use probability theory to describe the accuracy of this estimator: for each possible value of p we can compute, for every number $\varepsilon>0$, the probability $P_p(|x/n-p|\leqslant \varepsilon)$. Or instead of this very detailed description, we can use expected absolute error $E_p|x/n-p|$ or root-mean-square error $\sqrt{(E_p|x/n-p|^2)}$ as a measure of "size" of the random error $(x/n-p)$. The subscript in P_p and E_p designates the P measure indexed by the parameter value p.

In this example the P was left completely unspecified, but in most applied problems the P is at least partially specified. For a process whose outcome is a numerical measurement x, it may be plausible to think of x as being the sum of a constant μ and a large number of unidentified random variables that are more or less independent, have expectations averaging to 0, and have more or less equal variances. Then the Central Limit Theorem (Halmos, section 12) shows the plausibility of the fairly explicit P model that x has a Normal distribution whose expectation μ and variance σ^2 are unknown and to be estimated from observations.

NEYMAN-PEARSON HYPOTHESIS TESTING. In this type of Statistics problem we consider two alternative P models, called *Hypothesis* and *Alternative Hypothesis*; each may or may not fully specify P. And we use observations to decide between the two. The performance of a *test* (a rule that specifies, for each possible set of observations, which of the two to accept) is described by giving its *power* P(Accepting Alternative) for every possible specific P. (Or equivalently by giving its *operating characteristic*, P(Accepting Hypothesis), which is just $1-P$(Accepting Alternative).) This kind of Hypothesis Testing amounts to a very simplified form of Estimation, and is very widely used in applications where a choice has to be made between two possible actions.

KARL PEARSON HYPOTHESIS TESTING. (This differs from Neyman-Pearson testing in being concerned with the acceptability of one hypothesis instead of the relative acceptability of two.) Even when P is fully specified, the use of Statistics is essential because it is always necessary to check from observations whether this model or hypothesis is an accurate enough description of the real process. A *test* (the Karl Pearson chi-square test is a commonly used one) is a rule for deciding whether to accept or reject the hypothesis. Acceptance is always tentative, meaning that the model is not ruled out by the available data; rejection is the firm decision that the model must be discarded as inconsistent with observations.

In the coin-tossing example apparent symmetry may lead us to take the model or Hypothesis $p=1/2$ or, MUCH better, $|p-1/2|\leqslant a$ with the specified number a an important feature of the real situation. Our test would be:

If $|x/n-1/2|<k$, accept the model,
If $|x/n-1/2|\geqslant k$, reject the model;

where k is a number chosen so that $P(|x/n-1/2|\geqslant k)$ is negligibly small, say no more than $\alpha=.05$ or .01, whenever the model is true. We reject a model when, if the model were true, there would be a negligibly small probability of getting observations so apparently discrepant from it as the observations we have in fact seen. This is just the universal basis upon which any model of reality is rejected: "if the model were accurate, then we wouldn't be seeing this."

The "physical meaning" of Probability. This heading is, of course, absurd. Probability has no "meaning" or "interpretation," physical or otherwise: it simply IS. Such headings indicate an intent to ask what it is in the real world that Probability is a model for. Because of the universal freedom to use the same mathematics as model for many different things, there can be no unique answer to this question. One fairly generally agreed-upon answer is this:

> Probability $P(A)$ is (a model for) the proportion of occurrences of A that would be seen if the process were repeated an indefinitely large number of times.

Of course, this answer is vague, and necessarily so. No answer can be made precise, because if we attempt a precise description of reality we find ourselves with, not reality, but an abstract model of reality; and must then describe what this corresponds to in reality, leaving the desired precise answer always just beyond our grasp. Foundational writings that attempt this impossible task are usually lengthy and complicated enough to thoroughly confuse both reader and writer; they fail to make clear from sentence to sentence whether they are talking about mathematics or reality; and they show an uncritical reliance on the meaning of language, which is, after all, only an abstract model of experience, and one that is very imperfectly shared by its users.

In the coin-tossing example, here is how we might attempt to make the above answer precise: (we must consider definite finite numbers of repetitions; an infinite sequence of tosses is not something of the real world).

> *First attempt*: "By $P(H)=p$ we mean physically that in a modestly large number n of real tosses, there is a modestly large probability $P_p(|x/n-p|\leqslant\varepsilon)=p'$ that the proportion x/n of occurrences of H will be fairly close (within ε) to p."

But here we notice that "probability p'" is mathematics, not reality, so we have to say what it is in reality that this corresponds to.

> *Second attempt*: To the first attempt add "where by 'probability p'' we mean physically that in a modestly large number n' of repetitions of the whole n tosses, there is a modestly large probability p'' that the proportion of these repetitions in which the $|x/n-p|$ doesn't exceed ε, will be fairly close (within ε') to p'."

But now we notice that "probability p''" is mathematics, and we would have to specify what it corresponds to in reality, and so on and on, with success always one step away.

The above attempt was made here because in statistics we need that second-stage description of what probability corresponds to in the real world. It is this, perhaps more than anything else, that makes the physical interpretation of statistical methods difficult for students and users.

Additional Bibliography

One of the most striking changes since 1944 is in the list of books available for further reading. There is now an embarrassment of riches, only a few of which can be included here.

1. N. E. Steenrod, *et al.*, How to write mathematics, American Mathematical Society, 1973.

2. K-L. Chung, A course in probability theory, Harcourt, Brace and World, New York, 1968.

3. ———, Elementary probability theory with stochastic processes, Springer-Verlag, New York, 1974.

4. J. L. Doob, Foundations of probability theory and its influence on the theory of statistics, in *On the history of statistics and probability*, D. B. Owen ed., (Proceedings of a symposium on the American Mathematical Heritage, to celebrate the Bicentennial of the U.S.A., held at Southern Methodist University May 27–29, 1974) pp. 195–204. Marcel Dekker, New York, 1976.

5. W. Feller, An introduction to probability theory and its applications, Vol. 1, 3rd ed., Wiley, New York, 1968.

6. ———, An introduction to probability theory and its applications, Vol. 2, 2nd. ed., Wiley, New York, 1971.

7. H. Freeman, Introduction to statistical inference, Addison-Wesley, Reading, Mass., 1963.

8. A. N. Kolmogorov, Foundations of probability (English translation, ed. N. Morrison), Chelsea, New York, 1950.

9. A. Wald, Contributions to the theory of statistical estimation and testing hypotheses, Ann. Math. Statist., 10 (1939) 299–326. (Also appears in *Selected papers in statistics and probability by Abraham Wald*, Inst. Math. Statist., 1955.)

9

MARK KAC

Professor Kac was born on August 3, 1914, in Krzemieniec, Poland. He received the Ph.D. degree from the John Casimir University in Lwów in 1937, after which he worked as an actuary at the Phoenix Company, a Polish insurance firm. Professor Kac came to the United States in 1938, and after a year at The Johns Hopkins University, he joined the faculty of mathematics at Cornell University. He became a naturalized citizen in 1943, and from 1943 to 1945 served as a member of the Office of Scientific Research and Development. Professor Kac was a member of the Institute for Advanced Study at Princeton in 1951–52. He received the Parnas Foundation Fellowship in Poland in 1938–39 and held a Guggenheim Fellowship in 1946–47. He was elected to the American Academy of Arts and Sciences in 1959 and to the National Academy of Sciences in 1965. Professor Kac has been a member of the Council of the American Mathematical Society, its Vice-President from 1964 to 1966, and Editor of the *Transactions of the American Mathematical Society* from 1955 to 1958. He was appointed Professor and became a member of the Rockefeller Institute on July 1, 1961. He was Lorentz Visiting Professor of Theoretical Physics in Leiden, the Netherlands, in 1963, and has been Chairman of the Division of Mathematical Sciences of the National Research Council of the National Academy of Sciences in 1965–67. He was Andrew D. White Professor-at-Large at Cornell, 1965–72, a Senior Visiting Fellow at Oxford in 1969, Visiting Fellow, Brasenose College, Oxford, and Solvay Lecturer, University of Brussels, 1971. He is a member of The American Philosophical Society, The Royal Norwegian Academy of Trondheim, and the Institute of Mathematics.

Professor Kac's significant contributions to so many branches of mathematics and its applications, including the theory of probability, statistics, analysis, and number theory, are contained in his numerous papers—already more than 80 in number—which have appeared in many scientific publications throughout the world.

He is the author of Carus Monograph No. 12, *Statistical Independence in Probability, Analysis, and Number Theory*; *Probability and Related Topics in Physical Science*, 1959; and *Mathematics and Logic* (with S. Ulam), 1968. The article for which he has received the Chauvenet Prize has been reprinted as a CEM Film Manual, since it is the script of his film with the same title.

In accepting the 1968 Chauvenet Prize, Professor Kac indicated that he felt both honored and flattered by having been voted for the second time the recipient of the Chauvenet Prize.

RANDOM WALK AND THE THEORY OF BROWNIAN MOTION*

MARK KAC,† Cornell University

1. Introduction. In 1827 an English botanist, Robert Brown, noticed that small particles suspended in fluids perform peculiarly erratic movements. This phenomenon, which can also be observed in gases, is referred to as Brownian motion. Although it soon became clear that Brownian motion is an outward manifestation of the molecular motion postulated by the kinetic theory of matter, it was not until 1905 that Albert Einstein first advanced a satisfactory theory.

The theory was then considerably generalized and extended by the Polish physicist Marjan Smoluchowski, and further important contributions were made by Fokker, Planck, Burger, Fürth, Ornstein, Uhlenbeck, Chandrasekhar, Kramers and others [1]. On the purely mathematical side various aspects of the theory were analyzed by Wiener, Kolomgoroff, Feller, Lévy, Doob, and Fortet [2]. Einstein considered the case of the *free* particle that is, a particle on which no forces other than those due to the molecules of the surrounding medium are acting. His results can be briefly summarized as follows.

Consider the motion of the projection of the free particle‡ on a straight line which we shall call the x-axis. What one wants is the probability

$$\int_{x_1}^{x_2} P(x_0 \mid x; t)dx$$

that at time t the particle will be between x_1 and x_2 if it was at x_0 at time $t=0$. Einstein was then able to show that the "probability density" $P(x_0 \mid x; t)$§ must satisfy the partial differential equation

$$\frac{\partial P}{\partial t} = D\frac{\partial^2 P}{\partial x^2}, \tag{1}$$

where D is a certain physical constant. The conditions imposed on P are

$$(2)\qquad \begin{aligned} &(a) && P \geqq 0 \\ &(b) && \int_{-\infty}^{\infty} P(x_0 \mid x; t)dx = 1 \\ &(c) && \lim_{t\to 0} P(x_0 \mid x; t) = 0, \qquad \text{for } x \neq x_0. \end{aligned}$$

* This is an extended version of an address delivered at the annual meeting of the Association at Swarthmore, Pennsylvania, December 26–27, 1946.

† John Simon Guggenheim Memorial Fellow.

‡ In what follows we shall identify this projection with the particle itself and hence consider the so-called one-dimensional Brownian motion.

§ The notation $P(x_0 \mid x; t)$ and $P(n \mid m; s)$ for conditional probabilities is that used by Wang and Uhlenbeck [1]. It does not conform with the notation adopted in the statistical literature. Had we adopted the latter notation we would write $P(x; t \mid x_0)$ and $P(m; s \mid n)$.

Conditions (a) and (b) are the usual ones imposed upon a probability density and condition (c) expresses the *certainty* that at $t=0$ the particle was at x_0. It is well known that (1) and (2) imply that

$$P(x_0 \mid x; t) = \frac{1}{2\sqrt{\pi D t}} e^{-(x-x_0)^2/4Dt} \tag{3}$$

and that the solution (3) is unique.

The greatness of Einstein's contribution was not, however, solely due to the derivation of (1), and hence (3). From the point of view of physical applications it was equally, or perhaps even more, important that he was able to show that

$$D = \frac{2RT}{Nf}, \tag{4}$$

where R is the universal gas constant, T the absolute temperature, N the Avogadro number, and f the friction coefficient. The friction coefficient f, in the case the medium is a liquid or a gas at ordinary pressure, can in turn be expressed in terms of viscosity and the size of the particle [**3**].

It was relation (4) that made possible the determination of the Avogadro number from Brownian motion experiments, an achievement for which Perrin was awarded the Nobel prize in 1926. However, the derivation of (4) belongs to physics proper, and presents no particular mathematical interest; we shall therefore not be concerned with it in the sequel.

As soon as the theory for the free particle was established, a natural question arose as to how it should be modified in order to take into account outside forces as, for example, gravity. Assuming that the outside force acts in the direction of the x-axis and is given by an expression $F(x)$, Smoluchowski has shown that (1) should in this case be replaced by

$$\frac{\partial P}{\partial t} = -\frac{1}{f}\frac{\partial}{\partial x}(PF) + D\frac{\partial^2 P}{\partial x^2}. \tag{5}$$

Two cases of special interest and importance are:

$F(x) = -a$; field of constant force (for example, gravity).

$F(x) = -bx$; elastically bound particle (for example, small pendulum).

At this point it must be strongly emphasized that theories based on (1) and (5) are only approximate. They are valid only for relatively large t and, in the case of an elastically bound particle, only in the overdamped case, that is, when the friction coefficient is sufficiently large. These limitations of the theory were already recognized by Einstein and Smoluchowski but are often disregarded by writers who stress that in Brownian motion the velocity of the particle is infinite. This paradoxical conclusion is a result of stretching the theory beyond the bounds of its applicability. An improved theory (known as "exact") was advanced by Uhlenbeck and Ornstein and by Kramers. The Uhlenbeck-Ornstein approach was further elaborated by Chandrasekhar and Doob.

In what follows we shall be concerned with a discrete approach to the Einstein-Smoluchowski (approximate) theory. This approach was first suggested by Smoluchowski himself; it consists in treating Brownian motion as a discrete random walk. Smoluchowski used this approach only in connection with a free particle but we shall also treat other classical cases. Moreover, a re-interpretation of one of the discrete models will allow us to discuss the important question of recurrence and irreversibility in thermodynamics.

The main advantages of a discrete approach are pedagogical, inasmuch as one is able to circumvent various conceptual difficulties inherent to the continuous approach. It is also not without a purely scientific interest and it is hoped that it may suggest various generalizations which will contribute to the development of the Calculus of Probability.

2. The free particle. Imagine a particle which moves along the x-axis in such a way that in each step it can move either Δ to the right or Δ to the left, the duration of each step being τ. The fact that we are dealing with a free particle is interpreted by assuming that the probabilities of moving to the right or to the left are equal, and hence each equal $\frac{1}{2}$. Instead of $P(x_0|x; t)$ we now consider $P(n\Delta|m\Delta; s\tau) = P(n|m; s)$ which is the probability that the particle is at $m\Delta$ at time $s\tau$, if at the beginning it was at $n\Delta$. Noticing that $P(n|m; s)$ is also the probability that after s games of "heads or tails" the gain of a player is $\nu = m - n$, we can write

$$(6)\quad P(n \mid m; s) = \begin{cases} \dfrac{1}{2^s} \dfrac{s!}{\left(\dfrac{s+|\nu|}{2}\right)!\left(\dfrac{s-|\nu|}{2}\right)!} & \text{if } |\nu| \leqq s \text{ and } |\nu| + s \text{ is even,} \\ 0 & \text{otherwise.} \end{cases}$$

Suppose now that Δ and τ approach 0 in such a way that

$$(7)\qquad \frac{\Delta^2}{2\tau} = D, \qquad n\Delta \to x_0, \qquad s\tau = t.$$

It then follows from the classical Laplace-De Moivre theorem [4] that

$$(8)\qquad \lim \sum_{x_1 < m\Delta < x_2} P(n \mid m; s) = \frac{1}{2\sqrt{\pi D t}} \int_{x_1}^{x_2} e^{-(x-x_0)^2/4Dt} dx,$$

and hence the fundamental result of Einstein emerges as a consequence of what in probability theory we call a "limit theorem."

It is both important and instructive to point out a striking formal connection between the discrete (random walk) and the continuous (Einstein) approaches. We notice that $P(n|m; s)$ satisfies the difference equation

$$(9)\qquad P(n \mid m; s+1) = \tfrac{1}{2}P(n \mid m-1; s) + \tfrac{1}{2}P(n \mid m+1; s),$$

which we write in the equivalent form

$$\frac{P(n\Delta \mid m\Delta; (s+1)\tau) - P(n\Delta \mid m\Delta; s\tau)}{\tau}$$

$$(10) \qquad = \frac{\Delta^2}{2\tau}\left\{\frac{P(n\Delta \mid (m+1)\Delta; s\tau) - 2P(n\Delta \mid m\Delta; s\tau) + P(n\Delta \mid (m-1)\Delta; s\tau)}{\Delta^2}\right\}.$$

In the limit (7) this difference equation goes over formally into the differential equation

$$(11) \qquad \frac{\partial P}{\partial t} = D\frac{\partial^2 P}{\partial x^2},$$

which as noted before was the basis of Einstein's theory. This formal connection between the two approaches can be made rigorous, but we shall not go into this. However, we shall use it as a guiding heuristic principle in constructing models of Brownian motion when outside forces are taken into account.

Finally, let us mention that it is the relation

$$\frac{\Delta^2}{2\tau} = D,$$

which is responsible for the conclusion that the velocity of a Brownian particle is infinite. In fact, in our model, the ratio Δ/τ plays the role of the instantaneous velocity and it obviously approaches infinity as $\Delta \to 0$.

3. Particle in a field of constant force and in the presence of a reflecting barrier. We again consider random walk along the x-axis in which a particle can move Δ to the right or Δ to the left, the duration of each step being τ. We now introduce the following new assumptions:

(a) The probability of a move to the right is $q = \frac{1}{2} - \beta\Delta$, and consequently the probability of a move to the left is $p = \frac{1}{2} + \beta\Delta$. Here β is a certain physical constant, and Δ must be chosen sufficiently small so that $q > 0$.

(b) When the particle reaches the point $x = 0$ (*reflecting barrier*) it must, in the next step, move Δ to the right.

Without the assumption (b) the problem would be quite simple and of no great physical interest. In actual experiments with heavy Brownian particles, like those of Perrin, the bottom of the container acts as a reflecting barrier and the elucidation of its influence on the Brownian motion is of considerable theoretical interest.

This problem has been solved by Smoluchowski, on the basis of his equation (5) but we shall show that one can solve the discrete problem and obtain Smoluchowski's result by passing to a limit.

Assuming that the particle starts from $n\Delta \geq 0$ (n an integer) we seek an explicit expression for $P(n \mid m; s)$. We first notice that $P(n \mid m; s)$ satisfies, for

$m \geq 2$, the difference equation

$$P(n \mid m; s+1) = qP(n \mid m-1; s) + pP(n \mid m+1; s), \tag{12}$$

and that for $m=1$ and $m=0$ we have

$$P(n \mid 1; s+1) = P(n \mid 0; s) + pP(n \mid 2; s), \tag{12a}$$

$$P(n \mid 0; s+1) = pP(n \mid 1; s). \tag{12b}$$

We also have the initial condition

$$P(n \mid m; 0) = \delta(m, n), \tag{13}$$

where $\delta(m, n)$ denotes, as usual, the Kronecker delta.

The difference equation (12) when rewritten in the form analogous to (10) can be shown to go over formally (in the limit $\Delta \to 0$, $\tau \to 0$, $\Delta^2/2\tau = D$, $n\Delta \to x_0$, $m\Delta \to x$, $s\tau = t$) into the differential equation

$$\frac{\partial P}{\partial t} = D \frac{\partial^2 P}{\partial x^2} + 4\beta D \frac{\partial P}{\partial x}, \tag{14}$$

which is of the form (5) with $F(x) = -a = -4\beta Df$.

To find $P(n \mid m; s)$ we use a method which is basic in the study of the so-called Markoff chains, of which our problem is but a particular example, and which in its essentials goes back to Poincaré [5]. Let $(p)_s$ be the (infinite) vector

$$(p)_s = \begin{bmatrix} P(n \mid 0; s) \\ P(n \mid 1; s) \\ P(n \mid 2; s) \\ \vdots \end{bmatrix} \tag{15}$$

and A the infinite matrix

$$A = \begin{Vmatrix} 0 & p & 0 & 0 & 0 \cdots \\ 1 & 0 & p & 0 & 0 \cdots \\ 0 & q & 0 & p & 0 \cdots \\ 0 & 0 & q & 0 & p \cdots \\ \cdots & \cdots & \cdots & \cdots & \cdots \end{Vmatrix}. \tag{16}$$

Then, the difference equation (12), (12*a*) and (12*b*), can be written in the matrix form as

$$(p)_{s+1} = A(p)_s. \tag{17}$$

Thus it follows immediately that

$$(p)_s = A^s(p)_0, \tag{18}$$

where $(p)_0$ is the vector

$$(p)_0 = \begin{pmatrix} 0 \\ \cdot \\ \cdot \\ 0 \\ 1 \\ 0 \\ \cdot \\ \cdot \\ \cdot \end{pmatrix},$$

1 being the nth component, the components being numbered from zero on. Interpreting (18), we see that

$$P(n \mid m; s) = \text{the } (m, n) \text{ element of } A^s. \tag{19}$$

To make use of (19), we notice that if $R > n+s$ and we consider the finite matrix A_R, which is the upper left R by R submatrix of A, then for $m < R$

$$\text{the } (m, n) \text{ element of } A^s = \text{the } (m, n) \text{ element of } A_R^s, \tag{20}$$

or equivalently,

$$\text{the } (m, n) \text{ element of } A^s = \lim_{R \to \infty} \text{ of the } (m, n) \text{ element of } A_R^s. \tag{21}$$

For each R there exist matrices P_R and Q_R such that

$$P_R Q_R = I \tag{22}$$

and

$$A_R = P_R \begin{bmatrix} \lambda_0(R) & & & 0 \\ & \lambda_1(R) & & \\ & & \ddots & \\ 0 & & & \lambda_{R-1}(R) \end{bmatrix} Q_R, \tag{23}$$

$\lambda_0(R), \lambda_1(R), \cdots, \lambda_{R-1}(R)$ being the eigenvalues of the matrix A_R. To simplify the notation we write λ_j for $\lambda_j(R)$.

Multiplying the matrix A_R s times by itself and making use of (22), we obtain

$$A_R^s = P_R \begin{bmatrix} \lambda_0^s & & & 0 \\ & \lambda_1^s & & \\ & & \ddots & \\ 0 & & & \lambda_{R-1}^s \end{bmatrix} Q_R \tag{24}$$

and one can calculate the (m, n) element of A_R^s explicitly provided the diagonalization (23) can be performed explicitly. This indeed is the case.

Let $(x)_0, (x)_1, \cdots, (x)_{R-1}$ be the "right" and $(y)_0, (y)_1, \cdots, (y)_{R-1}$ the "left" eigenvectors belonging respectively to the eigenvalues $\lambda_0, \lambda_1, \cdots, \lambda_{R-1}$. In other words, for $k=0, 1, \cdots, R-1$, we have

$$A_R(x)_k = \lambda_k(x)_k,$$
$$A_R'(y)_k = \lambda_k(y)_k,$$

where A_R' is the transpose of A_R.

Suppose furthermore that the vectors can be so normalized that

$$(x)_k\cdot(y)_k = 1, \qquad k = 0, 1, \cdots, R-1, \tag{25}$$

where $(x)_k\cdot(y)_k$ is the inner (dot) product of the vectors. Since it is well known that

$$(x)_j\cdot(y)_k = 0 \qquad \text{for } \lambda_j \neq \lambda_k$$

we see that, in the case when all the eigenvalues are distinct, we can take as P_R the matrix whose columns are the vectors $(x)_k$ and for Q_R the matrix whose rows are the vectors $(y)_k$.

In order to determine the eigenvalues and the right eigenvectors we consider the system of linear equations

$$\begin{aligned} px_1 &= \lambda x_0 \\ x_0 + px_2 &= \lambda x_1 \\ qx_1 + px_3 &= \lambda x_2 \\ &\cdots\cdots\cdots \\ qx_{R-2} \qquad &= \lambda x_{R-1}, \end{aligned} \tag{26}$$

and the extended infinite system

$$\begin{aligned} px_1 &= \lambda x_0 \\ x_0 + px_2 &= \lambda x_1 \\ &\cdots\cdots\cdots \\ qx_{R-1} + px_{R+1} &= \lambda x_R \\ &\cdots\cdots\cdots. \end{aligned} \tag{27}$$

If we can find non-trivial solutions of (27), for which

$$x_R = 0, \tag{28}$$

we will have found solutions of (26).

It turns out that (28) will yield an equation in λ which has R distinct roots and thus our procedure gives us all eigenvalues, and consequently all right eigenvectors. Multiplying the members of the equations of (27) by 1,

$z, z^2, \dots$, and adding, we obtain formally

$$x_0 z + q\sum_1^\infty x_k z^{k+1} + p\sum_1^\infty x_k z^{k-1} = \lambda \sum_0^\infty x_k z^k$$

or, upon introducing the abbreviation,

$$f(z) = \sum_0^\infty x_k z^k, \tag{29}$$

we have

$$x_0 z + qz[f(z) - x_0] + \frac{p}{z}[f(z) - x_0] = \lambda f(z). \tag{30}$$

From (30) we obtain

$$f(z) = \frac{p}{q} x_0 \left\{ -1 + \frac{1 - \lambda z}{qz^2 - \lambda z + p} \right\}, \tag{31}$$

and since this function is analytic in the neighborhood of zero the formal procedure used above can be justified.

Let ρ_1 and ρ_2 be the *reciprocals* of the roots of

$$qz^2 - \lambda z + p = 0. \tag{32}$$

We have then

$$f(z) = \frac{p}{q} x_0 \left\{ -1 + \frac{1 - \lambda z}{p(1 - \rho_1 z)(1 - \rho_2 z)} \right\}, \tag{33}$$

and introducing partial fractions,

$$\frac{1 - \lambda z}{p(1 - \rho_1 z)(1 - \rho_2 z)} = \frac{1}{p}\frac{\lambda - \rho_1}{\rho_2 - \rho_1}\frac{1}{1 - \rho_1 z} + \frac{1}{p}\frac{\rho_2 - \lambda}{\rho_2 - \rho_1}\frac{1}{1 - \rho_2 z}. \tag{34}$$

Thus

$$x_k = \frac{x_0}{q}\left(\frac{\lambda - \rho_1}{\rho_2 - \rho_1}\rho_1^k + \frac{\rho_2 - \lambda}{\rho_2 - \rho_1}\rho_2^k \right) \qquad \text{for } k \geqq 1, \tag{35}$$

and, in particular, the equation $x_R = 0$ assumes the form

$$\frac{\lambda - \rho_1}{\rho_2 - \rho_1}\rho_1^R + \frac{\rho_2 - \lambda}{\rho_2 - \rho_1}\rho_2^R = 0. \tag{36}$$

Equation (36) must now be solved for λ. Assuming R to be even, and seeking solutions in the form

$$\lambda = 2\sqrt{pq}\cos\Theta, \qquad 0 \leqq \Theta \leqq \pi,$$

we are led to the equation

$$\frac{\tan R\Theta}{\tan \Theta} = \frac{1}{2p-1}.$$

For $R>(2p-1)^{-1}$ this equation is seen to have $R-2$ distinct roots, $\Theta_1, \Theta_2, \cdots, \Theta_{R-2}$, which lie in the subintervals

$$\left(\frac{j\pi}{R} - \frac{\pi}{2R}, \frac{j\pi}{R} + \frac{\pi}{2R}\right),$$

where j ranges through the integers from 1 to $R-1$ with the exception of $j=R/2$. Corresponding to $\Theta_1, \Theta_2, \cdots, \Theta_{R-2}$ we have $R-2$ distinct eigenvalues,

$$\lambda_k = 2\sqrt{pq}\cos\Theta_k, \qquad k = 1, 2, \cdots, R-2,$$

and the components of the right eigenvector belonging to λ_k can be written in the form

$$\overset{(m)}{x_k} = a_k\left(\frac{q}{p}\right)_*^{m/2}\left(\cos m\Theta_k - 2\beta\Delta\frac{\sin m\Theta_k}{\sin\Theta_k}\right),$$

where

$$\left(\frac{q}{p}\right)_*^{\mu} = \begin{cases} \left(\frac{q}{p}\right)^{\mu} & \text{if } \mu > 0, \\ q & \text{if } \mu = 0, \end{cases}$$

and a_k is a normalizing constant which will be fixed later. For sufficiently large R the remaining eigenvalues λ_0 and λ_{R-1} can be shown to be given by the formulas

$$\lambda_0 = 2\sqrt{pq}\cosh\theta_0, \qquad \lambda_{R-1} = -\lambda_0,$$

where θ_0 is the only positive root of the equation

$$\frac{\tanh R\theta}{\tanh\theta} = \frac{1}{2p-1}.$$

The components of the corresponding right eigenvectors are given by the expressions

$$\overset{(m)}{x_0} = a_0\left(\frac{q}{p}\right)_*^{m/2}\left(\cosh m\theta_0 - 2\beta\Delta\frac{\sinh m\theta_0}{\sinh\theta_0}\right)$$

$$\overset{(m)}{x_{R-1}} = a_{R-1}(-1)^m\left(\frac{q}{p}\right)_*^{m/2}\left(\cosh m\theta_0 - 2\beta\Delta\frac{\sinh m\theta_0}{\sinh\theta_0}\right).$$

It remains now to find the left eigenvectors. This can be accomplished in exactly the same manner and we merely quote the results. We obtain

$$y_k^{(m)} = b_k\left(\frac{p}{q}\right)^{m/2}\left(\cos m\Theta_k - 2\beta\Delta\frac{\sin m\Theta_k}{\sin\Theta_k}\right)$$

for $m=0, 1,\ldots, R-1$; $k=1, 2, \cdots, R-2$, and

$$y_0^{(m)} = b_0\left(\frac{p}{q}\right)^{m/2}\left(\cosh m\theta_0 - 2\beta\Delta\frac{\sinh m\theta_0}{\sinh\theta_0}\right)$$

$$y_{R-1}^{(m)} = b_{R-1}(-1)^m\left(\frac{p}{q}\right)^{m/2}\left(\cosh m\theta_0 - 2\beta\Delta\frac{\sinh m\theta_0}{\sinh\theta_0}\right).$$

To satisfy the normalization conditions (25) we must have

$$a_k b_k\left(q + \sum_{m=1}^{R-1} f_m^2(\Theta_k)\right) = 1, \qquad k = 1, 2, \cdots, R-2, \tag{37}$$

$$a_k b_k\left(q + \sum_{m=1}^{R-1} F_m^2(\theta_0)\right) = 1, \qquad k = 0, R-1, \tag{38}$$

where

$$f_m(\Theta) = \cos m\Theta - 2\beta\Delta\frac{\sin m\Theta}{\sin\Theta}$$

and

$$F_m(\theta) = \cosh m\theta - 2\beta\Delta\frac{\sinh m\theta}{\sinh\theta}.$$

We can, of course, put $a_0=a_1=\cdots=a_{R-1}=1$, and determine the b's from (37) and (38). Referring back to (19), (20), and (24), and recalling that columns of P_R are the right eigenvectors $(x)_k$, and the rows of Q_R are the left eigenvectors $(y)_k$, we obtain

$$P(n \mid m; s) = \sum_{k=0}^{R-1} \lambda_k^s x_k^{(m)} y_k^{(n)}, \tag{39}$$

or, more explicitly,

$$\begin{aligned} P(n \mid m; s) = {} & b_0(2\sqrt{pq}\cosh\theta_0)^s\left(\frac{p}{q}\right)^{n/2}\left(\frac{q}{p}\right)_*^{m/2} F_m(\theta_0)F_n(\theta_0)\left[1 + (-1)^{m+n+s}\right] \\ & + \left(\frac{p}{q}\right)^{n/2}\left(\frac{q}{p}\right)_*^{m/2}(2\sqrt{pq})^s \sum_{k=1}^{R-2} b_k \cos^s\Theta_k f_m(\Theta_k)f_n(\Theta_k). \end{aligned} \tag{40}$$

Making use of (21), we can achieve considerable simplification by letting $R\to\infty$. In fact, it can be shown that

$$P(n \mid m; s) = \frac{p-q}{2pq}\left(\frac{q}{p}\right)_*^m\left[1 + (-1)^{m+n+s}\right] \tag{41}$$

$$+\frac{2}{\pi}\left(\frac{p}{q}\right)^{n/2}\left(\frac{q}{p}\right)_{*}^{m/2}(2\sqrt{pq})^{s}\int_{0}^{\pi}\cos^{s}\Theta\frac{\tan^{2}\Theta}{(p-q)^{2}+\tan^{2}\Theta}f_{n}(\Theta)f_{m}(\Theta)d\Theta.$$

Although in various places we have tacitly assumed that p and q are different from $\frac{1}{2}$, the final formula (41) can easily be seen to be valid also for the case $p=q=\frac{1}{2}$. In this case (free particle in the presence of a reflecting barrier) the formula assumes the remarkably simple form

$$P(n\mid m;s)=\frac{2}{\pi}(1)_{*}^{m/2}\int_{0}^{\pi}\cos^{s}\Theta\cos m\Theta\cos n\Theta d\Theta, \tag{42}$$

and the right member can be expressed in terms of binomial coefficients. This formula can also be derived in a much simpler way using, for instance, the classical method of images.

In the limit

$$\Delta\to 0,\qquad \tau\to 0,\qquad \frac{\Delta^{2}}{2\tau}=D,\qquad n\Delta\to x_{0},\qquad s\tau=t,$$

one can show that

$$\lim\sum_{x_{1}<m\Delta<x_{2}}P(n\mid m;s)=\int_{x_{1}}^{x_{2}}P(x_{0}\mid x;t)dx,$$

where

$$P(x_{0}\mid x;t)=4\beta e^{-4\beta x}+e^{-2\beta(x-x_{0})}e^{-4\beta^{2}Dt}\frac{2}{\pi}\int_{0}^{\infty}e^{-Dy^{2}t}\frac{y^{2}}{4\beta^{2}+y^{2}}g(x,y)g(x_{0},y)dy, \tag{43}$$

and

$$g(x,y)=\cos xy-2\beta(\sin xy)/y.$$

The proof of this theorem is not elementary inasmuch as it utilizes the so called "continuity theorem for Fourier-Stieltjes transforms" [6]. Formula (43) can be shown to be equivalent with Smoluchowski's formula given in [1].

4. An elastically bound particle. Again the particle can move either Δ to the right or Δ to the left, and the duration of each step is τ. However, the probability of moving in either direction depends on the position of the particle. More precisely, if the particle is at $k\Delta$ the probabilities of moving right or left are

$$\frac{1}{2}\left(1-\frac{k}{R}\right)\quad\text{or}\quad\frac{1}{2}\left(1+\frac{k}{R}\right),$$

respectively. R is a certain integer, and possible positions of the particle are limited by the condition $-R\leqq k\leqq R$. The basic probabilities $P(n\mid m;s)$ now satisfy the difference equation

$$(44)\quad P(n \mid m; s+1) = \frac{R+m+1}{2R} P(n \mid m+1; s) + \frac{R-m+1}{2R} P(n \mid m-1; s),$$

which must be solved with the initial condition

$$(45)\qquad P(n \mid m; 0) = \delta(m, n).$$

In the limit

$$\Delta \to 0, \quad \tau \to 0, \quad R \to \infty, \quad \frac{\Delta^2}{2\tau} = D, \quad \frac{1}{R\tau} \to \gamma,$$

$$s\tau = t, \quad n\Delta \to x_0, \quad m\Delta \to x,$$

the difference equation (44) is seen to go over formally into the differential equation

$$(46)\qquad \frac{\partial P}{\partial t} = \gamma \frac{\partial (xP)}{\partial x} + D \frac{\partial^2 P}{\partial x^2}$$

which is Smoluchowski's equation (5) with $F(x) = -x/\gamma f$.

The discrete problem in a different form and in a different connection was first proposed and discussed by P. and T. Ehrenfest in 1907 [7]. In the next section we shall discuss their original formulation. A fairly detailed treatment was given by Schrödinger and Kohlrausch in 1926 [8] and a brief exposition can be found in the review article of Wang and Uhlenbeck [1]. It seems that Schrödinger and Kohlrausch were the first to point out the connection between the Ehrenfest model and Brownian motion of an elastically bound particle. However, an explicit solution of (44) with the initial condition (45) was apparently not known. I have recently found such a solution using the matrix method described in Section 3 [9]. Instead of the infinite matrix of that section we must now consider the finite matrix

$$(47)\qquad B = \begin{pmatrix} 0 & \frac{1}{2R} & 0 & 0 & 0 \cdots 0 \\ 1 & 0 & \frac{2}{2R} & 0 & 0 \cdots 0 \\ 0 & 1 - \frac{1}{2R} & 0 & \frac{3}{2R} & 0 \cdots 0 \\ \cdot & \cdot & \cdot & \cdot & \cdot \\ 0 & 0 & \cdots & \frac{1}{2R} & 0 \end{pmatrix}$$

and the problem is again reduced to finding the eigenvalues $\lambda_{-R}, \lambda_{-R+1}, \cdots, \lambda_0, \cdots, \lambda_{R-1}, \lambda_R$ of B and matrices P and Q such that

$$PQ = I \tag{48}$$

and

$$B = P \begin{bmatrix} \lambda_{-R} & & & & \\ & \lambda_{-R+1} & & 0 & \\ & & \ddots & & \\ & 0 & & \lambda_{R-1} & \\ & & & & \lambda_R \end{bmatrix} Q. \tag{49}$$

As before, $P(n|m; s)$ is the (m, n) element of B^s, where

$$B^s = P \begin{bmatrix} \lambda_{-R}^s & & & & \\ & \lambda_{-R+1}^s & & 0 & \\ & & \ddots & & \\ & 0 & & \lambda_{R-1}^s & \\ & & & & \lambda_R^s \end{bmatrix} Q. \tag{50}$$

In order to perform the diagonalization (49) explicitly we start (following the procedure of Section 3) by trying to find the eigenvalues and the right eigenvectors of B. For this purpose we consider the system of linear equations

$$\begin{aligned} \frac{1}{2R} x_1 &= \lambda x_0 \\ x_0 + \frac{2}{2R} x_2 &= \lambda x_1 \\ \left(1 - \frac{1}{2R}\right) x_1 + \frac{3}{2R} x_3 &= \lambda x_2 \\ \cdots\cdots\cdots\cdots\cdots \\ \frac{1}{2R} x_{2R-1} &= \lambda x_{2R}, \end{aligned} \tag{51}$$

and the auxiliary infinite system

$$\begin{aligned} \frac{1}{2R} x_1 &= \lambda x_0 \\ x_0 + \frac{2}{2R} x_2 &= \lambda x_1 \\ \left(1 - \frac{1}{2R}\right) x_1 + \frac{3}{2R} x_3 &= \lambda x_2 \\ \cdots\cdots\cdots\cdots\cdots \end{aligned} \tag{52}$$

$$\frac{1}{2R} x_{2R-1} + \frac{2R+1}{2R} x_{2R+1} = \lambda x_{2R}$$

$$\frac{2R+2}{2R} x_{2R+2} = \lambda x_{2R+1}$$

$$\cdots\cdots\cdots\cdots\cdots\cdots\cdots\cdots\cdots.$$

If we can find non-trivial solutions of (52) for which

$$x_{2R+1} = 0 \tag{53}$$

we will have found solutions of (51). It will turn out that this procedure will again yield all eigenvalues and right eigenvectors. Multiplying the members of the equations of (52) by $1, z, z^2, \cdots$, and adding, we obtain formally

$$\sum_{k=0}^{\infty}\left(1 - \frac{k}{2R}\right) x_k z^{k+1} + \sum_{k=0}^{\infty} \frac{k}{2R} x_k z^{k-1} = \lambda \sum_{k=0}^{\infty} x_k z^k,$$

or, introducing the abbreviation

$$f(z) = \sum_{k=0}^{\infty} x_k z^k,$$

$$z f(z) - \frac{z^2}{2R} f'(z) + \frac{1}{2R} f'(z) = \lambda f(z).$$

We thus get the differential equation

$$f'(z) = 2R \frac{\lambda - z}{1 - z^2} f(z), \tag{54}$$

whose solution satisfying $f(0) = x_0$ is easily found to be

$$f(z) = x_0 (1 - z)^{R(1-\lambda)} (1 + z)^{R(1+\lambda)}. \tag{55}$$

Since $f(z)$ is analytic in the neighborhood of $z=0$ the formal procedure can be justified.

We now notice that if

$$\lambda = \frac{j}{R}, \qquad j = -R, -R+1, \cdots, 0, \cdots, R-1, R, \tag{56}$$

$f(z)$ is a polynomial of degree $2R$, and hence $x_{2R+1} = 0$. The numbers (56) are thus seen to be eigenvalues of B and, since there are $2R+1$ of them, we see that we have found *all* the eigenvalues. It also follows that the components of the right eigenvector belonging to the eigenvalue $\lambda_j = j/R$ can be taken as

$$C_0^{(j)} = 1,\ C_1^{(j)},\ C_2^{(j)}, \cdots, C_{2R}^{(j)},$$

where the C's are defined by the identity

$$(1-z)^{R-j}(1+z)^{R+j} \equiv \sum_{k=0}^{2R} C_k^{(j)} z^k. \tag{57}$$

So far we have followed very closely the procedure described in Section 3. Surprisingly enough, we encounter unexpected difficulties in trying to carry out the analogy still further and determine by similar means the left eigenvectors.

To find the matrix Q we resort to a different method. Let us first recall that P can be taken as the matrix whose jth column (for convenience columns and rows are numbered from $-R$ to R) is

$$\begin{bmatrix} 1 \\ C_1^{(j)} \\ C_2^{(j)} \\ \vdots \\ C_{2R}^{(j)} \end{bmatrix}.$$

Matrix Q must satisfy the equation

$$P'Q' = I,$$

which is an immediate consequence of the equation $PQ=I$, and hence denoting by $\alpha_{-R}, \cdots, \alpha_0, \cdots, \alpha_R$, the consecutive elements of the jth column of Q', we must have

$$\sum_{k=-R}^{R} C_{R+r}^{(k)}\alpha_k = \delta(j, r), \qquad r = -R, \cdots, R. \tag{58}$$

From (58) it follows that

$$z^{R+j} = \sum_{r=-R}^{R} \delta(j, r) z^{R+r} = \sum_{r=-R}^{R} z^{R+r} \sum_{k=-R}^{R} C_{R+r}^{(k)}\alpha_k = \sum_{k=-R}^{R} \alpha_k \sum_{r=-R}^{R} C_{R+r}^{(k)} z^{R+r}$$

$$= \sum_{k=-R}^{R} \alpha_k \sum_{s=0}^{2R} C_s^{(k)} z^s,$$

or, by virtue of (57),

$$z^{R+j} = \sum_{k=-R}^{R} \alpha_k (1-z)^{R-k}(1+z)^{R+k} = (1-z)^{2R} \sum_{l=0}^{2R} \alpha_{l-R}\left(\frac{1+z}{1-z}\right)^l.$$

Thus

$$\frac{z^{R+j}}{(1-z)^{2R}} = \sum_{l=0}^{2R} \alpha_{l-R}\left(\frac{1+z}{1-z}\right)^l. \tag{59}$$

Let

$$\zeta = \frac{1+z}{1-z},$$

so that

$$z = -\frac{1-\zeta}{1+\zeta} \quad \text{and} \quad 1 - z = \frac{2}{1+\zeta}.$$

In terms of ζ (59) assumes the form

$$\frac{(-1)^{R+j}}{2^{2R}}(1-\zeta)^{R+j}(1+\zeta)^{R-j} = \sum_{l=0}^{2R} \alpha_{l-R}\zeta^l, \tag{60}$$

and since by (57)

$$(1-\zeta)^{R+j}(1+\zeta)^{R-j} = \sum_{l=0}^{2R} C_l^{(-j)}\zeta^l,$$

we obtain, by comparing coefficients of corresponding powers of ζ,

$$\alpha_{l-R} = \frac{(-1)^{R+j}}{2^{2R}} C_l^{(-j)},$$

or finally,

$$\alpha_s = \frac{(-1)^{R+j}}{2^{2R}} C_{R+s}^{(-j)}. \tag{61}$$

Formula (61) determines explicitly the elements of Q' (and hence of Q), and it is now possible to write an explicit expression for $P(n|m; s)$. In fact, making use of (50), we obtain

$$P(n \mid m; s) = \frac{(-1)^{R+n}}{2^{2R}} \sum_{j=-R}^{R} \left(\frac{j}{R}\right)^s C_{R+j}^{(-n)} C_{R+m}^{(j)}. \tag{62}$$

In the limit

$$\Delta \to 0, \qquad \tau \to 0, \qquad \frac{\Delta^2}{2\tau} = D, \qquad \frac{1}{R\tau} \to \gamma, \qquad s\tau = t, \qquad n\Delta \to x_0,$$

we have

$$\lim \sum_{x_1 < m\Delta < x_2} P(n \mid m; s) = \int_{x_1}^{x_2} P(x_0 \mid x; t)\,dx,$$

where

$$P(x_0 \mid x; t) = \frac{\sqrt{\gamma}}{\sqrt{2\pi D(1 - e^{-2\gamma t})}} e^{-\gamma(x - x_0 e^{-\gamma t})^2 / 2\gamma(1 - e^{-2\gamma t})}. \tag{63}$$

The proof is again made to depend on the continuity theorm for Fourier-Stieltjes transforms.

The frequency function (63) was first discovered by Lord Raleigh [10]. Its connection with Brownian motion of an elastically bound particle, in the strongly overdamped case, was established by Smoluchowski who arrived at it quite independently.

5. The Ehrenfest model. Irreversibility and recurrence. Imagine $2R$ balls numbered consecutively from 1 to $2R$, distributed in two boxes (I and II) so that at the beginning there are $R+n$, $-R \leqq n \leqq R$, balls in box I. We chose at random an integer between 1 and $2R$ (all these integers are assumed to be equiprobable) and move the ball, whose number has been drawn from the box in which it is, to the other box. This process is then repeated s times and we ask for the probability $Q(R+n|R+m;\ s)$ that after s drawings there should be $R+m$ balls in box I.

A moment's reflection will persuade one that this formulation (originally proposed by P. and T. Ehrenfest) [7] is equivalent to the random walk formulation of Section 4, if one interprets the excess over R of balls in box I as the displacement of the particle ($\Delta = 1$). Thus

$$Q(R + n \,|\, R + m;\ ) = P(n \,|\, m;\ s),$$

where $P(n|m;\ s)$ has the meaning of Section 4.

In the present formulation we have a simple and convenient model of heat exchange between two isolated bodies of unequal temperatures. The temperatures are symbolized by the numbers of balls in the boxes and the heat exchange is not an orderly process, as in classical thermodynamics, but a random one like in the kinetic theory of matter. The realistic value of the model is greatly enhanced by the fact that the average excess over R of the number of balls in box I, namely, the quantity

$$\sum_{m=-R}^{R} mP(n \,|\, m;\ s)$$

can easily be shown to be equal to

$$n\left(1 - \frac{1}{R}\right)^{s} \tag{64}$$

which in the limit $R \to \infty$, $1/R\tau \to \gamma$, $s\tau = t$, gives

$$ne^{-\gamma t},$$

or the Newton law of cooling.

There are several proofs of (64) [11]. The most straightforward one, which is not however the simplest, is based on formula (62).

The Ehrenfest model is also particularly suited for the discussion of a famous paradox which at the turn of this century nearly wrecked Boltzmann's inspired efforts to explain thermodynamics on the basis of kinetic theory. In classical thermodynamics the process of heat exchange of two isolated bodies of unequal temperatures is irreversible. On the other hand, if the bodies are treated as a dynamical system the famed "Wiederkehrsatz" of Poincaré asserts that "almost every" state (except for a set of states which, when interpreted as points in phase space, form a set of Lebesgue measure 0) of the system will be, to an arbitrarily prescribed degree of accuracy, again approximately achieved. Thus, argued Zermelo, the irreversibility postulated in thermodynamics and the "recurrence" properties of dynamical systems are irreconcilable. Boltzmann then replied that the "Poincaré cycles" (time intervals after which states "nearly recur" for the first time,—the word "nearly" requiring further specification) are so long compared to time intervals involved in ordinary experiences that predictions based on classical thermodynamics can be fully trusted. This explanation, though correct in principle, was set forth in a manner which was not quite convincing and the controversy raged on. It was mainly through the efforts of Ehrenfest and Smoluchowski that the situation became completely clarified, and the irreversibility interpreted in a proper statistical manner.

It will now be easy to discuss this explanation by appealing to the Ehrenfest model. Let $P'(n|m; s)$ denote the probability that after s drawings (the duration of each drawing is τ) $R+m$ balls will be observed *for the first time* in box I if there were $R+n$ balls in that box at the beginning. In particular, $P'(n|n; s)$ is the probability that the recurrence time of the state "n" (defined by the presence of $R+n$ balls in box I) is $s\tau$. One can then show that

$$\sum_{s=1}^{\infty} P'(n \mid n; s) = 1, \tag{65}$$

or, in other words: *each state is bound to recur with probability 1.* This is the statistical analogue of the "Wiederkehrsatz." One can show furthermore that the mean recurrence time, namely, the quantity

$$\theta_n = \sum_{s=1}^{\infty} s\tau P'(n \mid n; s)$$

is equal to

$$\tau \frac{(R+n)!(R-n)!}{(2R)!} 2^{2R}. \tag{66}$$

This is the statistical analogue of a "Poincaré cycle," and it tells us, roughly speaking, how long, on the average, one will have to wait for the state "n" to recur.

If $R+n$ and $R-n$ differ considerably, θ_n is enormous. For example, if $R=10000$, $n=10000$, $\tau=1$ second, we get

$$\theta = 2^{20000} \text{ seconds (of the order of } 10^{6000} \text{ years!).}$$

If on the other hand, $R+n$ and $R-n$ are nearly equal, θ_n is quite short. If in the above example we set $n=0$ we get (using Stirling's formula)

$$\theta \sim 100\sqrt{\pi} \text{ seconds} \sim 175 \text{ seconds}.$$

It was Smoluchowski who advanced the rule [12] that if one starts in a state with a long recurrence time the process will appear as irreversible. In our example if one starts with 20000 balls in one box and none in the other, one should observe, for a long time, an essentially irreversible flow of balls. On the other hand, if the mean recurrence time is short, there is no sense to speak about irreversibility.

We now give the proofs of (65) and (66). We shall base our considerations on a formula which Professor Uhlenbeck used for similar purposes in some of his unpublished notes. The formula in question is:

$$P(n\,|\,m;\,s) = P'(n\,|\,m;\,s) + \sum_{k=1}^{s-1} P'(n\,|\,m;\,k)P(m\,|\,m;\,s-k). \tag{67}$$

To convince oneself of the validity of this formula we divide all possible ways of reaching "m" from "n" in s steps into classes according to when "m" has been reached for the first time. We then observe that starting from "n" one can reach "m" in s steps in the following s mutually exclusive ways:

(1) "m" is reached for the first time after s steps.

(2) "m" is reached for the first time in 1 step and then, starting from "m" it is again reached in $s-1$ steps.

(3) "m" is reached for the first time in 2 steps and then, starting from "m", it is reached again in $s-2$ steps, and so forth. We note furthermore that the probability that "m" will be reached for the first time in k steps and then, starting from "m," it will be reached again in $s-k$ steps, is

$$P'(n\,|\,m;\,k)P(m\,|\,m;\,s-k). \tag{68}$$

This completes the proof of (67).

It should be emphasized that the justification of using the product of probabilities in (68) rests upon the fact that in our process the past is independent of the future. In other words, once we know that the system starts, say, from "m," its subsequent behavior is independent of the way in which "m" was reached in the first place.

We introduce now the generating functions

$$h(n\,|\,m;\,z) = \sum_{s=1}^{\infty} P(n\,|\,m;\,s)z^s \tag{69}$$

$$g(n\,|\,m;\,z) = \sum_{s=1}^{\infty} P'(n\,|\,m;\,s)z^s, \tag{70}$$

and note that (67) is equivalent to

$$h(n\,|\,m;\,z) = g(n\,|\,m;\,z) + h(m\,|\,m;\,z)g(n\,|\,m;\,z),$$

or

$$g(n \mid m; z) = \frac{h(n \mid m; z)}{1 + h(m \mid m; z)}. \tag{71}$$

In particular,

$$g(n \mid n; z) = \frac{h(n \mid n; z)}{1 + h(n \mid n; z)} = 1 - \frac{1}{1 + h(n \mid n; z)}, \tag{72}$$

and we also note that

$$\frac{dg(n \mid n; z)}{dz} = \frac{\dfrac{dh(n \mid n; z)}{dz}}{(1 + h(n \mid n; z))^2}. \tag{73}$$

From the definition of $g(n \mid n; z)$, we obtain

$$\lim_{z \to 1} g(n \mid n; z) = \sum_{s=1}^{\infty} P'(n \mid n; s) \tag{74}$$

$$\tau \lim_{z \to 1} \frac{dg(n \mid n; z)}{dz} = \sum_{s=1}^{\infty} s\tau P'(n \mid n; s). \tag{75}$$

It is from these formulas that we shall derive (65) and (66). We have, using (62)

$$h(n \mid n; z) = \frac{(-1)^{R+n}}{2^{2R}} \sum_{j=-R}^{R} \sum_{s=1}^{\infty} \left(\frac{jz}{R}\right)^s C_{R+j}^{(-n)} C_{R+n}^{(j)},$$

and since

$$1 = \frac{(-1)^{R+n}}{2^{2R}} \sum_{j=-R}^{R} C_{R+j}^{(-n)} C_{R+n}^{(j)},$$

we obtain

$$1 + h(n \mid n; z) = \frac{(-1)^{R+n}}{2^{2R}} \sum_{j=-R}^{R} \frac{1}{1 - \dfrac{j}{R} z} C_{R+j}^{(-n)} C_{R+n}^{(j)}. \tag{76}$$

All terms in the sum on the right hand side of (76) are regular at $z=1$ except the term corresponding to $j=R$, which has a simple pole at that point. Thus we can write

$$1 + h(n \mid n; z) = p(z) + \frac{(-1)^{R+n}}{2^{2R}} C_{2R}^{(-n)} C_{R+n}^{(R)} \frac{1}{1-z},$$

where $p(z)$ is regular at $z=1$. We see that

$$\lim_{z \to 1} (1 + h(n \mid n; z)) = \infty$$

and hence, using (72) and (74)

$$\sum_{s=1}^{\infty} P'(n \mid n; s) = 1.$$

It is easy to see that

$$\frac{(-1)^{R+n}}{2^{2R}} C_{2R}^{(-n)} C_{R+n}^{(R)} = \frac{1}{2^{2R}} \frac{(2R)!}{(R+n)!(R-n)!},$$

and, denoting this expression by ω, we have (for $|z| > 1$)

$$\frac{dg(n \mid n; z)}{dz} = \frac{(1-z)^2 p'(z) + \omega}{[(1-z)p(z) + \omega]^2},$$

and hence

$$\lim_{z \to 1} \frac{dg(n \mid n; z)}{dz} = \frac{1}{\omega}.$$

This together with (75) yields (66).

The above considerations can be extended to more general processes. However, Markoffian processes (*i.e.*, processes for which (68) is valid) are still the only ones for which one can also calculate the "fluctuation" of the recurrence time, namely, the quantity

$$\sum_{s=1}^{\infty} s^2 \tau^2 P'(n \mid n; s) - \theta_n^2. \tag{77}$$

Without going into the details, let us mention that (77) can be calculated in terms of

$$\lim_{z \to 1} \frac{d^2 g(n \mid n; z)}{dz^2}.$$

The fluctuation (77) gives us a measure of stability of the mean recurrence time inasmuch as it permits us to estimate how likely (or unlikely) it is to get a specified deviation of the actual recurrence time from the mean. It may seem that since the generating function $g(n|n; z)$ is known explictly it should be easy to get an explicit expression for $P'(n|n; s)$. This, however, is not the case. We have not succeeded in finding such an expression, except for $P'(0|0; s)$, and even then we had to use a different method. We shall give a brief description of this method. Let

$$P(n \mid m; 1) = p_{nm}.$$

Then,

$$P'(n \mid n; s) = \sum_{m_1, \cdots, m_{s-1}}' p_{nm_1} p_{m_1 m_2} \cdots p_{m_{s-1} n},$$

where the accent on the summation sign indicates that $m_j \neq n, j=1, 2, \cdots, s-1$. Now let

$$\epsilon_i = \begin{cases} 0 & \text{if } i = n \\ 1 & \text{if } i \neq n. \end{cases}$$

Noticing that

$$\epsilon_i^2 = \epsilon_i,$$

we can write

$$P'(n \mid n; s) = \sum_{m_1, \cdots, m_{s-1}} p_{nm_1}\epsilon_{m_1}p_{m_1m_2}\epsilon_{m_2}p_{m_2m_3} \cdots \epsilon_{m_{s-2}}p_{m_{s-2}m_{s-1}}\epsilon_{m_{s-1}}p_{m_{s-1}n},$$

where the summation is now extended over all m_j. If B is the matrix

$$((p_{nm})),$$

and B_1 the matrix

$$((\epsilon_n p_{nm} \epsilon_m)),$$

we see that

$$P'(n \mid n; s) = (n, n) \text{ element of } BB_1^{s-2}B. \tag{78}$$

We may note that B_1 is obtained from B by crossing out the nth row and the nth column of the latter, and replacing them by a row and column consisting entirely of zeros. If B_1 can be explicitly diagonalized, that is, written in the form

$$B_1 = P_1 \begin{pmatrix} \mu_1 & & 0 \\ & \mu_2 & \\ & & \ddots \\ 0 & & & \cdot \end{pmatrix} Q_1,$$

where

$$P_1 Q_1 = I$$

one can calculate $P'(n \mid n; s)$, explicitly using (78).

We have applied this method to the Ehrenfest model, but only in the case when the middle (zeroth) row and column of B are replaced by a row and column consisting entirely of zeros have we been able to diagonalize explicitly the resulting matrix B_1. The diagonalization proceeds very much as in Section 4, but it has proved necessary to distinguish between the cases when R is even or odd. In case R is even, we were able to derive the formula

$$P'(0 \mid 0; s) = -\frac{1}{2^{2R-1}} \cdot \frac{R+1}{2R} \sum \left(\frac{j}{R}\right)^{s-2} C_{R-j}^{(-1)} C_{R-1}^{(j)}, \qquad s \geqq 2, \tag{79}$$

where the summation is extended over all odd integers j between $-R$ and R. The details of the derivation are somewhat tedious and will not be reproduced here. Formula (79) furnishes a partial solution to a question left open by Wang and Uhlenbeck [1].

References

1. An extensive list of references to the physical literature can be found in the following articles: G. E. Uhlenbeck and L. S. Ornstein, On the theory of Brownian motion, Phys. Rev. 36 (1930) pp. 823–841, M. C. Wang and G. E. Uhlenbeck, On the theory of Brownian motion II, Rev. Mod. Phys. 17 (1945) pp. 323–342. For a very complete summary of earlier results see the important paper of Smoluchowski, Drei Vortrage über Diffusion, Brownsche Molekularbewegung und Koagulation von Kolloidteilchen, Phys. Zeit. 17 (1916) pp. 557–571 and 585–599.

2. References to the work of Wiener, Kolmogoroff, Doob and Feller are given in the second article in [1]. See also P. Lévy, Sur certains processus stochastiques homogènes, Comp. Math. 7 (1939) pp. 283–339, R. Fortet, Les fonctions aléatoires du type de Markoff associées à certain équations paraboliques, Jour. de Math. 22 (1943) pp. 177–243.

3. For a brief discussion of the nature of the friction coefficient f see the first article in [1].

4. For proofs of this theorem the reader is referred to H. Cramér, Mathematical Methods of Statistics, Princeton University Press (1946), in particular, pp. 198–203.

5. For a complete presentation of the matrix method as applied to Markoff chains see M. Fréchet, Traité du Calcul des Probabilités, Tome I, Fasc. III Second Livre, Paris, Gauthier-Villars (1936).

6. For the proof of the continuity theorem in its most general form see Cramér's book [4] pp. 96–100.

7. Über zwei bekannte Einwände gegen das Boltzmannsche H-Theorem, Phys. Zeit. 8 (1907) pp. 311–314.

8. Das Ehrenfestsche Model der H-Kurve, Phys. Zeit. 27 (1926), pp. 306–313.

9. M. Kac, Bull. Am. Math. Soc. 52 (1946) p. 621 (abstract).

10. See footnote 27 in the second paper of [1].

11. See the paper [8] and for a derivation based on an entirely different principle H. Steinhaus, La théorie and les applications des fonctions indépendantes au sens stochastique, Actualités Scientifiques et Industrielles 738 (1938) Paris, Hermann et Cie, pp. 57–73, in particular, pp. 61–64.

12. See the third article in [1] p. 568.

APPENDIX TO "RANDOM WALK AND THE THEORY OF BROWNIAN MOTION"

MARK KAC, Rockefeller University, New York

"Random Walk and the Theory of Brownian Motion" now almost thirty years old generated a substantial follow-up literature which would be nearly impossible to review even briefly. Most of it is of technical nature directed toward modification of formulation or treatment of the Ehrenfest Model.

Only the paper by the late F. G. Hess *Alternative solution to the Ehrenfest problem*, Amer. Math. Monthly, 61 (1954) 323–327, contains in my opinion a genuinely novel idea (also to be found in a somewhat less explicit form in A. J. F. Siegert, *On the approach to statistical equilibrium*, Phys. Rev., 76 (1949) 1708–1714).

The idea is the following. In my original treatment the state of the system is defined by the number m of balls in one of the two boxes (chosen and fixed once and for all). With this definition the process is a Markoff chain and the problem is to find the eigenvalues and the eigenvectors (both right and left) of the $(2R+1)\times(2R+1)$ matrix of transition probabilities $P(m|n;1)$. Hess defines the state as $(\varepsilon_1,\varepsilon_2,\ldots,\varepsilon_{2R})$ where $\varepsilon_i=1$ if ball i is in box I and $\varepsilon_i=0$ if ball i is in box II. With this definition of the state the process is again a Markoff chain but the problem is to diagonalize a much larger $(2^{2R}\times 2^{2R})$ matrix of transition probabilities $P(\varepsilon_1,\varepsilon_2,\ldots,\varepsilon_{2R}|\eta_1,\eta_2,\ldots,\eta_{2R},1)$ $(\eta_i=0,1)$. Remarkably enough this is much easier owing to the fact that the matrix is a linear combination of tensor products of $2R$ simple 2×2 matrices.

In physical terminology $(\varepsilon_1,\varepsilon_2,\ldots,\varepsilon_{2R})$ is a "microstate" while $m=\varepsilon_1+\varepsilon_2+\cdots+\varepsilon_{2R}$ is a "macrostate."

Hess's treatment is well worth studying, for it may be applicable to other problems.

Finally, an oddity. From a paper by Istvan Vincze, *Über das Ehrenfestsche Modell des Wärmeübertragung*, Archiv der Mathematik, XV (1964) 394–400, I have learned that the eigenvalues of $P(m|n;1)$ (actually $2RP(m|n;1)$) were surmised by Sylvester.

Sylvester's paper—a record in brevity—is reproduced below in its entirety:

THÉORÈME SUR LES DÉTERMINANTS.

[*Nouvelles Annales de Mathématiques*, XIII. (1854), p. 305.]

SOIENT les déterminants

$$|\lambda|,\ \begin{vmatrix}\lambda,1\\1,\lambda\end{vmatrix},\ \begin{vmatrix}\lambda,1,0\\2,\lambda,2\\0,1,\lambda\end{vmatrix},\ \begin{vmatrix}\lambda,1,0,0\\3,\lambda,2,0\\0,2,\lambda,3\\0,0,1,\lambda\end{vmatrix},\ \begin{vmatrix}\lambda,1,0,0,0\\4,\lambda,2,0,0\\0,3,\lambda,3,0\\0,0,2,\lambda,4\\0,0,0,1,\lambda\end{vmatrix},\ \begin{vmatrix}\lambda,1,0,0,0,0\\5,\lambda,2,0,0,0\\0,4,\lambda,3,0,0\\0,0,3,\lambda,4,0\\0,0,0,2,\lambda,5\\0,0,0,0,1,\lambda\end{vmatrix};$$

la loi de formation est évidente; effectuant, on trouve

$$\lambda,\ \lambda^2-1^2,\ \lambda(\lambda^2-2^2),\ (\lambda^2-1^2)(\lambda^2-3^2),\ \lambda(\lambda^2-2^2)(\lambda^2-4^2),$$
$$(\lambda^2-1^2)(\lambda^2-3^2)(\lambda^2-5^2),\ \lambda(\lambda^2-2^2)(\lambda^2-4^2)(\lambda^2-6^2),$$

et ainsi de suite.

There is no indication that Sylvester had a general proof and the mystery is why did he publish the note. Perhaps a sleuth with a historical bent will find a solution.

10

EDWARD JAMES McSHANE

Edward James McShane was born in New Orleans on May 10, 1904. He received the degrees of B.E. and B.S. from Tulane where he majored in physics. After completing his M.S. in 1925 he went to Chicago where he received his Ph.D. under Bliss. In 1947 he received an honorary Sc.D. from Tulane. He taught at Wichita, 1928–29, and was a National Research Council Fellow at Chicago, Princeton, Ohio State, and Harvard during the period 1930–32. In 1932–33 he was *Hilfsassistent* at the University of Göttingen. Joining the Princeton University faculty in 1933 as an instructor, he became an assistant professor in 1934. He was appointed a full professor at the University of Virginia in 1935 and in 1957 he was named Alumni Professor, a position he held until his retirement in 1973.

He was Head Mathematician at the Ballistics Research Lab at the Aberdeen Proving Grounds from 1942–45. He spent the year 1949–50 at the Institute of Advanced Study. In 1955–56 he held a Fulbright Award and spent the year at the University of Utrecht. From 1956 to 1968 he was a member of the National Science Board. In 1963–64 he was Visiting Professor at the Rockefeller University, and for one semester (autumn, 1969) he was a visiting research professor at the Research Institute for Mathematical Analysis at the University of Kyoto. Since his retirement he has been visiting professor for one semester (Jan.–May, 1975) at the Virginia Military Institute and for two months (Sept., Oct., 1975) at the Colorado State University. He was "starred" in American Men of Science in 1938.

Professor McShane was a member of the National Academy of Sciences, the American Mathematical Society (editor, *The Transactions*, 1944–46, President 1958–59), the American Philosophical Society, and the Mathematical Association of America (President, 1953–54).

His early mathematical work was in the field of the calculus of variations as a member of the Chicago school under G. A. Bliss. He also worked in the general theory of integration in function spaces. His interests and his publications have made him a leader in functional analysis and the general theory of limits. His work during the period at Aberdeen was in the area of exterior ballistics where he collaborated with J. L. Kelley and F. V. Reno in producing the book *Exterior Ballistics*. In addition, he wrote the books *Integration* (1944), *Order-Preserving Maps and Integration Processes* (1953), *Real Analysis*, with T. A. Botts (1959), and *Stochastic Calculus and Stochastic Models* (1974). Since his retirement he has been writing a new book on integration theory.

Professor McShane has had a highly successful career as a research mathematician. In addition, he has always shown extreme concern for excellence in the teaching of mathematics and the proper exposition of research. He strove constantly to make research results more accessible to the general mathematical public. "In thousands of different incidents which have gone unrecorded he has been helpful to graduate students, sympathetic to the problems of teachers in many little-known schools and colleges he has visited, generous in bringing his mathematical intellect to the aid of some young mathematician grappling with a problem, wise in counsel, trusted in difficulty, and inspiring in his readily-sensed ideals of human behavior." (*Professor Edward James McShane*, The Monthly, 71 (1964), pp. 1–2.) His own words in his retiring address to the Association typify the spirit of the Chauvenet awards, "Everyone of us is touched in some way or other by the problems of mathematical communication. Every one of us can make some contribution, great or small, within his own proper sphere of activity. And every contribution is needed if mathematics is to grow healthily and usefully and beautifully."

PARTIAL ORDERINGS AND MOORE-SMITH LIMITS*

E. J. McSHANE, University of Virginia

1. Introduction. Everybody knows that the concept of limit is fundamental in analysis. But somewhere about the time he reaches advanced calculus, a student may perhaps begin to wonder how many disguises this concept of limit can assume. Apart from the fact that the dependent variable which is doing the converging may be a real number or a complex number or a vector or a function or what not else, the independent variable may also take any one of a multitude of forms. Thus if the dependent variable is real, we still have to consider convergence of sequences of reals, of multiple sequences, of functions of a real variable x as x tends to some x_0 or as x tends to infinity, of functions of several real variables, and so on. The saving feature is of course that all these assorted definitions have strong resemblances. Nevertheless, the student may be forgiven if he wishes that somebody would put all these different but related ideas into fewer packages, so that it wouldn't be necessary to do almost the same thing over and over again for the slightly different kinds of limit. This may occur to him even if he has never heard of E. H. Moore's dictum, that the existence of strong similarities in the central part of different theories indicates the existence of a more general theory of which these different theories are all special cases. But in such a situation as this there is something to be considered beside generality. What we really want is a treatment of the subject which is not only unified, but is also elegant and easily understood. As a matter of fact, such a theory has been in existence for twenty-nine years. The treatment of limits† devised by E. H. Moore and his then student H. L. Smith appeals to me as one which can readily fit into a course in the theory of functions of a real variable or a course in advanced calculus, not only strengthening the content but also providing a continuous thread tying the subject matter together. And besides this, their treatment has another virtue which is important to a beginner and by no means to be despised by those who are no longer beginners; it follows closely the lines of the simplest of all limit theories, that is the theory of convergent sequences, follows them indeed so closely that many of the proofs can be taken over with only trivial notational changes. Thus the theorems for the more complicated limiting processes are obtained along with the simpler ones, and at a cost in effort little greater than that demanded by the simplest of all.

Let us then look over some of the definitions of limit that an undergraduate might be expected to know, and try to separate the essential parts shared by the different examples. From one point of view, we could classify all the defini-

* Presented as an invited address before the Mathematical Association of America Dec. 30, 1950.

† Moore, E. H. and Smith, H. L., A general theory of limits. Amer. J. Math., vol. 44, 1922, p. 102. Here we should also mention the very inclusive theory of limits based on filters, devised by H. Cartan, Theorie des filtres, C. R. Acad. Sci. Paris, vol. 205, 1937, pp. 595–598, Filtres et ultrafiltres, ibid., pp. 777–779. This same theory was also expounded by H. L. Smith, apparently independently (A general theory of limits, Nat. Math. Mag., vol. 12, 1938, pp. 371–379).

tions into two principal varieties. The first type is exemplified by the limit of a function $f(x)$, defined for all real x, as x approaches x_0. Roughly stated, the limit of f is l if $f(x)$ is near l whenever x is near x_0. The other type is exemplified by the limit of a sequence. The limit of f_n ($n=1, 2, \cdots$) as "n tends to infinity" is l if f_n is near l for all n after a certain n_0. If we would generalize the first type, we should study the idea "x is near x_0" and seek to give it a meaning both precise and general. Thus we are led to study neighborhoods, and eventually arrive at the idea of a topological space. In this day it would be utterly superfluous to stress the importance of the study of topological spaces. But at the moment we are not going in that direction. It is the other type of limit that is to engage our attention.

2. Directed sets and nets. What then are the essential elements of the definition of the limit of a sequence? To begin with, the property that n is an integer is quite irrelevant since the same kind of definition applies to the limit of $f(x)$ as x tends to ∞ over the whole real number system. But the relation "n is after n_0," or in symbols $n > n_0$, is important. To begin with, it is a binary relation on the integers. That is, if we write an integer n, and then write the name $>$ of the relation, and then write another integer m, we obtain a meaningful sentence "$n > m$"; the subject is "n," the verb is "$>$," and the object is "m." The sentence may be true or it may be false, but it is not nonsense. This is what we shall mean by a binary relation on a set A; R is a binary relation on A if whenever a and b are members of A, "aRb" is a meaningful sentence, either true or false.

But a relation that could in any reasonable sense be thought of as "after" must have another property. If a is after b and b is after c, we surely can reasonably ask that it should also be true that a is after c. This is the property called "transitivity." Formally, a binary relation R on A is transitive if whenever a, b and c are elements of A such that aRb and bRc, then it is true that aRc. We shall consider that binary relations R with this property are important enough to be dignified by the name of "partial orderings"; so henceforth a "partial ordering" of A shall be a transitive binary relation on A. (But not all mathematicians agree on this meaning, or on any other, for the expression "partial ordering.") Also, we shall make a slight notational change as a crutch to memory; instead of the non-committal name R for the partial ordering, we shall usually use $\succ$, as a reminder of its kinship with $>$. The statement "$a \succ b$" could be read "a follows b."

However, we have not yet completed our task of finding all the essential properties used in the definition of a sequence. To show this, we observe that if we define $n \succ m$ to mean that $n-m$ is a positive even integer, $\succ$ is a partial ordering. We could repeat the definition of limit of a sequence with the single change of replacing $>$ by $\succ$, and obtain an intelligible definition. But if we define f_n to be 0 if n is even and 1 if n is odd, we find that $f_n=0$ if $n \succ 2$, so $\lim f_n$ is 0; while $f_n=1$ if $n \succ 3$, so $\lim f_n$ is 1. This is most undesirable. Let us therefore look back at the proof that a sequence cannot have two different limits. Suppose that h and $k \neq h$ are both limits of f_n; let ϵ be a positive number less than the difference between h and k. For all n after a certain n' we have f_n within $\epsilon/2$ of h; for all n after a certain n'' we have f_n within $\epsilon/2$ of k. *Now choose an n which is after both n' and n''.* For this n, f_n differs by less than $\epsilon/2$ from both h and k, which therefore must be within ϵ of each other. But this contradicts the definition of ϵ, and so the sequence cannot have two different limits. If we try to carry this through with the relation $\succ$ in place of $>$, we find that the trouble occurs at the italicized sentence; if n' is even and n'' is odd, there is no n such that $n \succ n'$ and $n \succ n''$. This partial ordering has the peculiarity that if $n \succ m$, then if m is even so is n, and if m is odd so is n.

Accordingly, we avoid the trouble of non-unique limits by limiting our attention to those partial orderings in which this situation never arises. That is, we assume that our partial orderings have the property that for each pair of members of the set (not necessarily distinct) there is an element of the set which follows both of them. Moore and Smith gave this property a name* which has fallen into disuse, probably because the property has no importance except when combined with transitivity. When a relation $\succ$ partially orders a set A and also has the property just described, it is now customary to say that the relation $\succ$ "directs" the set A. To state it in full:

A set A is directed by a relation $\succ$ if $\succ$ is a binary relation on A with the properties:

(i) *if a, b, and c are elements of A such that $a \succ b$ and $b \succ c$, then $a \succ c$;*

(ii) *If a and b are elements of A, there exists an element c of A such that $c \succ a$ and $c \succ b$.*

One more bit of terminology is called for. A function which assigns to each positive integer n a real number f_n as functional value has long been known as a sequence of real numbers; more generally, a function which assigns to each n a functional value f_n in a set M is called a sequence of elements of M. We wish to have a name for a function which assigns to each element a of a directed set A a real number $f(a)$ as functional value. For this J. L. Kelley† proposes the name "net"‡ of real numbers; more generally, if f is a function which assigns to each element a of a directed set A a functional value $f(a)$ in a set M, we shall call the function f a "net" of elements of M.

3. Definition of convergence. Now we are in a position to attempt to take over the whole theory of convergent sequences, replacing the word "sequence"

* Compositive.

† J. L. Kelley, *Convergence in topology*, Duke Math. J., vol. 17, 1950, p. 277.

‡ Kelley writes me that this was suggested by Norman Steenrod in a conversation between Kelley, Steenrod and Paul Halmos. Kelley's own inclination was to the name "way"; the analogue of a subsequence would then be a "subway"! Since a stream with its tributaries is a good example of a system directed by the relation "downstream from," I incline toward "stream" rather than "net." But "net" has the great advantage of having seen print first.

by "net," the sign $>$ by $\succ$, and the integers by the elements of the directed set A. For example, the definition of limit becomes:

DEFINITION. *Let $f(a)$, a in A, be a net of real numbers, and let k be a real number. Then $\lim_{a \text{ in } A, \succ} f(a) = k$ means that for every positive ϵ there is an element a_ϵ of A such that $|f(a)-k| < \epsilon$ whenever $a \succ a_\epsilon$.*

For simplicity of notation we shall omit such parts of the symbolism under lim as can be left out without danger of confusion. For example, if $\succ$ is the only partial order of A that we are considering, we omit mentioning it; if all the functions we are interested in at a given moment are defined on the same directed set A, we condense the symbol to $\lim_a f(a)$ or even to $\lim f$. But this is a device familiar to all of us. By such simple devices we take over all the essential parts of the theory of limits. For example, if $f(a)$ and $g(a)$ are both defined and real valued for all a in a directed set A, and have the respective finite limits h and k, then $f+g$ has the limit $h+k$. As another important theorem, let f be a net of real numbers defined on a directed set A and having the property that $f(a) \geqq f(a')$ whenever $a \succ a'$. (Such nets may be called "monotone nondecreasing," but I think that Garrett Birkhoff's terminology "isotone" is much to be preferred. The corresponding name for nets such that $f(a) \leqq f(a')$ when ever $a \succ a'$ is "antitone," to replace the older expression "monotone non-increasing.") The ordinary proof for sequences carries over to show that if f has this property and also has a finite upper bound, it has a limit, and the limit is the same as the least upper bound.

The proof of the Cauchy criterion for convergence could be taken over too, but we shall sketch one which has the virtue of applying in any metric space by merely replacing absolute differences by distances. Suppose it known that the Cauchy condition is necessary and sufficient for the convergence of sequences. Let $f(a)$, a in A, be a net which satisfies the Cauchy condition for nets; that is, for each positive ϵ there is an a_ϵ in A such that $|f(a)-f(a')| < \epsilon$ whenever both a and a' are $\succ a_\epsilon$. We choose successively $a_1, a_2, \cdots$ in A such that $a_n \succ a_{n-1}$ and $|f(a)-f(a')| < 1/n$ whenever a and a' are $\succ a_n$. Then the numbers $f(a_n)$, $n=1, 2, \cdots$ form a Cauchy sequence, so they have a limit k. Now to show that the net has k as limit, let ϵ be a positive number, and pick $n > 2/\epsilon$, so that $|f(a_{n+1})-k| < \epsilon/2$. If $a \succ a_n$, by definition of a_n we have $|f(a_{n+1})-f(a)| < 1/n < \epsilon/2$, so $|f(a)-k| < \epsilon$, completing the proof. This particular theorem is quite convenient, since it provides at once all the assorted forms of the Cauchy criterion for all the various limit processes occurring in advanced calculus and theory of functions of a real variable.

There is, however, one portion of the general theory on which we ask deferment for a few moments. This is the idea of subsequence and the theorems connected with the idea. Before we take this up, we wish to stop to look at some special cases. For we have now generalized the theory of limits from the simple cases of ordinary sequences to something more general, and it is natural that we should pause to see how much new territory we have taken in.

4. Examples. In showing that some special kind of limit process is covered by the Moore-Smith theory there is a simple pattern that we often use. The symbol for the limit usually has some notation under the letters lim that indicates, pictorially speaking, in which direction the independent variable is "going." For instance, the limit of a sequence $\lim_{n\to\infty} f_n$ is a limit "as n goes to infinity." To change over to the notation of nets we try to put a partial ordering on the independent variable in such a way that $a \succ b$ shall means that a has "gone further" than b. Thus in the case of sequences, $m \succ n$ should mean that m has "gone further toward infinity than n has," that is $m > n$. In the case of a limit $\lim_{x\to c} f(x)$, $x \succ x'$ should mean that x has approached closer to c than x' has, that is $|x-c| < |x'-c|$. In this case, though, we add as usual the requirement that x and x' should not equal c.

As a first trial, we consider a real valued function f defined on a set A of real numbers having no finite upper bound, and look at $\lim_{x\to\infty} f(x)$, wherein we of course understand that x is restricted to A. According to the pattern described in the preceding paragraph, we define $x \succ x'$ to mean $x > x'$. It is trivially easy to show that A is directed by $\succ$, and that the Moore-Smith limit $\lim_{x \text{ in } A, \succ} f(x)$ is the same as $\lim_{x\to\infty} f(x)$. For a slightly less trivial example we take a double sequence, $f_{m,n}$ $(m, n = 1, 2, \cdots)$ of reals. Since we want the limit as both m and n tend to infinity, we define $(m, n) \succ (m', n')$ to mean that $m > m'$ and $n > n'$. We easily prove that the set of all pairs of positive integers is directed by this relation and that the Moore-Smith limit in this case reduces to $\lim_{m,n\to\infty} f_{m,n}$. The same device applies to sequences with more than two subscripts.

Suppose that A is a set of real numbers, and p a real number such that arbitrarily near it are numbers $a \neq p$ in A. Let f be a real valued function on A. We wish to show that $\lim_{x\to p} f(x)$ is a special case of the Moore-Smith limit. So we define $x \succ x'$ to mean that neither x nor x' is p and that $|x-p| < |x'-p|$. The Moore-Smith definition reduces to the usual $\epsilon-\delta$-definition except that in it the δ's are restricted to the form $|x-p|$, x in A. This restriction is easily seen to be without effect, so $\lim_{x\to p} f(x)$ is another example of the Moore-Smith theory. The same device can equally well be applied in spaces of any number of dimensions, and in fact in any metric space.

We shall now discuss unordered sums, historically the first step toward the Moore-Smith theory.* Let X be any set, and f a real-valued function on X. Even when X has an order-relation, for instance when X is the set of positive integers, we wish to form partial sums with more and more elements, chosen without regard to order. So we let A consist of all finite subsets of X, and for each a_1 and a_2 in A we define $a_1 \succ a_2$ to mean that a_1 contains a_2, that is all points belonging to a_2 belong to a_1. This set is clearly directed, since for any a and b in A the set c consisting of all points in a and all points in b satisfies $c \succ a$, $c \succ b$. If a is in A and consists say of $x_1, \cdots, x_k$, we define $S(a)$ to mean $f(x_1) + \cdots + f(x_k)$. This is a net; and $\lim S = k$ means that to each positive

* E. H. Moore, *Definition of limit in general analysis*, Proc. Nat. Acad. Sci., vol. 1, 1915, pp 628–632.

ϵ corresponds a finite set a_ϵ such that whenever a is a finite subset of A containing a_ϵ, consisting say of $x_1, \cdots, x_h$, then $|f(x_1)+\cdots+f(x_h)-k|<\epsilon$. It is not difficult to see that when X consists of the positive integers, lim S exists if and only if the series $f(1)+f(2)+\cdots$ is unconditionally convergent, in which case lim $S=\sum_{n=1}^{\infty} f(n)$.

Next we consider a still less traditional looking kind of net. Suppose that C is a plane curve defined by equations $x=x(t)$; $y=y(t)$, $a\leqq t\leqq b$. We wish to give precision to the idea that the length of C is the limit, in some sense, of the lengths of polygons inscribed in C as they acquire more and more vertices. So for A we choose the set of all polygons inscribed in C, the word "inscribed" being understood in the following sense. In $[a, b]$ we choose points $t_0=a<t_1<t_2<\cdots<t_{n-1}<t_n=b$, and we join the points $\{x(t_0), y(t_0)\}$, $\{x(t_1), y(t_1)\}, \cdots, \{x(t_n), y(t_n)\}$ in that order by line segments. The result is a polygon inscribed in C. If P and P' are in A, we say that $P\succ P'$ if and only if all the vertices of P' are also vertices of P. This is easily seen to satisfy (i) of the definition of direction; and if P' and P'' are in A, we can construct an inscribed polygon whose vertices are all the vertices of P' and all those of P'', so that $P\succ P'$ and $P\succ P''$, and part (ii) of the definition is also satisfied. Let $L(P)$ be the length of P. This is a net of real numbers. Its limit, if it exists, is called the length of the curve C. Since we easily see that $L(P')\geqq L(P'')$ whenever $P'\succ P''$, the general theorem on isotone (or non-decreasing) nets tells us that the limit is the same as the least upper bound, so the length of C is also the least upper bound of the lengths of all polygons inscribed in C.

5. Application to the definition of the integral. Our next example is more important, because it is one which is often unsatisfactorily treated in text-books. Let f be defined and bounded on an interval $[a, b]$ of real numbers; we wish to define its (Riemann) integral. First we choose numbers $a=x_0<x_1<\cdots<x_n=n$, and other numbers $\xi_1, \cdots, \xi_n$ such that ξ_i is between x_{i-1} and x_i. Then we form the sum $\sum_{i=1}^{n}(x_i-x_{i-1})f(\xi_i)$, and we take some kind of limit of this sum. The question is, though, what kind of limit. Sometimes one hears "the limit as the number of subdivisions tends to infinity," with some qualifying remark about the lengths of the subdivisions. But this sum is not a single-valued function of the number of subdivisions. Some kind of extension of the concept of limit is called for. One alternative is to look sternly at the student and say "That's perfectly clear, isn't it?" This is what Mark Kac calls "proof by intimidation." A better way is to widen the theory to cover multiple-valued functions, treating the value of the sum as a multiple-valued function of the length of the longest subinterval (x_{i-1}, x_i). A third way, which we now discuss, is to use the Moore-Smith limit. Let us use the name "partition" for a system of division points $a=x_0<x_1<\cdots<x_n=b$ together with the intermediate points $\xi_1, \cdots, \xi_n$ satisfying $x_{i-1}\leqq\xi_i\leqq x_i$; the "norm" of this partition shall be the greatest of the numbers $x_1-x_0, x_2-x_1, \cdots, x_n-x_{n-1}$. If P is the partition just described, we use it to determine a sum $S(P)=\sum_i f(\xi_i)(x_i-x_{i-1})$. This is a single-valued function

of the partition P; in fact, it was just for the purpose of making the sum single-valued that we defined P to consist of both division and intermediate points. Now we wish to express the integral as the limit of $S(P)$ as something happens. This something may be thought of in either of two ways. We may say that we want the longest of the subintervals in the partition to approach zero, or we may say that we want the limit as we cut up the interval (a, b) finer and finer. If we choose the first of these points of view, we would say that for two partitions P and P', the statement $P \succ P'$ should mean that the norm of P is less than the norm of P'. If we choose the other point of view, we would say that the statement $P \succ P'$ means that all the points of division $x_0', x_1', \cdots, x_m'$ occurring in P' are among the points $x_0, x_1, \cdots, x_n$ occurring in P. In either case, the intermediate points ξ_i are disregarded in defining the order. For the kind of integral we are here discussing it does not matter which of the two definitions of order we choose; if the finite sum has a limit when one of the two definitions of $\succ$ is used it has the same limit when the other is used, and this limit is by definition the (Riemann) integral of f. The choice of limiting process is in this case a matter of taste, and likewise in the case of the Riemann integral in higher dimensional spaces. However, it should be mentioned that when we go deeper into analysis and study the Stieltjes integral it does make a difference which definition of $\succ$ we use. It is not that one is "right" and the other "wrong"; the two definitions lead to different integrals, both of which have been investigated.*

6. Subnets. Now we come back to the question of subnets, postponed a few pages ago. The ordinary definition of subsequence has a straightforward generalization, but this generalization does not prove perfectly satisfactory. J. L. Kelley has proposed an alternative which may at first seem drastic, since it is not merely another scheme for generalizing the idea of subsequence with which we are all familiar, but proposes that we replace that old and familiar idea with a new one. Suppose that f_n, $n = 1, 2, \cdots$ is a sequence. We are used to saying that a subsequence of this is a sequence f_{n_j}, $n = 1, 2, \cdots$, wherein the subscripts n_j are positive integers such that $n_1 < n_2 < n_3 < \cdots$. But if we look at the proofs of the theorems on subsequences, we find that this last condition is not used in its full strength. What is used is the consequence that as j tends to ∞, so does n_j. Accordingly, Kelley defined a sequence f_{n_j} to be a subsequence of f_n if the n_j are positive integers such that as j tends to ∞ so does n_j. The generalization to nets is obvious. If $f(a)$, a in A is a net, another net $g(b)$, b in B is a subnet of the first if there is a function $a = a(b)$, b in B such that for all b in B, $g(b) = f\{a(b)\}$, and such also that for each a' in A there is a b' in B such that whenever $b \succ b'$ it is also true that $a(b) \succ a'$.

The principal uses of subsequences come by way of the theorem that if a sequence converges, every subsequence converges to the same limit. This gen-

* See, for example, L. M. Graves, *Theory of functions of real variables*, McGraw-Hill, 1946, pp. 260–261.

eralizes at once to nets. As an application, which in spite of its simplicity may serve to hint that our nets may have caught some strange fish in addition to what we wanted and expected, we consider a function $f(x)$, x in A defined on a set A of real numbers, and we assume that $\lim_{x\to c} f(x)$ exists and is equal to k. Let $x_1, x_2, x_3, \cdots$ be a sequence of points of A all different from c and having c as limit when $n\to\infty$. We *could* follow the familiar proof that then $f(x_n)$ has k as limit. But we need not; for $f(x_n)$, $n=1, 2, \cdots$ is by our definition a subnet of the net $f(x)$, x in A (wherein as before we define $a \succ a'$ to mean $0<|a-c| <|a'-c|$), so the subnet must have k as limit.

7. Relations to topology. So far our discussion has been restricted to topics that an undergraduate might encounter. Now we shall touch briefly on some less elementary uses of Moore-Smith convergence. But first we introduce an abbreviation. If a statement involving members a of a directed set A is true for all a which follow some a' in A, we say that the statement is "eventually" true. Thus we have a clear concept to replace the idea of "becomes and remains" often mentioned in elementary texts. It is in fact convenient to use this idea of "eventually" even in the beginning of the study of Moore-Smith limits. If a statement P is eventually true (say for $a\succ a'$) and another statement Q is eventually true (say for $a\succ a''$), we choose a^* such that $a^*\succ a'$ and $a^*\succ a''$; then P and Q are both true for $a\succ a^*$, so the joint statement "P and Q" is eventually true.

A topological space is a set X together with a specified collection of subsets of X, called the "open sets," such that the union of arbitrarily many open sets is open, the intersection of finitely many open sets is open, and the empty set and X itself are open. A neighborhood of a point x of X is an open set containing x. A net $x(a)$, a in A of points of X converges to a point x' of X if for every neighborhood of x', $x(a)$ is eventually in that neighborhood. A point x' is a *cluster point* of the net $x(a)$, a in A if, crudely stated, $x(a)$ "keeps coming back" to every neighborhood of x'; precisely,† for each neighborhood U of x' and each a' in A, $x(a)$ is in U for some $a\succ a'$.

It is well known that if X is a subset of a euclidean space, either it has all three of the following properties or else it has none of them:

(I) *From every covering of X by open sets it is possible to extract finitely many sets which cover X.*

(II) *Every sequence of points of X has a cluster point in X.*

(III) *Every sequence of points of X contains a subsequence which converges to a point of X.*

These three properties continue to be equivalent for some spaces more general than subsets of euclidean spaces, for example they are equivalent if X is per-

† For this relationship between U and the net $x(a)$, a in A, Halmos proposes the terminology "x is frequently in A U."

fectly separable; but they are not equivalent for all topological spaces. However, as soon as we replace sequences by nets the equivalence is restored. That is, if we write

(II′) *Every net of points of X has a cluster point in X,*

(III′) *For every net $x(a)$, a in A, of points of X there is a subnet converging to a point of X,*

then every topological space X either has all three properties (I), (II′), (III′), or it lacks all three. Suppose that X has property (I), and let $x(a)$, a in A be a net of points of X. If this had no cluster point, for each x in X we could find a neighborhood $U(x)$ such that $x(a)$ is eventually out of $U(x)$. Finitely many of these cover X, say $U(x_1), \cdots, U(x_n)$. For $a \succ$ a certain a_{n_j}, $x(a)$ is not in $U(x_j)$. Choose $a \succ a_1, \cdots, a_n$; then $x(a)$ is out of all the $U(x_j)$ and yet is in X, which is impossible. So X has property (II′). Conversely, suppose X lacks property (I). Then there is a collection K of open sets covering X but such that no finite subcollection of K covers X. Let A consist of all finite subsets of K, and order A by defining $a \succ a'$ to mean that all the sets which belong to a' also belong to a. A is directed by $\succ$, since if a' and a'' are in A, the set a consisting of all sets in a' together with all sets in a'' satisfies $a \succ a'$ and $a \succ a''$. For each such a, there is a point of X not in any of the sets which constitute a, since these sets form a finite subcollection of K and by hypothesis do not cover X. Pick such a point and call it $x(a)$. Because A is directed, these $x(a)$, a in A form a net. Now for any x' in X and any set U of the family K which contains x', we let a' be the subset of K consisting of U alone. If $a \succ a'$, then $x(a)$ is outside all the sets in a, in particular is outside U, so x' is not a cluster point of the net. This holds for all x' in X, so X lacks property (II′).

Property (III′) plainly implies (II′). To show the converse,* let $x(a)$, a in A be a net of points of X having x' as cluster point. Let B consist of all the pairs (U, a) in which U is a neighborhood of x' and a is a point of A such that $x(a)$ is in U. We order these by defining $(U, a) \gg (U', a')$ to mean that U is contained in U' and $a \succ a'$. This is obviously transitive. If (U', a') and (U'', a'') are in B, let U be the intersection of U' and U''. This is a neighborhood of x', so there is an a in A such that $a \succ a'$ and $a \succ a''$ and $x(a)$ is in U. So by definition $(U, a) \gg (U', a')$ and $(U, a) \gg (U'', a'')$, and B is directed by $\gg$. For each $b = (U, a)$ in B we define $a(b)$ to be the second component of b, that is $a(b) = a$. We must prove, first, that $x\{a(b)\}$, b in B is a subnet of $x(a)$, a in A, and second, that it converges to x'. We do both of these at once, as follows. Let a' be an arbitrary member of A and U' an arbitrary neighborhood of x'. There is an a'' in A such that $a'' \succ a'$ and $x(a'')$ is in U'. Define $b' = (U', a'')$. For every $b = (U, a)$ such that $b \gg b'$, we have first $a \succ a'' \succ a'$, that is $a(b) \succ a'$, whence $x\{a(b)\}$, b in B is a subnet of $x(a)$, a in A. And second we have U contained in

* J. L. Kelley, *loc. cit.*

U', so that $x\{a(b)\}$, being in U, is also in U', whenever $b \gg b'$; and therefore x' is the limit of $x\{a(b)\}$, b in B. This completes the proof of the theorem. It will be noticed that Kelley's definition of subnet is exactly what is needed; under previous definitions of subnet the theorem cannot be established.

When we are dealing with subspaces of finite dimensional spaces, it is quite convenient to be able to maneuver back and forth between property (I), in "Heine-Borel" arguments, and properties (II) and (III), in "Bolzano-Weierstrass" arguments. The proof just completed shows that the same kind of convenience is available even in the most general topological spaces, provided that we agree to use nets instead of sequences and to adopt Kelley's definition of subsequence.

8. Convergence in partially ordered spaces. In the earlier pages of this paper we were concerned with sets of real numbers and nets of real numbers. The real numbers have a natural topology, and if we think of the reals as a space of points equipped with this topology the natural generalization is to topological spaces. On the other hand, the reals also constitute a partially ordered, in fact a directed, system. It is thus also reasonable to wonder how much we can deduce from the order relation, and to investigate partially ordered sets and nets of points of partially ordered sets. Suppose then that X is a set of points partially ordered by a relation $\succ$. To save trouble we shall assume that $\succ$ is a *proper* partial ordering, which by definition means that if x and x' are two different points of X, the relation $x \succ x'$ and $x' \succ x$ cannot both be true. For example, when in discussing integrals we partially ordered partitions by defining $P \succ P'$ to mean that norm $P \leqq$ norm P', we thereby introduced an improper partial ordering. But if we had instead defined $P \succ P'$ to mean norm $P <$ norm P' the partial ordering would have been proper.

The definitions of upper and lower bounds and of least upper bounds (or suprema) and greatest lower bounds (or infima) can be taken over at once from the real numbers. Among the reals, there are two equivalent ways of defining completeness. The reals are complete in the Cauchy sense—every sequence of reals which satisfies the Cauchy condition is convergent to a real limit. They are also complete in the Dedekind sense—every non-empty set of real numbers which has an upper bound has a least upper bound, and likewise for lower bounds. The former of these is of course much used in metric spaces; it is the other which we now consider. We could of course take it over verbatim, and say that a properly partially ordered set is Dedekind-complete if every non-empty subset which has an upper bound also has a supremum. But a directed set with this property would then have a supremum and an infimum for each two points, and by definition such a set is a lattice. Now lattices are interesting objects to study, as anyone knows who has looked into Birkhoff's book about them.*

---

* Garrett Birkhoff, *Lattice theory*, Amer. Math. Soc. Colloquium Publications, vol. XXV, 1948.

But at this moment we do not wish to confine our attention to them. So we adopt another formulation which makes no difference when we are discussing the reals, but makes just the difference we want in other cases. We say that X is Dedekind-complete if every non-empty subset S of X which *is directed by* $\succ$ and has an upper bound also has a supremum, and every non-empty subset S of X which is directed by $\succ$ and has a lower bound also has an infimum. Thus for example, the set of circular regions of the plane $(x-x_0)^2+(y-y_0)^2\leqq r^2$, $r\geqq 0$ is Dedekind-complete if $\succ$ means $\supset$, "contains"; but these circles do not form a lattice.

If $f(a)$, a in A is a net of real numbers, the statement that the limit of $f(a)$ is k can be thus phrased in terms of order: For every real number $m<k$, it is eventually true that $f(a)>m$; and for every real number $n>k$, it is eventually true that $f(a)<n$. This suggests the following definition of convergence in partially ordered sets. If $f(a)$, a in A is a net of points of a partially ordered set X, it is convergent if there exist sets M and N in X, directed by $\succ$ and $\prec$ respectively, such that the supremum of M is the same as the infimum of N, and for every m in M and every n in N it is eventually true that $n\succ f(a)\succ m$. It is easy to show that the limit, if it exists, is unique. From the definition of limit we can proceed to the definition of continuity and to the study of continuous functions. Just where this leads us cannot yet be stated. I have done some studying of partially ordered sets and continuous functions on them, and expect to publish the results soon. But much remains to be done.

APPENDIX TO "PARTIAL ORDERINGS AND MOORE-SMITH LIMITS"

E. J. McSHANE, University of Virginia, Charlottesville

In 1954 E. J. McShane proposed (Canadian J. Math., 6 (1954) 161–168) a theory of limits that is an outgrowth of the Moore-Smith theory and is equivalent to it, but is in some respects more convenient. Let f be a function on a domain D, the values of f lying in a topological space Y. All the customary definitions have this in common: there is a family $\mathcal{A}$ of subsets of D, and the statement that f has limit y_0, or that $(f, \mathcal{A})$ converges to y_0, means that for each neighborhood U of y_0 in Y, there is a set A in family $\mathcal{A}$ such that for all x in A, $f(x)$ is in U. But with no restrictions on $\mathcal{A}$ we can prove no significant theorems. At the very least, when f_1 and f_2 are real-valued on a common domain D and $\mathcal{A}$ is a family of subsets of D, we wish to be able to prove that if f_1 is identically 0, $f(x)$ converges to 0 but not to 1, and if $f_1(x)$ and $f_2(x)$ both converge to 1, $f_1(x)+f_2(x)$ converges to 2. To do this we must assume that $\mathcal{A}$ is a non-empty family of non-empty sets directed by $\subset$; if

A_1 and A_2 are in $\mathcal{A}$, there is an A_3 that belongs to $\mathcal{A}$ and is contained in both. Such a family of sets is called a **direction** in D. If $\mathcal{A}$ and $\mathcal{B}$ are directions, B is a **subdirection** of A if each set A in $\mathcal{A}$ contains a set B in $\mathcal{B}$. The pair $(f,\mathcal{A})$ is called a **directed function** if f is a function and $\mathcal{A}$ is a direction in the domain of f.

The elementary theory of directed functions is studied in detail in *A Theory of Limits* (E. J. McShane, Studies in Modern Analysis, MAA Studies in Mathematics, 1 (1962) 7–29). We shall here add some examples and some applications to topology.

If f is a function on a domain D and D is directed by $\succ$, the "final sections" of D are the sets $D_x = \{x' \text{ in } D : x' \succ x\}$. The family $\mathcal{A}$ of all final sections of D is a non-empty family of non-empty sets. If $D_{x'}$ and $D_{x''}$ belong to it, then there is a member c of D such that $c \succ x'$ and $c \succ x''$. Then $D_c \subset D_{x'} \cap D_{x''}$, so $\mathcal{A}$ is a direction in D. The limit of $(f,\mathcal{A})$ is the Moore-Smith limit of the net consisting of the function f with the partial ordering $\succ$ of D. So every example of a net converts to an example of a directed function, and if the net has a limit, the directed function has the same limit.

We now present some further examples, and some applications to topology.

If f is defined on a subset D of a topological space X and takes values in a topological space Y, and x_0 is a point of X such that for every neighborhood U of x_0 the intersection $[D \backslash \{x_0\}] \cap U$ is non-empty, the family of all such intersections is a direction $\mathcal{A}$ in D. The limit of $(f,\mathcal{A})$ is then the limit of $f(x)$ as x tends to x_0, in the usual sense.

Let f be real-valued on an interval $[a,b]$ in the one-dimensional space R^1. Define partitions P as in §5, and for each P in the set D of all partitions define $S(P)$ as in §5. By a **gauge** on $[a,b]$ we shall mean a function γ on $[a,b]$ such that for each x in $[a,b]$, $\gamma(x)$ is a neighborhood of x in R^1. For each gauge γ define $D[\gamma]$ to be the set of all partitions $P = \{x_0,\ldots,x_n,\xi_1,\ldots,\xi_n\}$ such that for $j = 1,\ldots,n, [x_{j-1},x_j] \subset \gamma(\xi_j)$. By the theorem that begins on the bottom of p. 40 of the paper by T. H. Hildebrandt in this volume, for each gauge γ on $[a,b]$ the set $D[\gamma]$ is not empty. The family $\mathfrak{D}$ of all $D[\gamma]$ is directed by $\subset$; if $D[\gamma_1]$ and $D[\gamma_2]$ are in $\mathfrak{D}$ and we define γ_3 by setting $\gamma_3(x) = \gamma_1(x) \cap \gamma_2(x)$ for all x in $[a,b]$, it is clear that $D[\gamma_3]$ is contained in $D[\gamma_1]$ and in $D[\gamma_2]$. The "Riemann-complete" integral of f over $[a,b]$ is the limit of $(S,\mathfrak{D})$ if it exists. This integral was introduced and studied by R. Henstock (*Theory of the Integral*, Butterworth, 1963).

If the class of partitions is enlarged to include all sets $\{x_0,x_1,\ldots,x_n,\xi_1,\ldots,\xi_n\}$ with $x_0 = a < x_1 < \cdots < x_n = b$ and with all ξ_j in $[a,b]$ and S is defined as above, and $\mathfrak{L}$ is the family of sets $L[\gamma]$ with the property that P is in $L[\gamma]$ if and only if $[x_{j-1},x_j] \subset \gamma(\xi_j)$ for $j = 1,\ldots,n$, we again find that $(S,\mathfrak{L})$ is a directed function. If it converges, its limit is called the "gauge-integral" of f over $[a,b]$. (Cf. *A unified theory of integration*, E. J. McShane, Amer. Math. Monthly, 80 (1973) 349–359.)

It is interesting that in spite of the resemblance of these integrals to the Riemann integral, they are much more powerful. The Riemann-complete integral can be proved to be equivalent to the Denjoy integral, and the gauge-integral to be equivalent to the Lebesgue integral.

Let $(f, \mathfrak{A})$ be a directed function with values in a topological space Y. A point y_0 of Y is a cluster point of $(f, \mathfrak{A})$ if for each neighborhood U of y_0 and each set A in $\mathfrak{A}$, there is a point x of A such that $f(x)$ is in U. This concept is connected with convergence by the following theorem.

THEOREM 1. *Let $(f, \mathfrak{A})$ be a directed function with values in a topological space Y, and let y_0 be a point of Y. Then* (i) *if $\mathfrak{A}$ has a subdirection $\mathfrak{B}$ such that $(f, \mathfrak{B})$ converges to y_0, y_0 is a cluster point of $(f, \mathfrak{A})$;* (ii) *if y_0 is a cluster point of $(f, \mathfrak{A})$, there is a subdirection $\mathfrak{B}$ of $\mathfrak{A}$ such that $\mathfrak{B} \supset \mathfrak{A}$ and $(f, \mathfrak{B})$ converges to y_0.*

For (i), let U be a neighborhood of y_0 and A a set in $\mathfrak{A}$. There is a set B_1 of $\mathfrak{B}$ such that $f(x) \in U$ for all x in B_1, and there is a set B_2 of $\mathfrak{B}$ contained in $\mathfrak{A}$. There is an x in $B_1 \cap B_2$; then x is in A and $f(x)$ in U.

For (ii), let $\mathfrak{U}$ be the set of all neighborhoods of y_0 in Y, including Y, and let $\mathfrak{B}$ consist of all intersections $A \cap f^{-1}(U)$ with A in $\mathfrak{A}$ and U in $\mathfrak{U}$. By hypothesis these are non-empty. If $A_1 \cap f^{-1}(U_1)$ and $A_2 \cap f^{-1}(U_2)$ are in $\mathfrak{B}$ we choose A_3 in $\mathfrak{A}$ contained in A_1 and in A_2, and we choose U_3 in $\mathfrak{U}$ contained in U_1 and in U_2. Then $A_3 \cap f^{-1}(U_3)$ is a member of $\mathfrak{B}$ contained in $A_1 \cap f^{-1}(U_1)$ and in $A_2 \cap f^{-1}(U_2)$, so $\mathfrak{B}$ is directed by $\subset$. Every set A in $\mathfrak{A}$ is in $\mathfrak{B}$ since $A = A \cap f^{-1}(Y)$; and every set A in $\mathfrak{A}$ contains a set in $\mathfrak{B}$, namely A itself, so $\mathfrak{B}$ is a subdirection of $\mathfrak{A}$. If U is any neighborhood of y_0, we choose any A in $\mathfrak{A}$ and define $B = A \cap f^{-1}(U)$. Then B is in $\mathfrak{B}$ and $f(x)$ is in U for all x in B. This completes the proof of (ii).

A **maximal direction** in a set D is a direction in D that is not a proper subset of any direction in D. A directed function $(f, \mathfrak{A})$ is a **maximally directed function** if A is a maximal direction in the domain of f.

THEOREM 2. *Let $(f, \mathfrak{A})$ be a maximally directed function with values in a topological space Y, and let y_0 be in Y. If y_0 is a cluster point of $(f, \mathfrak{A})$, $(f, \mathfrak{A})$ converges to y_0.*

By Theorem 1, there is a directed function $(f, \mathfrak{B})$ such that $(f, \mathfrak{B})$ converges to y_0 and $\mathfrak{B} \supset \mathfrak{A}$. Since $(f, \mathfrak{A})$ is maximally directed, $\mathfrak{B} = \mathfrak{A}$.

The next theorem originated in H. Cartan's theory of filters.

THEOREM 3. *If D is any non-empty set, every direction in D is contained in a maximal direction in D.*

Let $\mathfrak{A}$ be a direction in D, and let $\mathfrak{D}$ be the collection of all directions in D that contain $\mathfrak{A}$. A subset $\mathfrak{C}$ of $\mathfrak{D}$ is said to be linearly ordered if whenever B_1 and B_2 are both in $\mathfrak{C}$, either $B_1 \subset B_2$ or $B_2 \subset B_1$. By the Hausdorff maximal principle (which is well known to be equivalent to the axiom of choice) there exists a linearly ordered subset $\mathfrak{C}$ of $\mathfrak{D}$ that is not contained in any linearly ordered subset $\mathfrak{C}'$ of $\mathfrak{D}$ except $\mathfrak{C}' = \mathfrak{C}$. We choose such a $\mathfrak{C}$, and define $\mathfrak{M}$ to be the union of all families (directions) that belong to $\mathfrak{C}$.

Obviously $\mathfrak{M} \supset \mathfrak{A}$, so it is not empty. If M belongs to $\mathfrak{M}$, it belongs to some direction $\mathfrak{B}$ in the linearly ordered collection $\mathfrak{C}$, so M is not empty. If M_1 and M_2

both belong to $\mathfrak{M}$, M_1 belongs to some direction $\mathfrak{B}_1$ in the collection $\mathfrak{C}$, and M_2 belongs to some direction $\mathfrak{B}_2$ in $\mathfrak{C}$. Since $\mathfrak{C}$ is linearly ordered, one of $\mathfrak{B}_1, \mathfrak{B}_2$ contains the other; say to be specific that $\mathfrak{B}_2 \supset \mathfrak{B}_1$. Then both M_1 and M_2 belong to the direction $\mathfrak{B}_2$, so there is a set B in the direction $\mathfrak{B}_2$ contained in $M_1 \cap M_2$. Since B belongs to the member $\mathfrak{B}_2$ of $\mathfrak{C}$, it is in $\mathfrak{M}$. This completes the proof that $\mathfrak{M}$ is a direction that contains A. It remains to prove that $\mathfrak{M}$ is maximal.

Suppose this is false. Then there exists a direction $\mathfrak{B}$ in D that contains $\mathfrak{M}$ but is different from $\mathfrak{M}$. This implies that $\mathfrak{B}$ contains some set B_1 that does not belong to $\mathfrak{M}$, hence does not belong to any direction in the linearly ordered collection $\mathfrak{C}$, and in particular $\mathfrak{B}$ cannot itself belong to $\mathfrak{C}$. Define $\mathfrak{C}_1 = \mathfrak{C} \cup \{\mathfrak{B}\}$. This is a collection of members of $\mathfrak{D}$. It contains $\mathfrak{C}$ and is different from $\mathfrak{C}$. Since $\mathfrak{B}$ contains $\mathfrak{M}$ it contains every direction $\mathfrak{B}_1$ in $\mathfrak{C}$, from which it follows readily that $\mathfrak{C}_1$ is linearly ordered. This contradicts the fact that $\mathfrak{C}$ is not a proper subset of any linearly ordered subset of $\mathfrak{D}$, and the proof is complete.

We can use this to extend and improve §7 of the paper.

THEOREM 4. *Let Y be a topological space. Then the following four statements are all true or all false.*

(I) *From every covering of Y by open sets it is possible to extract finitely many sets that cover Y.*

(II) *Every directed function with values in Y has a cluster point in Y.*

(III) *Every directed function with values in Y has a subdirected function that converges to a point of Y.*

(IV) *Every maximally directed function with values in Y converges to a point of Y.*

(I) implies (IV). For suppose (IV) false. Let $(f, \mathfrak{A})$ be a maximally directed function with values in Y that does not converge. By Theorem 2 it has no cluster point in Y. So for each y in Y there is a neighborhood $U[y]$ of y and a set $A[y]$ in $\mathfrak{A}$ such that $U[y] \cap f^{-1}(A[y])$ is empty. For each y we choose such a $U[y]$. These cover Y, but no finite subset of them covers Y. For let $U[y_1], \ldots, U[y_k]$ be a finite subset of them. For each y_j there is a set $A[y_j]$ in $\mathfrak{A}$ such that $A[y_j] \cap f^{-1}(U[y_j])$ is empty. The direction $\mathfrak{A}$ contains a member A that is contained in each $A[y_j]$, so for each x in A, $f(x)$ is not in any of the $U[y_j]$.

(IV) implies (III). Let $(f, \mathfrak{A})$ be a directed function with values in Y. By Theorem 3, $\mathfrak{A}$ is contained in a maximal direction $\mathfrak{B}$ in the domain of f. Then $(f, \mathfrak{B})$ is a subdirection of $(f, \mathfrak{A})$, and by (IV) it converges to a point of Y.

(III) implies (II), by Theorem 1.

(II) implies (I). For suppose (I) false. There exists a covering of Y by a collection $\mathcal{C}$ of open sets such that no finite subset of $\mathcal{C}$ covers Y. Let $\mathfrak{A}$ be the family of all sets $Y \setminus [C_1 \cup \cdots \cup C_k]$ with $C_1, \ldots, C_k$ in $\mathcal{C}$. By hypothesis the sets in $\mathfrak{A}$ are non-empty, and the intersection of two members of $\mathfrak{A}$ is a member of $\mathfrak{A}$, so $\mathfrak{A}$ is a direction. Let f be the identity function on Y. Then $(f, \mathfrak{A})$ is a directed function with values in Y. Each y in Y is contained in a member C of $\mathcal{C}$, and there is a neighborhood U of y contained in C. For no point x in the member $Y \setminus C$ of $\mathfrak{A}$ is $f(x)$ in U, so y is not a cluster point of $(f, \mathfrak{A})$.

11

RICHARD HUBERT BRUCK

R. H. Bruck was born in Pembroke, Ontario on December 26, 1914. He received his B.A. at Toronto in 1937, was a Fellow at Toronto, 1937–40, receiving his M.A. in 1938 and his doctorate in 1940 in the field of abstract algebra. He became instructor of mathematics at Alabama, 1940–42, and went to the University of Wisconsin at Madison in 1942, as instructor, 1942–44, assistant professor, 1944–47, associate professor, 1947–52, and professor since 1952, research professor since 1967. He was a Guggenheim Fellow and University Research Fellow at Wisconsin, 1946–47, a Fulbright Lecturer at the Research Institute, Australian Mathematical Society, 1963, Research Lecturer, Canadian Mathematical Congress, 1963, Visiting Professor, University of North Carolina at Chapel Hill, 1963–64. In addition, he was a consultant with the Rand Corporation at Santa Monica 1961–72, Assistant Editor of the *Bulletin*, 1945–47, and the *Proceedings*, 1955–57.

His principal mathematical interests lie in the area of representation theory, tensor algebra, linear non-associative algebra, the theory of loops, projective planes, and the general theory of graphs.

RECENT ADVANCES IN THE FOUNDATIONS OF EUCLIDEAN PLANE GEOMETRY*

R. H. BRUCK, University of Wisconsin

1. Introduction. A program of axiomatizing Euclidean plane geometry in a manner consistent with present standards of rigour was beautifully carried out by Hilbert [1].§ Upon reading Hilbert's book in its entirety one sees what is not at first evident—that Hilbert is alive to the interesting questions which arise when some of his geometric axioms are dropped or modified. And many mathematicians, both before and since the appearance of Hilbert's book, have investigated such problems.

Unfortunately for the wide audience whose interest in geometry was awakened in high school or university, answers to the deeper questions of the sort I have in mind, if they have been given at all, require a long excursion into abstract algebra. I know of no remedy for this situation. What can be done—or, at any rate, what the present paper attempts to do—is to give a pictorial account of some of the geometric axioms, a simple explanation of the algebraic problems which these pose and a brief account (with references rather than proofs) of the answers.

Think of this paper as an excursion from wherever you are towards a town named Cayley Numbers. As we saunter along, we pass by Planar Ternary Rings, Veblen-Wedderburn Systems, Division Rings with the Right Inverse Property, Right Alternative Division Rings, Alternative Division Rings—and down the hill we see Cayley Numbers.—Strange names they have for towns in these parts, but I believe you'll enjoy the scenery.

2. The axioms of incidence. We shall use geometric language quite informally. In particular we shall assume that everyone feels at home with phrases such as "point is on line," "line is through point" and with adjectives such as "collinear," "concurrent," "parallel." Such carelessness is a little dangerous (since the language may have unintended connotations) but saves a great many words.

A Euclidean (or affine) plane π is a system of undefined objects, called points and lines, subject to the following *axioms of incidence:*

(i) *If P, Q are distinct points of π, there is one and only one line, PQ, of π, through both of P and Q.*

(ii) *If a, b are distinct lines of π, there is at most one* (*and there may be no*) *point of π on both of a and b.*

* This paper had its origin in 1950 in my seminar at the University of Wisconsin. It has been presented since then by my students or myself in various forms at several universities. The present account was delivered by invitation to the Iowa Section of this Association in Ames, Iowa, April 30, 1954.

§ The book [1a] is included because it is in English and indicates the spirit of Hilbert's later approach but all specific references [1] are to [1b].

(iii) *If the point P of π is not on the line a of π, there is exactly one line of π which passes through P and is parallel to a.*

(iv) *There is at least one set, A, B, C, D, of four distinct points of π, no three of which are collinear.*

The axioms of incidence require so little of a Euclidean plane that very few theorems have been proved. Indeed, the main "theorem"—I like to call it Hall's Theorem—might be stated as follows: Any damn thing can happen. As a relatively respectable example of this we may note that the following system satisfies the axioms: π consists of four distinct points, A, B, C, D, and of six distinct lines, namely the following three pairs of parallels: AB, CD; AC, BD; AD, BC. Each line contains exactly two distinct points and each point lies on exactly three distinct lines.

Even so, the axioms of incidence do ensure a certain amount of regularity. Suppose a is a line of π and P is a point of π not on a. By (iii), there is exactly one line through P, say b, which does not meet a. The remaining lines through P are in one-to-one correspondence with the points of a, for,* by (i), (ii), (iii), each of these lines is a line PQ for exactly one point Q of a, and conversely. Hence we can say that there is "one more" line through P than there are points of a. Consequently, every two lines not through P have the same number of points. Then, considering the points A, B, C, D of (iv), we see that every line not through A has the same number of points as BC or BD or CD, and similarly for B, C, D. Thus, if some line of π has n points (where n is a positive integer or a transfinite cardinal) then every line has n points and every point lies on $n+1$ lines.

We can add: if a, b, c are distinct lines and if a is parallel to b and b parallel to c, then a is parallel to c—else the common point of a, c would have through it two parallels to b, namely a, c. Now consider two lines a, d, meeting in a point P. Each of the $n-1$ points of d, other than P, determines a unique parallel to a, and conversely. Hence the *parallel class* of a, consisting of a and the lines parallel to a, contains precisely n lines. Moreover the parallel class of a is the parallel class of each of the lines contained in it.

In a rigorous treatment of Euclidean planes it is necessary at various points to give special consideration to planes with a small number of points on each line. We shall be content to ignore this difficulty entirely.

3. Hall's planar ternary rings. How can we introduce coordinates into a Euclidean plane π subject only to the axioms of incidence? We cannot talk of rectangular axes—since we have no notion of angle aside from the special case of parallel lines. We cannot talk of lengths—we have no notion of distance. We cannot talk of the slope of a line—but, on the contrary, we shall do just that in a moment.

---

* A few simple diagrams, which we feel compelled to omit, make the following remarks quite obvious.

The method of Marshall Hall [2] seems as good as can be expected. We select an arbitrary point O of π and three distinct lines§ through O which we call the x-axis, the y-axis and the unit line (Fig. 1). On the unit line we select any point I (the unit point) distinct from O. The line through I parallel to the y-axis we call the slope line.

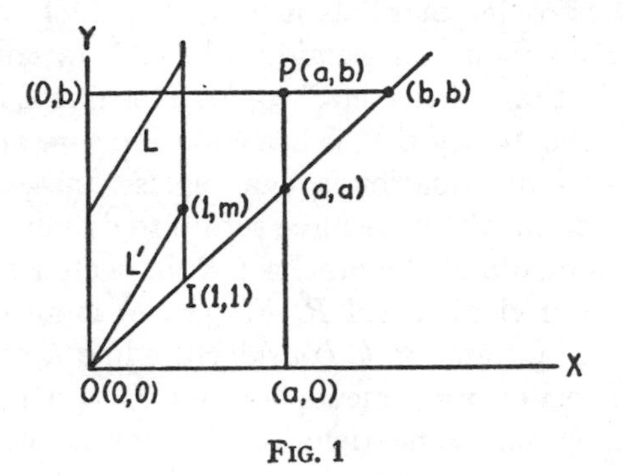

FIG. 1

Now we choose an arbitrary set R of elements or "labels" subject to two restrictions: (a) Among the elements of R are two distinct elements 0, 1. (b) The elements of R can be put into one-to-one correspondence with the points of the unit line OI (and hence with the points of any line of π). To each point of the unit line OI we assign a pair of coordinates (x, y) with $x=y$, where $x=y$ is an element of R. In particular we give the origin O the coordinates $(0, 0)$ and the unit point I the coordinates $(1, 1)$. We arrange that each point of OI has coordinates (a, a) for a uniquely determined element a of R and that, conversely, there is a unique point on OI with coordinates (b, b) for each element b of R.— Despite all this arbitrariness in assigning coordinates to the points of OI, we use the coordinates in such a way that any two ways of assigning them would be equally good or bad.

Next consider any point P of π. The line through P in the parallel class of the y-axis meets OI in a unique point; say the point with coordinates (a, a). And the line through P in the parallel class of the x-axis meets OI in a unique point, say (b, b). Then we assign to P the coordinates (a, b) (see Fig. 1). In particular, points of the x-axis have coordinates of form $(x, 0)$, points of the y-axis have the form $(0, y)$, and the four points $(0, 0)$, $(x, 0)$, $(0, y)$, (x, y) form the vertices of a parallelogram.

The line through the point (a, b) in the parallel class of the y-axis will naturally have the equation $x=a$. Similarly, it is clear what we mean by the line with equation $y=b$. We have yet to assign equations to the other lines, except that the unit line OI should certainly have the equation $y=x$.

Consider any line L. There is a unique line L' in the parallel class of L which

§ Note that, in Figure 1, OX, OY are perpendicular and angle XOY is bisected by the unit line OI. This is meaningless but somehow comforting.

passes through the origin O. If L' is the y-axis, we assign no slope to L or L'. Otherwise, L' must meet the slope line $x=1$ in a unique point, say the point $(1, m)$; in this case we assign to L (and L') the slope m (see Fig. 1). In particular, every line $y=b$ has slope 0, and every line in the parallel class of the unit line OI has slope 1.

At this stage every point of π has a unique pair of coordinates and every line, except for the lines $x=a$, has a unique slope. Now consider a line L which intersects the y-axis in the "y-intercept" $(0, b)$. This line L has a unique slope m. We should *like* to be able to say that L has equation $y=xm+b$; but, at the present stage,* at least, such an equation is meaningless. Instead, we use the plane π and the coordinate system which we have set up to define a *ternary operation* (or function) F on the elements of the label set R, in such a manner that, for each ordered triple a, m, b of elements of R, $F(a, m, b)$ is an element, say d, of R. This is done as follows: for any m, b, consider the line L of slope m through the y-intercept $(0, b)$. The line $x=a$ meets the line L in a unique point P whose coordinates are (a, d) for some definite element d of R. Then we define $F(a, m, b)=d$.

In view of the definition of F it is tautological to say (see Fig. 2) that the equation of the line with slope m, y-intercept $(0, b)$ is $y=F(x, m, b)$. As a particular case, the unit line OI has slope 1 and y-intercept $(0, 0)$. The equation of OI is surely $y=x$ and yet it is also $y=F(x, 1, 0)$. Consequently, $F(x, 1, 0)=x$ for every x in R.—There are other facts about F which arise, like this, directly from the definitions, and there are deeper facts which come by insisting upon the full import of the axioms of incidence.

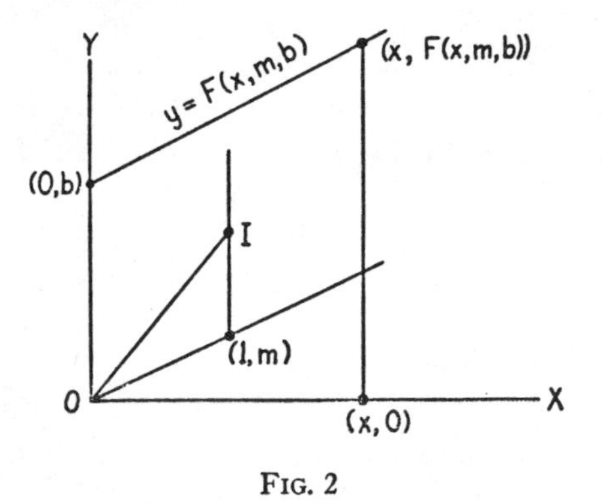

FIG. 2

The system (R, F), consisting of the label set R and the ternary operation F, is known as a *planar ternary ring*. Such systems are of course useless unless we have some way of singling out planar ternary rings from among all systems with ternary operations—for a suitable set of postulates see Appendix I. And even

* After we impose the vector axiom, equations of lines will indeed take this familiar form. (See §§ 6, 7, 8.)

then they are relatively useless until we find some way of handling them more easily than the Euclidean planes themselves.

4. Addition. For any planar ternary ring the operation of addition (+) is defined equationally by

$$a + b = F(a, 1, b), \qquad \text{all } a, b \text{ in } R. \tag{4.1}$$

The algebraic consequence of (4.1) is this: *the equation of the line with slope 1 and y-intercept* $(0, b)$ *can now be written* $y=x+b$.

The geometric counterpart of (4.1) is the notion of addition of points on the unit line OI. Every ordered pair A, B of points $A=(a, a)$, $B=(b, b)$ of OI uniquely determines a sum-point $S=(a+b, a+b)$ of OI as follows (see Fig. 3):

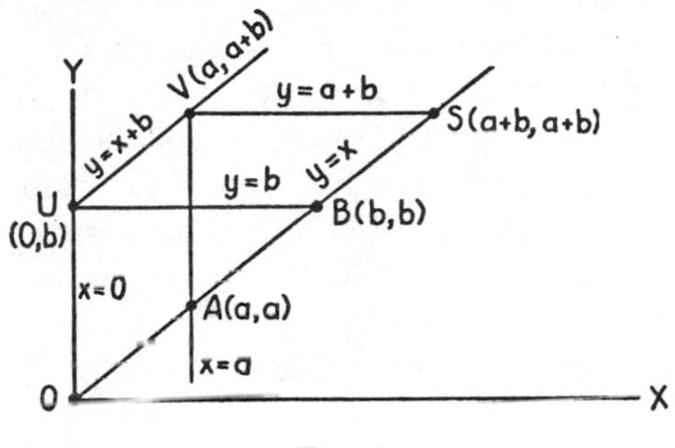

FIG. 3

the line $(y=b)$ of slope 0 through $B=(b, b)$ meets the y-axis in point $U=(0, b)$. The line $(y=x+b)$ of slope 1 through U and that line $(x=a)$ through A which is in the parallel class of the y-axis meet in a point $V=(a, a+b)$. The line $(y=a+b)$ of slope 0 through V meets the unit line $(y=x)$ in $S=(a+b, a+b)$. This geometric operation can be explained, without reference to coordinates or axes, in terms of three parallel classes: the lines in the parallel class of the x-axis (slope 0); the lines in the parallel class of the y-axis (no slope); the lines in the parallel class of the unit line OI (slope 1).

If we write $A+B=S$, it is easily verified from Figure 3 that if any two of A, B, S are arbitrarily assigned as points of OI, the third is uniquely determined by the equation. Moreover $A+0=A$ and $0+A=A$ for every point A of OI. This means (see Appendix I) that *the system* $(R, +)$ *is a loop.*

Now let us suppose, temporarily, that the Euclidean plane π satisfies all the usual axioms of high-school geometry, so that we can make use of line segments.* In Figure 3, from the parallelogram $AOUV$, $\underline{OA}=\underline{UV}$, and, from the parallelogram $UVBS$, $\underline{UV}=\underline{BS}$. Consequently, $\underline{OA}=\underline{BS}$, so that the sum of the line segments $\underline{OA}$, $\underline{OB}$ is equal to $\underline{OS}$. In this special case, then, the somewhat

* In order to distinguish between the line through the two points A, B and the line segment (or vector—see § 6) with initial point A, endpoint B, we underline the latter. Thus: line AB, line segment $\underline{AB}$.

arbitrary equation $A+B=S$ is illuminated by the equation $\underline{OA}+\underline{OB}=\underline{OS}$. The equation (4.1) is thus related to very familiar things indeed. We shall return to this subject in §6.

5. Multiplication. For any planar ternary ring (R, F) the operation of multiplication $(\cdot)$ is defined equationally by

$$ab = F(a, b, 0), \qquad \text{all } a, b \text{ in } R. \tag{5.1}$$

The algebraic consequence of (5.1) is this: *the equation of the line through* $O=(0, 0)$ *with slope* m *can now be written* $y=xm$.

The geometric counterpart of (5.1) is the notion of multiplication of points on the unit line OI. Every ordered pair A, B of points $A=(a, a)$, $B=(b, b)$ of OI uniquely determines a product-point $P=(ab, ab)$ of OI as follows (see Fig. 4): The line $(y=b)$ of slope 0 through B meets the slope line $(x=1)$ in a point $U=(1, b)$. The line $OU(y=xb)$ of slope b meets the line $(x=a)$ of no slope through A in a point $V=(a, ab)$. The line $(y=ab)$ of slope 0 through V meets the unit line $(y=x)$ in the point $P=(ab, ab)$. This geometric operation can be explained, without reference to coordinates or axes, in terms of three classes of lines: the parallel class of the x-axis (slope 0); the parallel class of the y-axis (no slope); the class consisting of all lines through O except the y-axis (one line for every slope).

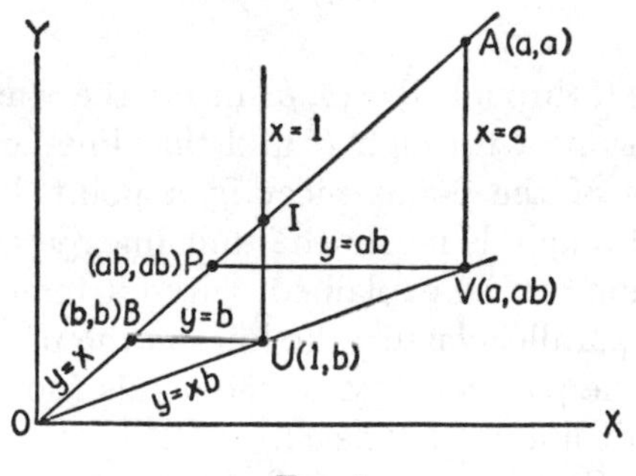

FIG. 4

If we write $A\cdot B=P$, we first note from Figure 4 that $A\cdot 0=0\cdot A=0$ for every point A of OI. Then, if we restrict attention to the set S consisting of the points of OI other than O, we find from Figure 4 that if any two of A, B, P are assigned in S, the equation $A\cdot B=P$ uniquely determines the third as a point in S. Moreover, $I\cdot A=A\cdot I=A$ for every A in S. This means (see Appendix I) that, if R^* denotes the set of elements of R exclusive of 0, *the system* $(R^*, \cdot)$ *is a loop.*

Again let us assume, temporarily, all the axioms of high school geometry. In Figure 4 we take $A\neq O$, $B\neq O$. From two sets of similar triangles,

$$\underline{OA}:\underline{OI} = \underline{OV}:\underline{OU} = \underline{OP}:\underline{OB},$$

so that

$$\underline{OA} \cdot \underline{OB} = \underline{OI} \cdot \underline{OP}.$$

Consequently, if we take $\underline{OI}$ to have unit length, we can parallel the abstract equation $A \cdot B = P$ with the familiar equation $\underline{OA} \cdot \underline{OB} = \underline{OP}$. (Although Hilbert introduces enough axioms to validate these calculations, we shall not quite do so.)

6. The vector axiom. At the end of §4 we temporarily made use of the notion of a line segment. In our usual thinking a line segment $\underline{AB}$ has both an inside and an outside. Such concepts require axioms of order (see Hilbert [1]) which we do not wish to introduce. We shall be content with a very mild notion of a *vector*. For present purposes, a vector $\underline{AB}$ consists merely of an ordered pair of distinct points A, B: an *initial point* A and an *endpoint* B. As before AB denotes the *line* through A and B.

We introduce a natural notion of equality of vectors. First of all, any vector is equal to itself: $\underline{AB} = \underline{AB}$. Next, if AB, $A'B'$ are distinct lines, the vectors $\underline{AB}$, $\underline{A'B'}$ will be called equal if and only if the lines AB, $A'B'$ are parallel and the lines AA', BB' are parallel. (Note that, in this case, if $\underline{AB} = \underline{A'B'}$ then, also, $\underline{BA} = \underline{B'A'}$, $\underline{AA'} = \underline{BB'}$, $\underline{A'A} = \underline{B'B}$.) Next suppose that AB, $A'B'$, $A''B''$ are distinct lines and that, according to our definition, $\underline{AB} = \underline{A'B'}$ and $\underline{A'B'} = \underline{A''B''}$. Our notion of equality will be useless unless we can be sure that $\underline{AB} = \underline{A''B''}$. It certainly is true that AB, $A''B''$ are parallel. Moreover, if A, A', A'' are collinear then B, B', B'' are collinear and the line $AA'' = AA'$ is parallel to the line $BB'' = BB'$; so that, in this case, it is true that $\underline{AB} = \underline{A''B''}$. But if A, A', A'' are not collinear the desired equality need not* hold. Therefore we force equality by imposing the vector axiom (see Fig. 5):

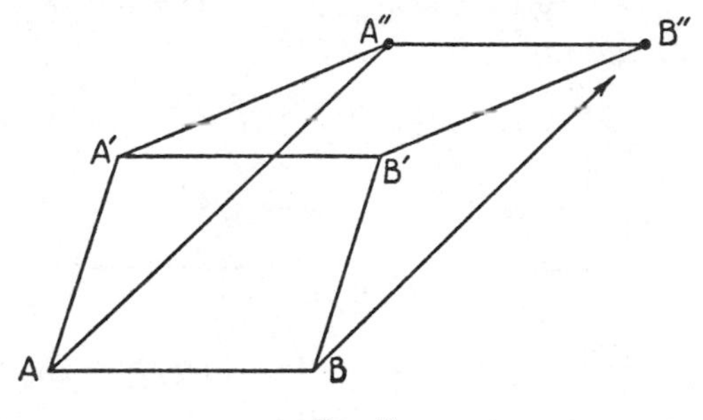

FIG. 5

THE VECTOR AXIOM. *If six distinct points of the Euclidean plane π form two triangles $AA'A''$, $BB'B''$, if the lines AB, $A'B'$, $A''B''$ are parallel and if the pairs of lines (AA', BB'), $(A'A'', B'B'')$ are parallel, then AA'', BB'' are also parallel.*

* There exist at least two essentially different Euclidean planes of order 9 (9 points on each line) in which the vector axiom fails. One type can be deduced from Hall [2] and another will be found in Carmichael [3].

The vector axiom is a special case of the axiom of Desargues and is one of the axioms used by Hilbert [1].

The vector axiom can be reached in still another way. According to the usual definition, the sum of two vectors $\underline{AA'}$, $\underline{A'A''}$ is the vector $\underline{AA''}$. If $\underline{BB'}=\underline{AA'}$ and $\underline{B'B''}=\underline{A'A''}$, we should like the sum $\underline{BB'}+\underline{B'B''}=\underline{BB''}$ to be equal to $\underline{AA''}$. If A, A', A'' are collinear, this is automatic, but in the case of Figure 5 we require the vector axiom.

Before we can go on, there is still one more aspect of vector equality which requires consideration. Suppose we have $\underline{AB}=\underline{CD}$ and $\underline{CD}=\underline{EF}$, where the lines AB, CD are distinct but AB, EF are identical. We would like to say that $\underline{AB}=\underline{EF}$, but in doing so we are in danger of serious trouble. For suppose that also $\underline{AB}=\underline{C'D'}$ and $\underline{C'D'}=\underline{EG}$; we need to be able to assert that $F=G$. Luckily no new axiom is needed. We will indicate how this is so by examining the case that the lines AB, CD, $C'D'$ are distinct. In this case, since $\underline{C'D'}=\underline{AB}$ and $\underline{AB}=\underline{CD}$, the vector axiom gives $\underline{C'D'}=\underline{CD}$. Then, since $\underline{C'D'}=\underline{CD}$ and $\underline{CD}=\underline{EF}$, the vector axiom gives $\underline{C'D'}=\underline{EF}$. However, $\underline{C'D'}=\underline{EG}$. Therefore the line through D' parallel to $C'E$ meets the line $ABEF$ in F and in G, and consequently $F=G$.—For the case that CD, $C'D'$ are the same line we simply introduce a new line $C''D''$ and argue as before.

At this stage, assuming the vector axiom, we can be confident that equality of vectors satisfies the usual reflexive, symmetric and transitive laws. There only remains to introduce zero vectors $\underline{AA}$ (with the same initial and final points) and to define $\underline{AA}=\underline{BB}$ for all points A, B; or, more conveniently, $\underline{AA}=0$ for all points A. We also give a symmetric definition of vector addition as follows: If $\underline{AB}$, $\underline{CD}$ are any two vectors, choose any point P, determine Q so that $\underline{PQ}=\underline{AB}$ and R so that $\underline{QR}=\underline{CD}$, and call $\underline{PR}$ the sum of the ordered pair of vectors $\underline{AB}$, $\underline{CD}$. It is easy to check that (in the sense of vector equality) the sum is independent of the point P.

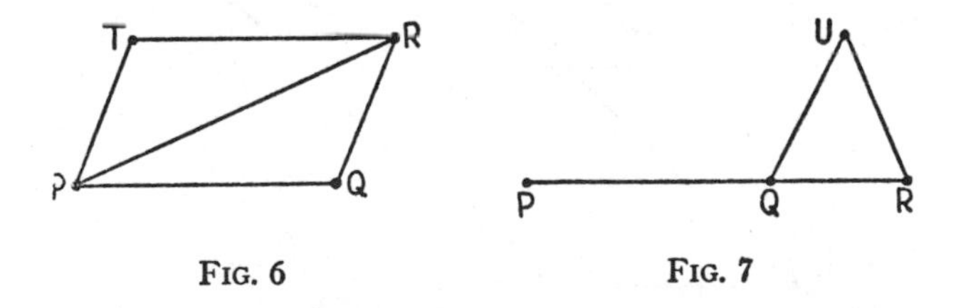

FIG. 6 FIG. 7

Now the associative law of vector addition is evident, since $(\underline{PQ}+\underline{QR})+\underline{RS}=\underline{PR}+\underline{RS}=\underline{PS}$ and $\underline{PQ}+(\underline{QR}+\underline{RS})=\underline{PQ}+\underline{QS}=\underline{PS}$. The commutative law of addition is equally evident for vectors $\underline{PQ}$, $\underline{QR}$ if P, Q, R are not collinear, since $\underline{PQ}+\underline{QR}=\underline{PR}$ while (see Fig. 6) $\underline{QR}+\underline{PQ}=\underline{PT}+\underline{TR}=\underline{PR}$. On the other hand, if P, Q, R are collinear, a simple device (see Fig. 7) allows us to use the non-collinear case along with associativity: $\underline{QR}+\underline{PQ}=(\underline{QU}+\underline{UR})+\underline{PQ}=\underline{QU}+(\underline{UR}+\underline{PQ})=\underline{QU}+(\underline{PQ}+\underline{UR})=(\underline{QU}+\underline{PQ})+\underline{UR}=(\underline{PQ}+\underline{QU})+\underline{UR}=\underline{PQ}+(\underline{QU}+\underline{UR})=\underline{PQ}+\underline{QR}$. Consequently, all the usual laws of vector addition are satisfied.

Now we are ready to consider the additive system $(R, +)$ defined by (4.1). More specifically, we consider Figure 3. In Figure 3, $\underline{OA} = \underline{UV}$ and $\underline{UV} = \underline{BS}$, so $\underline{OA} = \underline{BS}$. Therefore $\underline{OS} = \underline{OB} + \underline{BS} = \underline{OB} + \underline{OA} = \underline{OA} + \underline{OB}$. And, inasmuch as $S = (a+b, a+b)$, $A = (a, a)$, $B = (b, b)$, we can assert the following (see Appendix I):

In the presence of the vector axiom, the system $(R, +)$ is an abelian group isomorphic to the additive group of vectors.

7. Linearity. If the vector axiom had not already been thrust upon us in connection with equality of vectors, we could urge another reason for its adoption. Namely, we would like every planar ternary ring (R, F) of the Euclidean plane π to have the property of *linearity* embodied in

$$F(a, b, c) = ab + c \qquad \text{all } a, b, c, \text{ in } R. \tag{7.1}$$

where the addition and multiplication on the right hand side are as defined in (4.1), (5.1). The algebraic consequence of (7.1) is this: *the equation of the line with slope m, y-intercept $(0, b)$ can be written $y = xm + b$.*

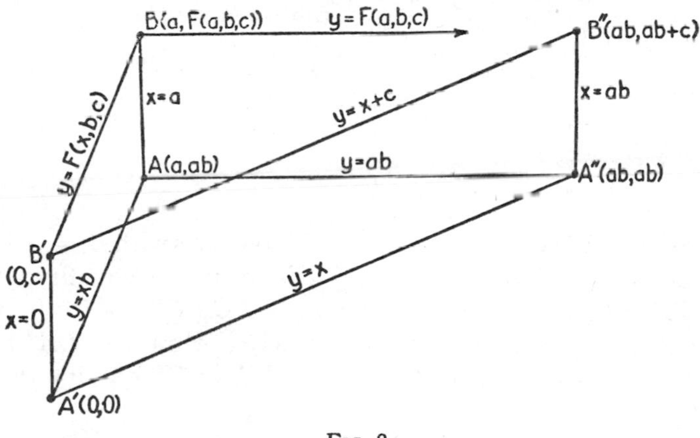

FIG. 8

In examining (7.1) we take $a \neq 0$, $c \neq 0$, $b \neq 0$, 1, since otherwise (7.1) holds trivially. Consider Figure 8, which emphasizes the essential nature of (7.1) by omitting irrelevant points (such as the unit point I) and lines (such as the x-axis). The y-axis appears as the line $x = 0$, and the unit line OI as the line $y = x$. (7.1) will hold if and only if the line $y = F(a, b, c)$ passes through the point $(ab, ab+c)$; that is (in Fig. 8) if and only if the lines AA'', BB'' are parallel. Hence, by comparison of Figure 8 with Figure 5, we see that, in the presence of the vector axiom, every planar ternary ring of π is linear. Now assume conversely that every planar ternary ring of π is linear, and consider Figure 5. With a little care we can construct a coordinate system in which the points and lines of Figure 5 have coordinates and equations of the forms indicated in Figure 8; then, by linearity, we can deduce that BB'' is parallel to AA''. To sum up:

A necessary and sufficient condition that every planar ternary ring of the Euclidean plane π be linear is that π satisfy the vector axiom.

8. Veblen-Wedderburn systems. The vector axiom has still another consequence, namely the right distributive law

$$(a + b)c = ac + bc, \qquad \text{all } a, b, c \text{ in } R, \tag{8.1}$$

of multiplication with respect to addition. We first note that (8.1) holds trivially if any one of a, b, c is zero or if $c=1$. Excluding these cases, consider Figure 9 below. The line $OA'(y=xc)$ meets the line $C'D'(x=a+b)$ in the point

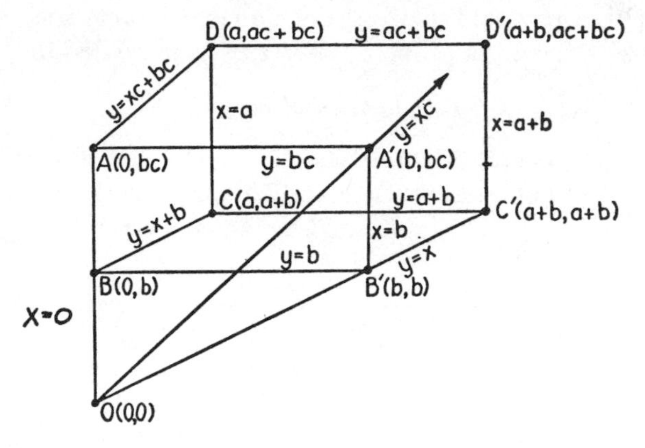

FIG. 9

$D''=(a+b, (a+b)c)$. We want to prove that D'' coincides with $D'=(a+b, ac+bc)$. By use of vectors, $\underline{A'D'}=\underline{A'B'}+\underline{B'C'}+\underline{C'D'}$. By vector equality, $\underline{A'B'}=\underline{AB}$, $\underline{B'C'}=\underline{BC}$, $\underline{C'D'}=\underline{CD}$. Therefore $\underline{A'D'}=\underline{AB}+\underline{BC}+\underline{CD}=\underline{AD}$. In particular, then, the line $A'D'$ has the slope of AD, namely c. However, OA' has slope c, so O, A', D' are collinear. That is, OA' meets $C'D'$ in D', proving that $D''=D'$. This completes the proof of (8.1).

To sum up, in the presence of the vector axiom each planar ternary ring (R, F) is linear, the additive system is an abelian group and the right distributive law (8.1) holds. A planar ternary ring with these properties is known as a Veblen-Wedderburn system, after O. Veblen and J. H. M. Wedderburn, who first studied* such systems in 1907. (For a complete set of postulates, see Appendix II.) It can be shown, conversely, that if one planar ternary ring of π is a Veblen-Wedderburn system then the vector axiom holds. Therefore:

THEOREM 1. *If the vector axiom holds in a Euclidean plane π, then every planar ternary ring of π is a Veblen-Wedderburn system. Conversely, if any one ternary ring of π is a Veblen-Wedderburn system, then the vector axiom holds in π.*

The theory of Veblen-Wedderburn systems is still relatively undeveloped—

* See [4]. The discussion in [2] is better suited to present purposes.

though I would hazard a guess that abstract algebra soon will be able to cope with these systems. In the meantime we are much hampered by the lack of the left distributive law, which must be paid for with an additional geometric axiom.

9. The distributive axiom. Our next axiom (see Fig. 10) is worded§ for ready comparison with one of Hilbert's:

THE DISTRIBUTIVE AXIOM. *Let seven distinct points of the Euclidean plane consist of two triangles $ABC, A'B'C'$ in perspective from a point O, with the pairs $(AB, A'B')$, $(BC, B'C')$ of corresponding sides parallel and with (*) BC parallel to OA. Then the third pair of sides, AC, $A'C'$, are also parallel.*

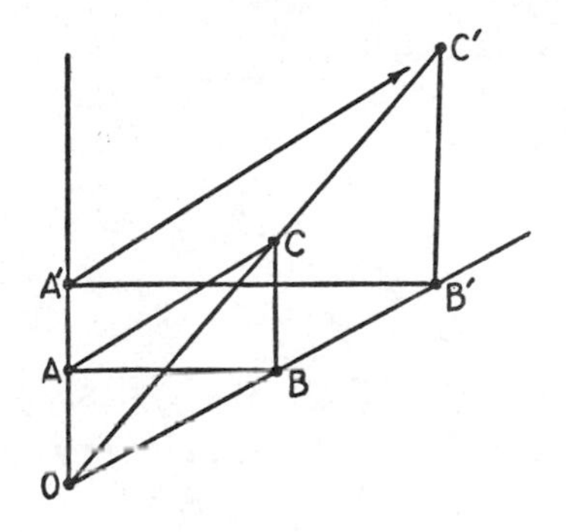

FIG. 10

The corresponding axiom of Hilbert [1] is stronger in that the restrictive hypothesis (*) is omitted. Both of these axioms are special cases of the axiom of Desargues.

Now consider the identity

$$F(a, b, ac) = aF(1, b, c), \qquad \text{all } a, b, c \text{ in } R. \tag{9.1}$$

In the presence of (7.1) (since $1b=b$ for every b) (9.1) is equivalent to the left distributive law

$$ab + ac = a(b + c), \qquad \text{all } a, b, c \text{ in } R. \tag{9.2}$$

Therefore we are interested in the geometric axiom which asserts (9.1) for every planar ternary ring (R, F) of π. Just as we showed that (7.1) was equivalent to the vector axiom, so we can show that (9.1) is equivalent to the distributive axiom. This is indicated by Figure 11 below.

10. Division rings with the right inverse property. Recall the common saying: "You get out of anything just what you put into it." Surely one would have to work hard to justify such a statement in mathematics. For example, if we put into the Euclidean plane π just enough to ensure that every coordinate ring of π is linear (the vector axiom) we get out all the properties of Veblen-

§ It would be neater to say that the three lines OAA', BC, $B'C'$ are parallel.

Wedderburn systems, including the right distributive law. When we insist further on the left distributive law (9.2) for every coordinate ring of π (that is, on the distributive axiom) the harvest is even more remarkable, as we shall see.

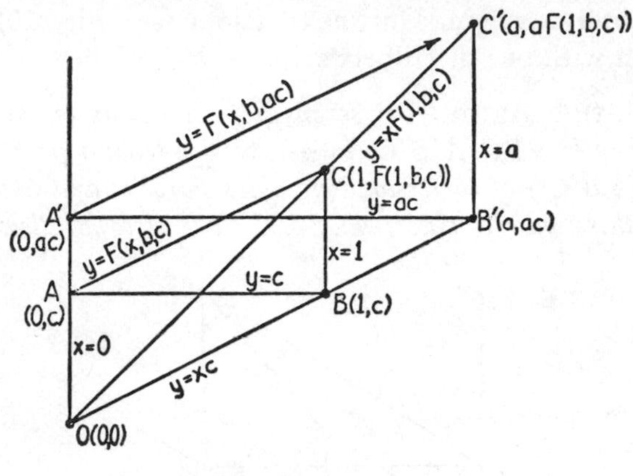

FIG. 11

A Veblen-Wedderburn system which also satisfies the left distributive law is much better known as a *division ring* (with identity element). (For a complete set of postulates see Appendix II.) Thus: *A necessary and sufficient condition that every planar ternary ring of a Euclidean plane π be a division ring is that π satisfy both the vector axiom and the distributive axiom.* In Theorem 1 it was stated that if any one coordinate ring of a Euclidean plane is a Veblen-Wedderburn system then (the vector axiom holds and) all are. A comparable statement would be false here: there exist Euclidean planes in which some but not all of the coordinate rings are division rings, the rest being merely Veblen-Wedderburn systems. For the correct theorem we need the notion of the right inverse property.

A division ring R (with identity element 1) is said to have the *right inverse property* if each nonzero element a of R has an inverse a^{-1} such that

$$(10.1)\qquad (ba)a^{-1} = b, \qquad \text{all } a, b \text{ in } R,\ a \neq 0.$$

(From (10.1) with $b=1$, $aa^{-1}=1$; thus (10.1) yields the weak associative law $(ba)a^{-1}=b(aa^{-1})$.) The correct theorem is as follows:

THEOREM 2. *The following properties are equivalent for a Euclidean plane π:*

(i) *π satisfies the vector axiom and the distributive axiom.*

(ii) *Every planar ternary ring of π is a division ring.*

(iii) *Every planar ternary ring of π is a division ring with the right inverse property.*

(iv) *Some planar ternary ring of π is a division ring with the right inverse property.*

We have already seen that (i) is equivalent to (ii) and we shall be content to show now that (i) implies (iii). Assuming (i), we have that every planar ternary ring (R, F) of π is a division ring. Consider Figure 12 below. If to Figure 12 we add another triangle $A'B'C'$ in such a manner that the hypotheses of the distributive axiom hold, $A'C'$ will be parallel to AC. We may phrase this more conveniently as follows: if the lines OB, OC remain fixed, the slope of AC is the same for every choice of $A\,(A \neq O)$ on the x-axis. We take $A=(b, 0)$ for any nonzero b. We assume that OB has the fixed slope $1+a$, distinct from 1, 0; so that $a\neq 0, -1$. The equation of OB is then $y=x(1+a)$. Since AB has equation $x=b$, the y-coordinate of B is $b(1+a)=b+ba$. BC, OC have equations $y=b+ba$, $y=x$ respectively, so that $C=(b+ba, b+ba)$. If AC has slope m, the equation of AC is $y=xm+k$ for some k, where, since A and C lie on the line, $0=bm+k$ and $b+ba=(b+ba)m+k=(bm+k)+(ba)m=(ba)m$. Therefore m satisfies

$$(ba)m = b + ba. \tag{10.2}$$

Since m is independent of b we may set $b=a^{-1}$ where a^{-1} is defined by $a^{-1}a=1$. Then (10.2) yields $m=a^{-1}+1$. Hence, for all $a\neq 0, -1$ and $b\neq 0$, $(ba)(a^{-1}+1)=b+ba$. From this we get the right inverse property (10.1).—Strictly speaking, we must examine (10.1) for the case $b=0$ and the case $a=-1$, but these give no trouble.

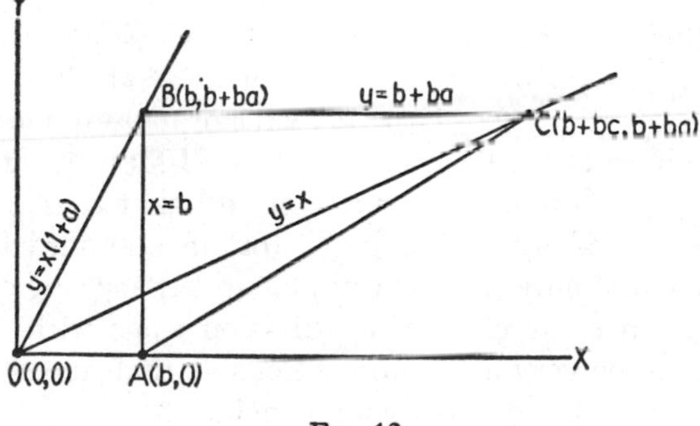

FIG. 12

11. The algebra begins in earnest. At this stage it becomes important to give an algebraic characterization of division rings with the right inverse property. The main theorems are as follows:

THEOREM 3. *Every division ring with the right inverse property is an alternative division ring, and conversely.*

THEOREM 4. *Every alternative division ring is either an associative division ring (that is, a field or skew-field) or a Cayley division algebra over its centre.*

We shall discuss these theorems briefly. It may be shown that a division ring R (with identity element 1) has the right inverse property if and only if it satisfies the identity

$$((ab)c)b = a((bc)b), \qquad \text{all } a, b, c \text{ in } R \tag{11.1}$$

From (11.1) with $c=1$ we derive

$$(ab)b = a(bb), \qquad \text{all } a, b \text{ in } R. \tag{11.2}$$

A ring satisfying (11.2) is called *right alternative;* and it is called *alternative* if it satisfies both (11.2) and

$$b(ba) = (bb)a, \qquad \text{all } a, b \text{ in } R. \tag{11.3}$$

In 1950, Skornyakov [5] proved Theorem 4 for characteristics other than 2, 3. Quite independently, Bruck and Kleinfeld [6] proved Theorem 4 for characteristic not 2 and later Kleinfeld [7], by combining the methods of the first two papers, removed the restriction as to characteristic. Since then, Kleinfeld [8] has given a definitive characterization of simple alternative rings and, incidentally, a new proof of Theorem 4.

In 1951, only a few months after I had become aware of Theorem 2, Skornyakov [9] proved that every right alternative division ring of characteristic not 2 is alternative. In the case of characteristic not 2, (11.2) implies (11.1); but this is false for characteristic 2. However, in 1953, San Soucie [10] showed that division rings of characteristic 2 which satisfy (11.1) are alternative. Thus Theorem 3 is true.

Instead of elaborating here the properties of Cayley division algebras, we refer the reader to an elementary discussion of these algebras from an entirely different point of view (Dickson [11]). It seems of more importance to indicate the geometric significance of Theorems 2, 3, 4. Hilbert [1] shows that if the Euclidean plane π satisfies the vector axiom and that strong form of the distributive axiom obtained by omitting (*) (or, in standard language, if π is Desarguesian) then, and only then, every planar ternary ring of π is an associative division ring. Since the class of all division rings with the right inverse property turns out to be very little more extensive than the class of all associative division rings, we draw the following conclusion:

If we intend to insist that every planar ternary ring of the Euclidean plane π be a division ring, we may as well go the whole way and require π to be Desarguesian.

12. Other points of view. It would be unjust to leave the present topic without some brief reference to a great mass of literature entirely neglected here. Most of this literature (including Hall [2], Veblen and Wedderburn [4]) is written in the language of projective rather than Euclidean planes, which did not suit my aims.* But of course the study of Desarguesian planes, for example, did not originate with Hilbert.

* I shall make no serious attempt to link the present discussion to projective geometry. The reader can discover how to do this for himself by first reading §2 of Hall [2] through the first three lines of p. 232 and then considering Hall's Figure 4 (p. 264) together with the statement (below the figure) of the projective axiom Theorem L. First, in Hall's Figure 4, delete line $AMNB$ and its points and then carefully redraw the figure in the resulting Euclidean plane so that parallel lines

More particularly, the geometric meaning of alternative division rings was first studied by Ruth Moufang (see the references in [2] or [6]) and characterized by the uniqueness of the projective construction of a fourth harmonic point. Later Hall [2] gave an independent characterization in terms of his Theorem L. Theorem L is a projective axiom which, in two of its Euclidean forms, becomes* respectively the vector axiom and the distributive axiom. The characterization of Veblen-Wedderburn systems is originally due to Hall.

One final remark. It is an amusing fact that Theorem 6.4 of Hall [2], although true, was not completely proved until 1953, just ten years after the publication date of Hall's paper. By a trivial slip, the theorem contains the word "two" where "three" would have been appropriate. In 1950 I had a rude awakening in this connection which led to Theorem 2. The record shows that a similar experience led Skornyakov to the study of right alternative division rings.—May there be more such slips!

Appendix I. Planar ternary rings. A planar ternary ring is a system (R, F) consisting of a set R and a ternary operation F subject to the following postulates:

(i) 0 and 1 are two distinct elements of R.

(ii) If a, b, c are in R, $F(a, b, c)$ is a uniquely defined element of R.

(iii) $F(0, b, c) = F(a, 0, c) = c$ for all a, b, c of R.

(iv) $F(a, 1, 0) = F(1, a, 0) = a$ for each a in R.

(v) If b, b', c, c' are in R, with $b \neq b'$, the equation $F(x, b, c) = F(x, b', c')$ has a unique solution x in R.

(vi) If a, a', b, b' are in R, with $a \neq a'$, the system of equations $F(a, x, y) = b$, $F(a', x, y) = b'$ has a unique solution x, y in R.

(vii) If a, b, c are in R, the equation $F(a, b, x) = c$ has a unique solution x in R.

A planar ternary ring (R, F) determines a unique Euclidean plane defined as follows: The points of the plane are the ordered pairs (x, y) of elements x, y of R. Each ordered pair $[m, b]$ of elements m, b of R is a line of the plane which passes through those points (x, y) such that $y = F(x, m, b)$. Each symbol $[a]$, a in R, is a line of the plane which passes through those points (x, y) such that $x = a$.

Addition is defined for a planar ternary ring (R, F) by $a + b = F(a, b, 0)$. The system $(R, +)$ is a loop. That is:

(viii) In the equation $x + y = z$, if any two of x, y, z are assigned as elements of R, the third is uniquely determined as an element of R.

(*e.g.*, RS, XY) appear parallel. You should recognize the figure for the vector axiom (triangles ZXY, TRS). Then note that Theorem L can be interpreted as the vector axiom: ZY is parallel to TS. Now begin afresh with Hall's Figure 4 (or, equivalently, restore the deleted line $AMNB$ and its points). This time delete line ARX and its points and apply the same process. Theorem L, as now interpreted, gives a statement about triangles NZT, MYS which is slightly different from but clearly equivalent to the distributive axiom.—As an alternative suggestion, the reader may prefer to consult a pamphlet by H. G. Forder [12] which (I am told—I have not yet seen it) covers much the same ground as the present paper with more emphasis on the projective formulation.

* See footnote pp. 15–16.

(ix) There exists an element 0 of R such that $0+a=a+0=a$ for every a in R.

A loop $(R, +)$ is a group provided:

(x) $(a+b)+c=a+(b+c)$ for all a, b, c of R, and is an abelian group if also

(xi) $a+b=b+a$ for all a, b of R.

Multiplication is defined for a planar ternary ring (R, F) by $ab=F(a, b, 0)$. In particular,

(xii) $0a=a0=0$ for all a in R.

If R^* consists of R with 0 removed, $(R^*, \cdot)$ is a loop; that is, (viii), (ix) hold with $R, +, 0$ replaced by $R^*, \cdot, 1$ respectively.

Appendix II. Special planar ternary rings. A Veblen-Wedderburn system is a system $(R, +, \cdot)$ consisting of a set R and two binary operations $+, \cdot$, subject to the following postulates:

(I) $(R, +)$ is an abelian group with zero 0.

(II.1) $(a+b)c=ac+bc$ for all a, b, c of R.

(III) $(R^*, \cdot)$ is a loop with identity 1.

(IV) $a0=0$ for each a of R.

(V) If a, a', b are in R, with $a\neq a'$, the equation $xa=xa'+b$ has a unique solution x in R.

A Veblen-Wedderburn system $(R, +, \cdot)$ becomes a planar ternary ring (R, F) when F is defined by $F(a, b, c)=ab+c$.

A division ring (with identity element 1) is a system $(R, +, \cdot)$ which satisfies (I), (II.1), (III) and

(II.2) $c(a+b)=ca+cb$ for all a, b, c of R.

Every division ring (with identity) is a Veblen-Wedderburn system, but not conversely.

Bibliography

1a. David Hilbert, Foundations of Geometry, translated by H. J. Townsend, Chicago, 1902.

1b. David Hilbert, Grundlagen der Geometrie, 7th edition, Berlin, 1930.

2. Marshall Hall, Projective planes, Trans. Amer. Math. Soc., vol. 54, 1943, pp. 229–277.

3. R. D. Carmichael, Groups of Finite Order, Ginn and Co., 1937.

4. O. Veblen and J. H. M. Wedderburn, Non-Desarguesian and non-Pascalian geometries, Trans. Amer. Math. Soc., vol. 8, 1907, pp. 379–383.

5. L. A. Skornyakov, Alternative fields, Ukrain. Math. Žurnal, vol. 2, 1950, pp. 70–85. (Russian.)

6. R. H. Bruck and Erwin Kleinfeld, The structure of alternative division rings, Proc. Amer. Math. Soc., vol. 2, 1951, pp. 878–890.

7. E. Kleinfeld, Alternative division rings of characteristic 2, Proc. Nat. Acad. Sci. (U. S. A.), vol. 37, 1951, pp. 818–820.

8. E. Kleinfeld, Simple alternative rings, Ann. of Math., vol. 58, 1953, pp. 544–547.

9. L. A. Skornyakov, Right-alternative fields, Izvestiya Akad. Nauk. SSSR Ser. Mat., vol. 15, 1951, pp. 177–184. (Russian.)

10. R. L. San Soucie, Right alternative division rings of characteristic two, Proc. Amer. Math. Soc., vol. 6, 1955, pp. 291–296.

11. L. E. Dickson, On quaternions and their generalizations and the history of the eight-square theorem, Ann. of Math., vol. 20, 1919, pp. 155–171.

12. H. G. Forder, Coordinates in Geometry, Auckland University College Math. Ser., no. 1, Auckland, New Zealand, 1954.

APPENDIX TO "RECENT ADVANCES IN THE FOUNDATIONS OF EUCLIDEAN GEOMETRY"

R. H. BRUCK, University of Wisconsin

In the twenty-two years since the appearance of my paper **[1]** the foundations of geometry have changed very markedly. It was possible in **[1]** to describe the situation clearly in simple terms, but this is true no longer. Perhaps it will take several decades before someone will be able to conceal the apparatus and reveal the essence for all to see.

One broad topic which has occupied many able mathematicians is the Lenz-Barlotti classification of projective planes. This has required a lot of group theory: some of the results had to await the discovery of the Suzuki groups; others use the Feit-Thompson theorem.

Another important topic is the construction of affine and projective planes. In particular, the construction of translation planes—which are the planes we discussed in **[1]**—has been reduced to a variety of problems concerning affine or projective spaces coordinatized by a field or skew-field. (Veblen-Wedderburn systems make a brief appearance for technical reasons and then can be abandoned.) An allied topic is Ostrom's theory of derivation of planes; this "quadratic" theory badly needs to be generalized to higher degrees.

Since 1949, no new information has been published concerning the following question: *For which positive integers n do there exist affine or projective planes of order n*? To emphasize the situation, consider an integer $n>1$ with the following two properties:

(I) *n is not a prime-power.*
(II) *Either* (II.1) $n\equiv 0$ *or* $3 \bmod 4$
or (II.2) $n\equiv 1$ *or* $2 \bmod 4$ *and* $n=a^2+b^2$ *for integers* a,b.

At the time of writing (January 1977) there is no integer $n>1$ satisfying (I), (II) for which it has been proved in a published paper whether or not there exists an affine or projective plane of order n. For every other integer $n>1$ the answer is known: There is a plane if (I) fails; there is no plane if I holds but II fails. Even one new result might set off an avalanche of research papers. And perhaps the avalanche will soon be upon us, since there is real hope for a decision soon in the case of order 10.

I would counsel the serious student of the foundations of geometry to acquire a deep knowledge of field theory including Galois theory of vector spaces in the roles of affine or projective or circle geometries, of curves and surfaces, of group theory and especially of the theory of linear groups, and of many peripheral subjects such as algebraic geometry, differential geometry, the theory of linear codes, computing.—It is not easy to guess what mathematical equipment is needed to solve the fundamental problems of geometry: Is it new techniques we lack, or merely new insights?

At present, Dembowski's book [2] is probably the best guide to the literature through 1967. At a more elementary level, I recommend Hughes and Piper [3]. Just where Manin [4] and Berlekamp [5] fit in I am not sure, but I sense that they should not be neglected. In addition to these books, the list of references contains a few papers which have appeared since 1967 and which emphasize the connections between construction problems and classical geometry. For further recent papers the reader must do his own digging in *Mathematical Reviews*.

References

1. R. H. Bruck, Recent advances in the foundations of Euclidean plane geometry, Slaught Memorial Paper No. 4, 2–17. Mathematical Association of America, 1955.

2. Peter Dembowski, Finite Geometries, Springer-Verlag, New York, 1968.

3. Daniel R. Hughes and F. C. Piper, Projective Planes, Springer-Verlag, New York, 1973.

4. Yu. I. Manin, Cubic Forms, American Elsevier, New York, 1974.

5. Elwyn R. Berlekamp, Algebraic Coding Theory, McGraw-Hill, New York, 1968.

6. R. H. Bruck, Construction problems of finite projective planes, Combinatorial Mathematics and its Applications (Proceedings of the 1967 Chapel Hill Conference) Chapter 27, 426–514, University of North Carolina Press, 1969.

7. ———, Some relatively unknown ruled surfaces in projective space, Archives, Nouvelle Série, Section des Sciences, Institut Grand-Ducal de Luxembourg, 34 (1970) 361–376.

8. ———, Circle geometry in higher dimensions, II, Geometriae Dedicata, 2 (1973) 133–188.

9. ———, Construction problems in finite projective spaces, Finite Geometric Structures and their Applications (Proceedings of the 1972 Conference at Bressanone) 107–188. Edizione Cremonese, Rome, 1973.

INDEX

Pages 1–312 refer to Volume I; pages 313–595 refer to Volume II.